**Butterworths Technical and Scientific Checkbooks**

# Mathematics 4 Checkbook

**J O Bird**
BSc(Hons), AFIMA, TEng(CEI), MITE

**A J C May**
BA, CEng, MIMechE, FITE, MBIM

**Butterworths**
London Boston Sydney Wellington Durban Toronto

First published 1981

© Butterworth & Co (Publishers) Ltd 1981

---

**British Library Cataloguing in Publication Data**

Bird, J. O.
    Mathematics 4 checkbook.
    1. Shop mathematics
    I. Title        II. May, A.J.C.
    510'.246    TJ1165

    ISBN 0-408-00660-9
    ISBN 0-408-00612-9 Pbk

---

Typeset by Scribe Design, Gillingham, Kent
Printed in Scotland by Thomson Litho Ltd., East Kilbride

# Contents

## SECTION III – FOURIER SERIES

# Note to Reader

As textbooks become more expensive, authors are often asked to reduce the number of worked and unworked problems, examples and case studies. This may reduce costs, but it can be at the expense of practical work which gives point to the theory.

Checkbooks if anything lean the other way. They let problem-solving establish and exemplify the theory contained in technician syllabuses. The Checkbook reader can gain *real* understanding through seeing problems solved and through solving problems himself.

Checkbooks do not supplant fuller textbooks, but rather supplement them with an alternative emphasis and an ample provision of worked and unworked problems. The brief outline of essential data—definitions, formulae, laws, regulations, codes of practice, standards, conventions, procedures, etc—will be a useful introduction to a course and a valuable aid to revision. Short-answer and multi-choice problems are a valuable feature of many Checkbooks, together with conventional problems and answers.

Checkbook authors are carefully selected. Most are experienced and successful technical writers; all are experts in their own subjects; but a more important qualification still is their ability to demonstrate and teach the solution of problems in their particular branch of technology, mathematics or science.

Authors, General Editors and Publishers are partners in this major low-priced series whose essence is captured by the Checkbook symbol of a question or problem 'checked' by a tick for correct solution.

# Preface

This textbook of worked problems provides coverage of the Technician Education Council level 4 units in Mathematics. However the text is not only for students studying on higher TEC courses for it contains material suitable for HNC, HND and for degree courses. Each topic considered in the text is presented in a way that assumes in the reader only the knowledge obtained in their TEC level 3 Mathematics units. This Checkbook provides a follow-up to the Checkbooks written for Mathematics 1, 2 and 3. Nearly 300 detailed worked problems, followed by some 520 further problems with answers are included.

The authors would like to express their appreciation for the friendly co-operation and helpful advice given to them by the publishers. Thanks are also due to Mrs Elaine Mayo for the excellent typing of the manuscript.

Finally, the authors would like to add a word of thanks to their wives, Elizabeth and Juliet, for their continued patience, help and encouragement during the preparation of this book.

J O Bird
A J C May
Highbury College of Technology
Portsmouth

# Butterworths Technical and Scientific Checkbooks

*General Editors for Science, Engineering and Mathematics titles:*
**J.O. Bird and A.J.C. May,** Highbury College of Technology, Portsmouth.

*General Editor for Building, Civil Engineering, Surveying and Architectural titles:*
**Colin R. Bassett,** lately of Guildford County College of Technology.

A comprehensive range of Checkbooks will be available to cover the major syllabus areas of the TEC, SCOTEC and similar examining authorities. A comprehensive list is given below and classified according to levels.

*Level 1 (Red covers)*
Mathematics
Physical Science
Physics
Construction Drawing
Construction Technology
Microelectronic Systems
Engineering Drawing
Workshop Processes & Materials

*Level 2 (Blue covers)*
Mathematics
Chemistry
Physics
Building Science and Materials
Construction Technology
Electrical & Electronic Applications
Electrical & Electronic Principles
Electronics
Microelectronic Systems
Engineering Drawing
Engineering Science
Manufacturing Technology
Digital Techniques
Motor Vehicle Science

*Level 3 (Yellow covers)*
Mathematics
Chemistry
Building Measurement
Construction Technology
Environmental Science
Electrical Principles
Electronics
Microelectronic Systems
Electrical Science
Mechanical Science
Engineering Mathematics & Science
Engineering Science
Engineering Design
Manufacturing Technology
Motor Vehicle Science
Light Current Applications

*Level 4 (Green covers)*
Mathematics
Building Law
Building Services & Equipment
Construction Technology
Construction Site Studies
Concrete Technology
Economics for the Construction Industry
Geotechnics
Engineering Instrumentation & Control

*Level 5*
Building Services & Equipment
Construction Technology
Manufacturing Technology

# 1 Solving equations by iterative methods

## A MAIN POINTS CONCERNED WITH SOLVING EQUATIONS BY ITERATIVE METHODS

1 Many equations can only be solved graphically or by methods of successive approximations to the roots, called **iterative methods**. Two methods of successive approximations are (i) an algebraic method, introduced in para. 3, and (ii) by using the Newton-Raphson formula, given in para. 5.

2 Both successive approximation methods rely on a reasonably good first estimate of the value of a root being made. One way of doing this is to sketch a graph of the function, say, $y = f(x)$, and determine the approximate values of roots from the points where the graph cuts the $x$-axis. Another way is by using a functional notation method. This method uses the property that the value of the graph of $f(x) = 0$ changes sign for values of $x$ just before and just after the value of a root. For example, one root of the equation $x^2 - x - 6 = 0$ is $x = 3$. Using functional notation:

$f(x) = x^2 - x - 6$
$f(2) = 2^2 - 2 - 6 = -4$
$f(4) = 4^2 - 4 - 6 = +6$

It can be seen from these results that the value of $f(x)$ changes from $-4$ at $f(2)$ to $+6$ at $f(4)$, indicating that a root lies between 2 and 4.

3 **An algebraic method of successive approximations**

This method can be used to solve equations of the form:

$a + bx + cx^2 + dx^3 + \ldots\ldots = 0$, where $a, b, c, d, \ldots\ldots$ are constants. Procedure:

**First approximation**

(a) Using a graphical or the functional notation method, (see para. 2), determine an approximate value of the root required, say $x_1$.

**Second approximation**

(b) Let the true value of the root be $(x_1 + \delta_1)$.

(c) Determine $x_2$ the approximate value of $(x_1 + \delta_1)$ by determining the value of $f(x_1 + \delta_1) = 0$, but neglecting terms containing products of $\delta_1$.

**Third approximation**

(d) Let the true value of the root be $(x_2 + \delta_2)$.

(e) Determine $x_3$, the approximate value of $(x_2 + \delta_2)$ by determining the value of $f(x_2 + \delta_2) = 0$, but neglecting terms containing products of $\delta_2$.

(f) The fourth and higher approximations are obtained in a similar way.

1

4   Using the techniques given in para. 3(b) to (f), it is possible to continue getting values nearer and nearer to the required root. The procedure is repeated until the value of the required root does not change on two consecutive approximations, when expressed to the required degree of accuracy. (See *Problems 1 and 2*.)

5   The Newton-Raphson formula, often just referred to as **Newton's method** may be stated as follows:

*if $r_1$ is the approximate value of a real root of the equation $f(x) = 0$, then a closer approximation to the root, $r_2$ is given by:*

$$r_2 = r_1 - \frac{f(r_1)}{f'(r_1)} \quad .$$

(If, as occasionally happens, the successive approximations of a root do not converge towards the value of the root, a new value of $r_1$ should be selected so that $f(r_1)$ has the same sign as $f''(r_1)$). The advantages of Newton's method over the algebraic method of successive approximations is that it can be used for any type of mathematical equation, (i.e. ones containing trigonometric, exponential, logarithmic, hyperbolic and algebraic functions), and it is usually easier to apply than the algebraic method. (See *Problems 3 to 5*.)

## B.  WORKED PROBLEMS ON SOLVING EQUATIONS BY ITERATIVE METHODS

*Problem 1*  Use an algebraic method of successive approximations to determine the value of the negative root of the quadratic equation: $4x^2 - 6x - 7 = 0$ correct to 3 significant figures. Check the value of the root by using the quadratic formula.

A first estimate of the values of the roots is made by using the functional notation method.

$f(x) \quad = 4x^2 - 6x - 7$

$f(0) \quad = 4(0)^2 - 6(0) - 7 = -7$

$f(-1) = 4(-1)^2 - 6(-1) - 7 = 3$

These results show that the negative root lies between 0 and $-1$, since the values of $f(x)$ change sign between $f(0)$ and $f(-1)$, (see para. 2). The procedure given in para. 3 for the root lying between 0 and $-1$ is followed.

**First approximation**

(a) Let a first approximation be such that it divides the interval 0 to $-1$ in the ratio of $-7$ to 3, i.e., let $x_1 = -0.7$.

**Second approximation**

(b) Let the true value of the root, $x_2$, be $(x_1 + \delta_1)$.

(c) Let $f(x_1 + \delta_1) = 0$, then, since $x_1 = -0.7$,

$$4(-0.7 + \delta_1)^2 - 6(-0.7 + \delta_1) - 7 = 0.$$

Hence,   $4[(-0.7)^2 + (2)(-0.7)(\delta_1) + \delta_1^2] - (6)(-0.7) - 6\delta_1 - 7 = 0.$

Neglecting terms containing products of $\delta_1$ gives:

$$1.96 - 5.6\delta_1 + 4.2 - 6\delta_1 - 7 \approx 0$$

i.e.      $\delta_1 \approx \dfrac{-1.96 - 4.2 + 7}{-5.6 - 6}$

$\approx \dfrac{0.84}{-11.6} \approx -0.072$

2

Thus, $x_2$, a second approximation to the root is $[-0.7+(-0.072)]$, i.e., $x_2 = -0.772$, correct to 3 significant figures.

The procedure given in (b) and (c) is now repeated for $x_2 = -0.772$.

**Third approximation**

(d) Let the true value of the root, $x_3$, be $(x_2+\delta_2)$.

(e) Let $f(x_2+\delta_2) = 0$, then, since $x_2 = -0.772$,

$4(-0.772+\delta_2)^2 - 6(-0.772+\delta_2) - 7 = 0$

$4[(-0.772)^2 + (2)(-0.772)(\delta_2) + \delta_2^2] - (6)(-0.772) - 6\delta_2 - 7 = 0$

Neglecting terms containing products of $\delta_2$ gives:

$2.384 - 6.176\delta_2 + 4.632 - 6\delta_2 - 7 \simeq 0$

i.e. $\delta_2 \simeq \dfrac{-2.384 - 4.632 + 7}{-6.176 - 6} \simeq \dfrac{-0.016}{-12.176} \simeq +0.0013$

Thus $x_3$, the third approximation to the root is $(-0.772+0.0013)$,

i.e. $x_3 = -0.771$, correct to 3 significant figures.

**Fourth approximation**

(f) The procedure given for the second and third approximations is now repeated for $x_3 = -0.771$.

Let the true value of the root, $x_4$, be $(x_3+\delta_3)$.

Let $f(x_3+\delta_3) = 0$, then since $x_3 = -0.771$,

$4(-0.771+\delta_3)^2 - 6(-0.771+\delta_3) - 7 = 0$

$4[(-0.771)^2 + (2)(-0.771)\delta_3 + \delta_3^2] - 6(-0.771) - 6\delta_3 - 7 = 0$

Neglecting terms containing products of $\delta_3$ gives:

$2.3778 - 6.168\delta_3 + 4.626 - 6\delta_3 - 7 \simeq 0$

i.e. $\delta_3 \simeq \dfrac{-2.3778 - 4.626 + 7}{-6.168 - 6} \simeq \dfrac{-0.0038}{-12.168} \simeq +0.0003$

Thus, $x_4$, the fourth approximation to the root is $(-0.771+0.0003)$

i.e. $x_4 = -0.771$, correct to 3 significant figures.

With reference to para. 4, since the values of the roots are the same on two consecutive approximations, when stated to the required degree of accuracy, then the negative root of $4x^2 - 6x - 7 = 0$ is **-0.771** correct to 3 significant figures. Checking, using the quadratic formula:

$x = \dfrac{-(-6) \pm \sqrt{[(-6)^2 - (4)(4)(-7)]}}{(2)(4)}$

$= \dfrac{6 \pm 12.166}{8} = -0.771$ and $2.27$, correct to 3 significant figures.

*Problem 2* Determine the value of the smallest positive root of the equation $3x^3 - 10x^2 + 4x + 7 = 0$, correct to 3 significant figures, using an algebraic method of successive approximations.

The functional notation method is used to find the value of the first approximation, (see para. 2).

$f(x) = 3x^3 - 10x^2 + 4x + 7$

$f(0) = 3(0)^3 - 10(0)^2 + 4(0) + 7 = 7$

$f(1) = 3(1)^3 - 10(1)^2 + 4(1) + 7 = 4$

$f(2) = 3(2)^3 - 10(2)^2 + 4(2) + 7 = -1$

Following the procedure given in para. 3:

**First approximation.**

(a) Let the first approximation be such that it divides the interval 1 to 2 in the ratio of 4 to $-1$, i.e., let $x_1$ be 1.8

3

**Second approximation**

(b) Let the true value of the root, $x_2$, be $(x_1 + \delta_1)$.

(c) Let $f(x_1 + \delta_1) = 0$, then since $x_1 = 1.8$,

$$3(1.8 + \delta_1)^3 - 10(1.8 + \delta_1)^2 + 4(1.8 + \delta_1) + 7 = 0$$

Neglecting terms containing products of $\delta_1$ and using the binomial series, gives:

$$3[1.8^3 + 3(1.8)^2\,\delta_1] - 10[1.8^2 + (2)(1.8)\delta_1] + 4(1.8 + \delta_1) + 7 \simeq 0$$

$$3(5.832 + 9.72\delta_1) - 32.4 - 36\delta_1 + 7.2 + 4\delta_1 + 7 \simeq 0$$

$$17.496 + 29.16\delta_1 - 32.4 - 36\delta_1 + 7.2 + 4\delta_1 + 7 \simeq 0$$

$$\delta_1 \simeq \frac{-17.496 + 32.4 - 7.2 - 7}{29.16 - 36 + 4} \simeq -\frac{0.704}{2.84} \simeq -0.25$$

Thus $x_2 \simeq 1.8 - 0.25 = 1.55$

**Third approximation**

(d) Let the true value of the root, $x_3$, be $(x_2 + \delta_2)$.

(e) Let $f(x_2 + \delta_2) = 0$, then since $x_2 = 1.55$,

$$3(1.55 + \delta_2)^3 - 10(1.55 + \delta_2)^2 + 4(1.55 + \delta_2) + 7 = 0$$

Neglecting terms containing products of $\delta_2$, gives:

$$11.17 + 21.62\delta_2 - 24.03 - 31\delta_2 + 6.2 + 4\delta_2 + 7 \simeq 0$$

$$\delta_2 \simeq \frac{-11.17 + 24.03 - 6.2 - 7}{21.62 - 31 + 4} \simeq \frac{-0.34}{-5.38} \simeq 0.063$$

Thus $x_3 \simeq 1.55 + 0.063 \simeq 1.613$

(f) Values of $x_4$ and $x_5$ are found in a similar way.

$$f(x_3 + \delta_3) = 3(1.613 + \delta_3)^3 - 10(1.613 + \delta_3)^2 + 4(1.613 + \delta_3) + 7 = 0$$

giving $\delta_3 \simeq 0.005$ and $x_4 \simeq 1.618$, i.e. 1.62 correct to 3 significant figures.

$$f(x_4 + \delta_4) = 3(1.618 + \delta_4)^3 - 10(1.618 + \delta_4)^2 + 4(1.618 + \delta_4) + 7 = 0$$

giving $\delta_4 \simeq 0$, correct to 4 significant figures and $x_5 \simeq 1.62$, correct to 3 significant figures.

Since $x_4$ and $x_5$ are the same when expressed to the required degree of accuracy, then, the required root is **1.62**, correct to 3 significant figures.

[**Note on accuracy and errors.** Depending on the accuracy of evaluating the $f(x + \delta)$ terms, one or two iterations (i.e. successive approximations) might be saved. However, it is not usual to work to more than about 4 significant figures accuracy in this type of calculation. If a small error is made in calculations, the only likely effect is to increase the number of iterations.]

*Problem 3* Use Newton's method to determine the positive root of the quadratic equation $5x^2 + 11x - 17 = 0$, correct to 3 significant figures. Check the value of the root by using the quadratic formula.

The functional notation method is used to determine the first approximation to the root, (see para. 2).

$$f(x) = 5x^2 + 11x - 17$$

$$f(0) = 5(0)^2 + 11(0) - 17 = -17$$

$$f(1) = 5(1)^2 + 11(1) - 17 = -1$$

$$f(2) = 5(2)^2 + 11(2) - 17 = 25$$

This shows that the value of the root is close to $x = 1$.

Let the first approximation to the root, $r_1$, be 1.

Newton's formula states that:

A closer approximation, $r_2 = r_1 - \dfrac{f(r_1)}{f'(r_1)}$

$f(x) = 5x^2 + 11x - 17$, thus, $f(r_1) = 5(r_1)^2 + 11(r_1) - 17$
$$= 5(1)^2 + 11(1) - 17 = -1$$

$f'(x)$ is the differential coefficient of $f(x)$, i.e.,
$f'(x) = 10x + 11$. Thus $f'(r_1) = 10(r_1) + 11 = 10(1) + 11 = 21$
By Newton's formula, a better approximation to the root is:

$$r_2 = 1 - \frac{-1}{21}$$

$= 1 - (-0.048) = 1.05$, correct to 3 significant figures.
A still better approximation to the root, $r_3$, is given by:

$$r_3 = r_2 - \frac{f(r_2)}{f'(r_2)} = 1.05 - \frac{[5(1.05)^2 + 11(1.05) - 17]}{[10(1.05) + 11]}$$

$= 1.05 - \dfrac{0.063}{21.5} = 1.05 - 0.003 = 1.047$, i.e. 1.05, correct to 3 significant figures.

Since the values of $r_2$ and $r_3$ are the same when expressed to the required degree of accuracy, the required root is **1.05**, correct to 3 significant figures.
Check, using the quadratic equation formula,

$$x = \frac{-11 \pm \sqrt{[121 - 4(5)(-17)]}}{(2)(5)} = \frac{-11 \pm 21.47}{10}.$$

The positive root is 1.047, i.e., **1.05**, correct to 3 significant figures.

*Problem 4* Taking the first approximation as 2, determine the root of the equation $x^2 - 3 \sin x + 2 \ln(x+1) = 3.5$, correct to 3 significant figures, by using Newton's method.

Newton's formula states that $r_2 = r_1 - \dfrac{f(r_1)}{f'(r_1)}$, where $r_1$ is a first approximation

to the root and $r_2$ is a better approximation to the root.
Since $f(x) = x^2 - 3 \sin x + 2 \ln(x+1) - 3.5$
$$f(r_1) = f(2) = 2^2 - 3 \sin 2 + 2 \ln 3 - 3.5, \text{ where } \sin 2 \text{ means the sine of 2 radians}$$
$$= 4 - 2.7279 + 2.1972 - 3.5 = -0.0307$$

$f'(x) = 2x - 3 \cos x + \dfrac{2}{x+1}$

$f'(r_1) = f'(2) = 2(2) - 3 \cos 2 + \dfrac{2}{3}$
$$= 4 + 1.2484 + 0.6667 = 5.9151$$

Hence, $r_2 = r_1 - \dfrac{f(r_1)}{f'(r_1)} = 2 - \dfrac{-0.0307}{5.9151}$

$= 2.005$ or 2.01, correct to 3 significant figures.
A still better approximation to the root, $r_3$, is given by:

$$r_3 = r_2 - \frac{f(r_2)}{f'(r_2)}$$

$$= 2.005 - \frac{[(2.005)^2 - 3 \sin 2.005 + 2 \ln 3.005 - 3.5]}{[2(2.005) - 3 \cos 2.005 + \dfrac{2}{2.005 + 1}]}$$

$= 2.005 - \dfrac{(-0.0010)}{5.938} = 2.005 + 0.00017$

i.e. $r_3 = 2.01$, correct to 3 significant figures.

Since the values of $r_2$ and $r_3$ are the same when expressed to the required degree of accuracy, then the required root is **2.01**, correct to 3 significant figures.

---

*Problem 5* Use Newton's method to find the root of:

$(x+4)^3 - e^{1.92x} + 5 \cos \frac{x}{3} = 9$, correct to 3 significant figures.

---

The functional notational method is used to determine the approximate value of the root.

$f(x) = (x+4)^3 - e^{1.92x} + 5 \cos \frac{x}{3} - 9$

$f(0) = (0+4)^3 - e^0 + 5 \cos 0 - 9 = 59$

$f(1) = 5^3 - e^{1.92} + 5 \cos \frac{1}{3} - 9 \simeq 114$

$f(2) = 6^3 - e^{3.84} + 5 \cos \frac{2}{3} - 9 \simeq 164$

$f(3) = 7^3 - e^{5.76} + 5 \cos 1 - 9 \simeq 19$

$f(4) = 8^3 - e^{7.68} + 5 \cos \frac{4}{3} - 9 \simeq -1660$

From these results, let a first approximation to the root be $r_1 = 3$.
Newton's formula states that a better approximation to the root,

$r_2 = r_1 - \dfrac{f(r_1)}{f'(r_1)}$

$f(r_1) = f(3) = 7^3 - e^{5.76} + 5 \cos 1 - 9 = 19.35$

$f'(x) = 3(x+4)^2 - 1.92e^{1.92x} - \dfrac{5}{3} \sin \dfrac{x}{3}$

$f'(r_1) = f'(3) = 3(7)^2 - 1.92e^{5.76} - \dfrac{5}{3} \sin 1 = -463.7$

Thus, $r_2 = 3 - \dfrac{19.35}{-463.7} = 3 + 0.042 = 3.042 = 3.04$, correct to 3 significant figures.

Similarly, $r_3 = 3.042 - \dfrac{f(3.042)}{f'(3.042)} = 3.042 - \dfrac{(-1.146)}{(-513.1)} = 3.042 - 0.0022$

$= 3.0398 = 3.04$, correct to 3 significant figures.

Since $r_2$ and $r_3$ are the same when expressed to the required degree of accuracy, then the required root is **3.04**, correct to 3 significant figures.

---

## C. FURTHER PROBLEMS ON SOLVING EQUATIONS BY ITERATIVE METHODS

In *Problems 1 to 5*, use an algebraic method of successive approximations to solve the equations given to the accuracy stated.

1  $3x^2 + 5x - 17 = 0$, correct to 3 significant figures.　　　　　　　　　$[-3.36; 1.69]$
2  $x^3 - 2x + 14 = 0$, correct to 3 decimal places.　　　　　　　　　　　$[-2.686]$
3  $2x^3 - 10x + 4 = 0$, correct to 3 significant figures.　　　　$[-2.41; 0.410; 2.00]$
4  $x^4 - 3x^3 + 7x - 5.5 = 0$, correct to 3 significant figures.　　　　　$[-1.53; 1.68]$
5  $x^4 + 12x^3 - 13 = 0$, correct to 4 significant figures.　　　　　$[-12.01; 1.000]$

In *Problems 6 to 12* use Newton's method to solve the equations given to the accuracy stated.

6

6  $x^2 - 2x - 13 = 0$, correct to 3 decimal places.            [−2.742; 4.742]

7  $3x^3 - 10x = 14$, correct to 4 significant figures.            [2.313]

8  $4x^3 - 16x^2 - 2x + 7 = 0$, correct to 3 significant figures.     [−0.668; 0.652; 4.02]

9  $x^4 - 3x^3 + 7x = 12$, correct to 3 decimal places.            [−1.721; 2.648]

10  $3x^4 - 4x^3 + 7x - 12 = 0$, correct to 3 decimal places.         [−1.386; 1.491]

11  $3 \ln x + 4x = 5$, correct to 3 decimal places.            [1.147]

12  $x^3 = 5 \cos 2x$, correct to 3 significant figures.            [0.744]

13  $300e^{-2\theta} + \dfrac{\theta}{2} = 6$, correct to 3 significant figures.         [2.05]

14  A Fourier analysis of the instantaneous value of a waveform can be represented by:

$$y = (t + \tfrac{\pi}{4}) + \sin t + \tfrac{1}{8} \sin 3t .$$

Use Newton's method to determine the value of $t$ near to 0.4, correct to 4 decimal places, when the amplitude, $y$, is 0.880.            [0.039 9]

15  A damped oscillation of a system is given by the equation: $y = -7.4e^{0.5t} \sin 3t$. Determine the value of $t$ near to 4.2, correct to 3 significant figures, when the magnitude $y$ of the oscillation is zero.            [4.19]

# 2 Hyperbolic functions

## A. MAIN POINTS CONCERNED WITH HYPERBOLIC FUNCTIONS

1 Functions which are associated with the geometry of the conic section called a hyperbola are called **hyperbolic functions**. Applications of hyperbolic functions include transmission line theory and catenary problems.

2 By definition:

(i) Hyperbolic sine of $x$, $\sinh x = \dfrac{e^x - e^{-x}}{2}$       (1)

'sinh $x$' is often abbreviated to 'sh $x$' and is pronounced as 'shine $x$'.

(ii) Hyperbolic cosine of $x$, $\cosh x = \dfrac{e^x + e^{-x}}{2}$       (2)

'cosh $x$' is often abbreviated to 'ch $x$' and is pronounced as 'kosh $x$'.

(iii) Hyperbolic tangent of $x$, $\tanh x = \dfrac{\sinh x}{\cosh x} = \dfrac{e^x - e^{-x}}{e^x + e^{-x}}$       (3)

'tanh $x$' is often abbreviated to 'th $x$' and is pronounced as 'than $x$'.

(iv) Hyperbolic cosecant of $x$, $\operatorname{cosech} x = \dfrac{1}{\sinh x} = \dfrac{2}{e^x - e^{-x}}$       (4)

'cosech $x$' is pronounced as 'coshec $x$'.

(v) Hyperbolic secant of $x$, $\operatorname{sech} x = \dfrac{1}{\cosh x} = \dfrac{2}{e^x + e^{-x}}$       (5)

'sech $x$' is pronounced as 'shec $x$'.

(vi) Hyperbolic cotangent of $x$, $\coth x = \dfrac{1}{\tanh x} = \dfrac{e^x + e^{-x}}{e^x - e^{-x}}$       (6)

'coth $x$' is pronounced as 'koth $x$'.

3 **Some properties of hyperbolic functions**

(i) Replacing $x$ by 0 in equation (1) gives: $\sinh 0 = \dfrac{e^0 - e^{-0}}{2} = \dfrac{1-1}{2} = 0$

(ii) Replacing $x$ by 0 in equation (2) gives: $\cosh 0 = \dfrac{e^0 + e^{-0}}{2} = \dfrac{1+1}{2} = 1$

(iii) If a function of $x$, $f(-x) = -f(x)$, then $f(x)$ is called an **odd function** of $x$. Replacing $x$ by $-x$ in equation (1) gives:

$$\sinh(-x) = \frac{e^{-x} - e^{-(-x)}}{2} = \frac{e^{-x} - e^{x}}{2} = -\left(\frac{e^{x} - e^{-x}}{2}\right) = -\sinh x$$

Replacing $x$ by $-x$ in equation (3) gives:

$$\tanh(-x) = \frac{e^{-x} - e^{-(-x)}}{e^{-x} + e^{-(-x)}} = \frac{e^{-x} - e^{x}}{e^{-x} + e^{x}} = -\left(\frac{e^{x} - e^{-x}}{e^{x} + e^{-x}}\right) = -\tanh x$$

Hence **sinh $x$ and tanh $x$ are both odd functions**, (see para. 5), as also are cosech $x = \dfrac{1}{\sinh x}$ and coth $x = \dfrac{1}{\tanh x}$.

(iv) If a function of $x$, $f(-x) = f(x)$, then $f(x)$ is called an **even function** of $x$. Replacing $x$ by $-x$ in equation (2) gives:

$$\cosh(-x) = \frac{e^{-x} + e^{-(-x)}}{2} = \frac{e^{-x} + e^{x}}{2} = \cosh x.$$

Hence **cosh $x$ is an even function**, (see para. 5), as also is sech $x = \dfrac{1}{\cosh x}$.

4 **Tables of exponential and hyperbolic functions** are available where values of sinh $x$ and cosh $x$ may be read directly, usually from an argument of $x = 0$ to $x = 6.0$. For example, sh $0.27 = 0.2733$, ch $0.60 = 1.1855$, sh $3.9 = 24.691$, ch $5.2 = 90.633$, and so on.

Values of hyperbolic functions for values of $x$ greater than 6.0 and for values of $x$ of greater accuracy than 2 significant figures are evaluated using the definitions of para. 2 or by using a calculator containing hyperbolic functions. (See *Problems 1 to 4*.)

5 **Graphs of hyperbolic functions**

(i) A graph of $y = \sinh x$ may be plotted using values from tables of hyperbolic functions. The curve is shown in *Fig 1*. Since the graph is symmetrical about the origin, sinh $x$ is an odd function (as stated in para. 2(iii)).

(ii) A graph of $y = \cosh x$ may be plotted using values from tables of hyperbolic functions. The curve is shown in *Fig 2*. Since the graph is symmetrical about the $y$-axis, cosh $x$ is an even function (as stated in para. 2(iv)). The shape of $y = \cosh x$ is that of a heavy rope or chain hanging freely under gravity and

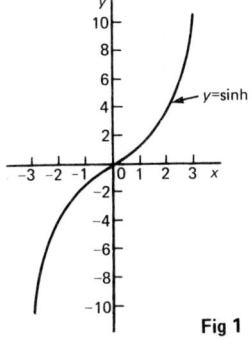

**Fig 1**

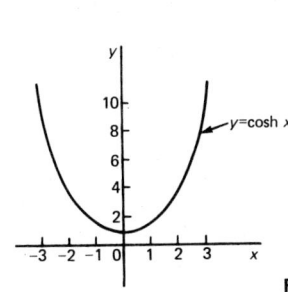

**Fig 2**

is called a **catenary**. Examples include, transmission lines, a telegraph wire or a fisherman's line.

(iii) Graphs of $y = \tanh x$, $y = \operatorname{cosech} x$, $y = \operatorname{sech} x$ and $y = \coth x$ are deduced in *Problems 5 and 6*.

6   For every trigonometric identity there is a corresponding hyperbolic identity. **Hyperbolic identities** may be proved by either

(i) replacing sh $x$ by $\dfrac{e^x - e^{-x}}{2}$   and ch $x$ by $\dfrac{e^x + e^{-x}}{2}$ , or

(ii) by using **Osborne's rule**, which states: *'the six trigonometric ratios used in trigonometrical identities relating general angles may be replaced by their corresponding hyperbolic functions, but the sign of any direct or implied product of two sines must be changed'.*
For example, since $\cos^2 x + \sin^2 x = 1$ then, by Osborne's rule, $\operatorname{ch}^2 x - \operatorname{sh}^2 x = 1$, i.e., the trigonometric functions have been changed to their corresponding hyperbolic functions and since $\sin^2 x$ is a product of two sines the sign is changed from $+$ to $-$. *Table 1* shows some trigonometrical identities and their corresponding hyperbolic identities. (See *Problems 7 to 13*)

TABLE 1

| Trigonometric identity | Corresponding hyperbolic identity |
|---|---|
| $\cos^2 x + \sin^2 x = 1$ | $\operatorname{ch}^2 x - \operatorname{sh}^2 x = 1$ |
| $1 + \tan^2 x = \sec^2 x$ | $1 - \operatorname{th}^2 x = \operatorname{sech}^2 x$ |
| $\cot^2 x + 1 = \operatorname{cosec}^2 x$ | $\coth^2 x - 1 = \operatorname{cosech}^2 x$ |
| *Compound angle formulae* | |
| $\sin(A \pm B) = \sin A \cos B \pm \cos A \sin B$ | $\operatorname{sh}(A \pm B) = \operatorname{sh} A \operatorname{ch} B \pm \operatorname{ch} A \operatorname{ch} B$ |
| $\cos(A \pm B) = \cos A \cos B \mp \sin A \sin B$ | $\operatorname{ch}(A \pm B) = \operatorname{ch} A \operatorname{ch} B \pm \operatorname{sh} A \operatorname{sh} B$ |
| $\tan(A \pm B) = \dfrac{\tan A \pm \tan B}{1 \mp \tan A \tan B}$ | $\operatorname{th}(A \pm B) = \dfrac{\operatorname{th} A \pm \operatorname{th} B}{1 \pm \operatorname{th} A \operatorname{th} B}$ |
| *Double angles* | |
| $\sin 2x = 2 \sin x \cos x$ | $\operatorname{sh} 2x = 2 \operatorname{sh} x \operatorname{ch} x$ |
| $\cos 2x = \cos^2 x - \sin^2 x$ | $\operatorname{ch} 2x = \operatorname{ch}^2 x + \operatorname{sh}^2 x$ |
| $\quad = 2 \cos^2 x - 1$ | $\quad = 2 \operatorname{ch}^2 x - 1$ |
| $\quad = 1 - 2 \sin^2 x$ | $\quad = 1 + 2 \operatorname{sh}^2 x$ |

7   **Differentiation of hyperbolic functions**

(i) $\dfrac{d}{dx}(\sinh x) = \dfrac{d}{dx}\left(\dfrac{e^x - e^{-x}}{2}\right) = \left[\dfrac{e^x - (-e^{-x})}{2}\right] = \left(\dfrac{e^x + e^{-x}}{2}\right) = \cosh x$

If $y = \sinh ax$, where '$a$' is a constant, then $\dfrac{dy}{dx} = a \cosh ax$.

(ii) $\dfrac{d}{dx}(\cosh x) = \dfrac{d}{dx}\left(\dfrac{e^x + e^{-x}}{2}\right) = \left[\dfrac{e^x + (-e^{-x})}{2}\right] = \left(\dfrac{e^x - e^{-x}}{2}\right) = \sinh x$

If $y = \cosh ax$, where '$a$' is a constant, then $\dfrac{dy}{dx} = a \sinh ax$.

(iii) Using the quotient rule of differentiation the derivatives of $\tanh x$, $\operatorname{sech} x$, $\operatorname{cosech} x$ and $\coth x$ may be determined using the results of (i) and (ii). (See *Problems 14 and 15*)

10

(iv) **Summary**

| $y$ or $f(x)$ | $\dfrac{dy}{dx}$ or $f'(x)$ |
|---|---|
| $\sinh ax$ | $a \cosh ax$ |
| $\cosh ax$ | $a \sinh ax$ |
| $\tanh ax$ | $a \operatorname{sech}^2 ax$ |
| $\operatorname{sech} ax$ | $-a \operatorname{sech} ax \tanh ax$ |
| $\operatorname{cosech} ax$ | $-a \operatorname{cosech} ax \coth ax$ |
| $\coth ax$ | $-a \operatorname{cosech}^2 ax$ |

(See *Problems 16 to 19*)

8   Equations of the form $a$ ch $x + b$ sh $x = c$, where $a$, $b$ and $c$ are constants may be solved either by:

   (a) plotting graphs of $y = a$ ch $x + b$ sh $x$ and $y = c$ and noting the points of intersection, or more accurately,

   (b) by adopting the following procedure:

     (i)   Change sh $x$ to $\left(\dfrac{e^x - e^{-x}}{2}\right)$ and ch $x$ to $\left(\dfrac{e^x + e^{-x}}{2}\right)$

     (ii)  Rearrange the equation into the form $pe^x + qe^{-x} + r = 0$, where $p$, $q$ and $r$ are constants.

     (iii) Multiply each term by $e^x$, which produces an equation of the form $p(e^x)^2 + re^x + q = 0$ (since $(e^{-x})(e^x) = e^0 = 1$).

     (iv) Solve the quadratic equation $p(e^x)^2 + re^x + q = 0$ for $e^x$ by factorising or by using the quadratic formula.

     (v)  Given $e^x = a$ constant, (obtained by solving the equation in (iv)), take Naperian logarithms of both sides to give $x = \ln$ (constant).

(See *Problems 20 to 23*)

9   **Series expansions for cosh $x$ and sinh $x$.**

   (i) By definition, $e^x = 1 + x + \dfrac{x^2}{2!} + \dfrac{x^3}{3!} + \dfrac{x^4}{4!} + \dfrac{x^5}{5!} + \ldots \ldots$

      Replacing $x$ by $-x$ gives: $e^{-x} = 1 - x + \dfrac{x^2}{2!} - \dfrac{x^3}{3!} + \dfrac{x^4}{4!} - \dfrac{x^5}{5!} + \ldots \ldots$

   (ii) $\cosh x = \dfrac{1}{2}(e^x + e^{-x})$

     $= \dfrac{1}{2}\left[\left(1 + x + \dfrac{x^2}{2!} + \dfrac{x^3}{3!} + \dfrac{x^4}{4!} + \dfrac{x^5}{5!} + \ldots \ldots\right) + \left(1 - x + \dfrac{x^2}{2!} - \dfrac{x^3}{3!} + \dfrac{x^4}{4!} - \dfrac{x^5}{5!} + \ldots\right)\right]$

     $= \dfrac{1}{2}\left[\left(2 + \dfrac{2x^2}{2!} + \dfrac{2x^4}{4!} + \ldots \ldots\right)\right]$, i.e. $\cosh x = 1 + \dfrac{x^2}{2!} + \dfrac{x^4}{4!} + \ldots \ldots$

(valid for all values of $x$)

     cosh $x$ is an even function and contains only even powers of $x$ in its series expansion.

   (iii) $\sinh x = \dfrac{1}{2}(e^x - e^{-x})$

     $= \dfrac{1}{2}\left[\left(1 + x + \dfrac{x^2}{2!} + \dfrac{x^3}{3!} + \dfrac{x^4}{4!} + \dfrac{x^5}{5!} + \ldots\right) - \left(1 - x + \dfrac{x^2}{2!} - \dfrac{x^3}{3!} + \dfrac{x^4}{4!} - \dfrac{x^5}{5!} + \ldots\right)\right]$

     $= \dfrac{1}{2}\left[2x + \dfrac{2x^3}{3!} + \dfrac{2x^5}{5!} + \ldots \ldots\right]$, i.e. $\sinh x = x + \dfrac{x^3}{3!} + \dfrac{x^5}{5!} + \ldots \ldots$

(valid for all values of $x$)

11

sinh $x$ is an odd function and contains only odd powers of $x$ in its series expansion. (See *Problems 24 to 26*)

## B. WORKED PROBLEMS ON HYPERBOLIC FUNCTIONS

*Problem 1* Using tables, evaluate sinh 5.4, correct to 4 significant figures.

$$\sinh 5.4 = \frac{1}{2}(e^{5.4}-e^{-5.4}) = \frac{1}{2}[(e^{5.0})(e^{0.4})-(e^{-5.0})(e^{-0.4})]$$

$$= \frac{1}{2}[(148.41)(1.4918)-(0.00674)(0.6703)] = \frac{1}{2}[221.398-0.004518]$$

i.e., sh **5.4 = 110.7**, correct to 4 significant figures, which may be checked by using a calculator.

*Problem 2* Determine the value of cosh 1.86, correct to 4 significant figures, using tables.

$$\cosh 1.86 = \frac{1}{2}(e^{1.86}+e^{-1.86}) = \frac{1}{2}[(e^{1.8})(e^{0.06})+(e^{-1.8})(e^{-0.06})]$$

$$= \frac{1}{2}[(6.0497)(1.0618)+(0.1653)(0.9418)] = \frac{1}{2}[6.4236+0.15568]$$

i.e., ch **1.86 = 3.290**, correct to 4 significant figures.

*Problem 3* Evaluate sinh 2.1926 correct to 4 significant figures.

$$\sinh 2.1926 = \frac{1}{2}(e^{2.1926}-e^{-2.1926}) = \frac{1}{2}(8.9585-0.11163),\text{ by calculator,}$$

$$= \textbf{4.423},\text{ correct to 4 significant figures.}$$

To calculate $e^{2.1926}$ using tables, let $e^{2.1926} = y$.
Taking Naperian logarithms of both sides gives: $2.1926 = \ln y$.
From Naperian logarithm tables, $y = 8.958$
Similarly, let $e^{-2.1926} = y$, then $-2.1926 = \ln y -2.1926 = -3+0.8074 = \bar{3}.8074$
Hence $\ln y = \bar{3}.8074 = \bar{3}.6974+(\bar{3}.8074-\bar{3}.6974) = \bar{3}.6974+0.1100$
Thus $y = 10^{-1} \times 1.116 = 0.1116$ from tables.

Thus $\sinh 2.1926 = \frac{1}{2}(8.958-0.1116) = \textbf{4.423}$, correct to 4 significant figures.

*Problem 4* Evaluate, correct to 4 significant figures, (a) th 0.52; (b) cosech 1.4; (c) sech 0.86; (d) coth 0.38.

(a) $\text{th } 0.52 = \dfrac{e^{0.52}-e^{-0.52}}{e^{0.52}+e^{-0.52}}$ . $e^{0.52} = (e^{0.5})(e^{0.02}) = (1.6487)(1.0202) = 1.6820$

$\qquad\qquad\qquad\qquad\qquad\qquad e^{-0.52} = (e^{-0.5})(e^{-0.02}) = (0.6065)(0.9802) = 0.5945$

Hence $\text{th } 0.52 = \dfrac{1.6820-0.5945}{1.6820+0.5945} = \dfrac{1.0875}{2.2765} = \textbf{0.4777}$

(b) $\text{cosech } 1.4 = \dfrac{1}{\sinh 1.4} = \dfrac{1}{1.9043}$ (from tables) $= \textbf{0.5251}$

(c) $\text{sech } 0.86 = \dfrac{1}{\cosh 0.86} = \dfrac{2}{e^{0.86}+e^{-0.86}}$

$$e^{0.86} = (e^{0.8})(e^{0.06}) = (2.2255)(1.0618) = 2.3630$$
$$e^{-0.86} = (e^{-0.8})(e^{-0.06}) = (0.4493)(0.9418) = 0.4232$$

Hence $\text{sech } 0.86 = \dfrac{2}{2.3630+0.4232} = \mathbf{0.7178}$

(d) $\coth 0.38 = \dfrac{1}{\text{th } 0.38} = \dfrac{\text{ch } 0.38}{\text{sh } 0.38} = \dfrac{1.0731}{0.3892} = \mathbf{2.757}$

*Problem 5* Sketch graphs of (a) $y = \tanh x$ and (b) $y = \coth x$ for values of $x$ between $-3$ and $3$.

A table of values is drawn up as shown in *Table 2*.

TABLE 2

| $x$ | $-3$ | $-2$ | $-1$ | $0$ | $1$ | $2$ | $3$ |
|---|---|---|---|---|---|---|---|
| sh $x$ | $-10.02$ | $-3.63$ | $-1.18$ | $0$ | $1.18$ | $3.63$ | $10.02$ |
| ch $x$ | $10.07$ | $3.76$ | $1.54$ | $1$ | $1.54$ | $3.76$ | $10.07$ |
| $y = \text{th } x = \dfrac{\text{sh } x}{\text{ch } x}$ | $-0.995$ | $-0.97$ | $-0.77$ | $0$ | $0.77$ | $0.97$ | $0.995$ |
| $y = \coth x = \dfrac{\text{ch } x}{\text{sh } x}$ | $-1.005$ | $-1.04$ | $-1.31$ | $\pm\infty$ | $1.31$ | $1.04$ | $1.005$ |

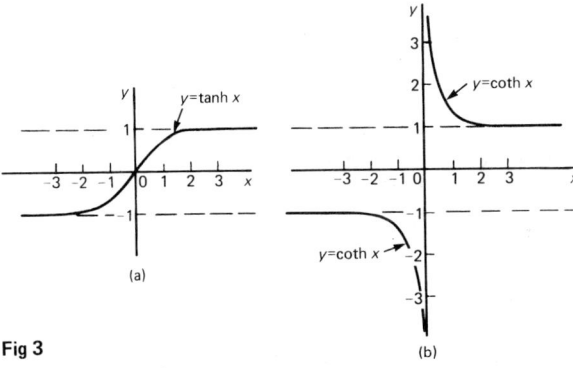

**Fig 3**

(a)  (b)

(a) A graph of $y = \tanh x$ is shown in *Fig 3(a)*.
(b) A graph of $y = \coth x$ is shown in *Fig 3(b)*.
Both graphs are symmetrical about the origin thus $\tanh x$ and $\coth x$ are odd functions.

*Problem 6* Sketch graphs of (a) $y = \text{cosech } x$ and (b) $y = \text{sech } x$ from $x = -4$ to $x = 4$, and, from the graphs, determine whether they are odd or even functions.

A table of values is drawn up as shown in *Table 3*.

13

TABLE 3

| $x$ | $-4$ | $-3$ | $-2$ | $-1$ | 0 | 1 | 2 | 3 | 4 |
|---|---|---|---|---|---|---|---|---|---|
| sh $x$ | $-27.29$ | $-10.02$ | $-3.63$ | $-1.18$ | 0 | 1.18 | 3.63 | 10.02 | 27.29 |
| cosech $x = \dfrac{1}{\text{sh}x}$ | $-0.04$ | $-0.10$ | $-0.28$ | $-0.85$ | $\pm\infty$ | 0.85 | 0.28 | 0.10 | 0.04 |
| ch $x$ | 27.31 | 10.07 | 3.76 | 1.54 | 1 | 1.54 | 3.76 | 10.07 | 27.31 |
| sech $x = \dfrac{1}{\text{ch }x}$ | 0.04 | 0.10 | 0.27 | 0.65 | 1 | 0.65 | 0.27 | 0.10 | 0.04 |

(a) A graph of $y = \text{cosech } x$ is shown in *Fig 4(a)*. The graph is symmetrical about the origin and is thus an **odd function**.

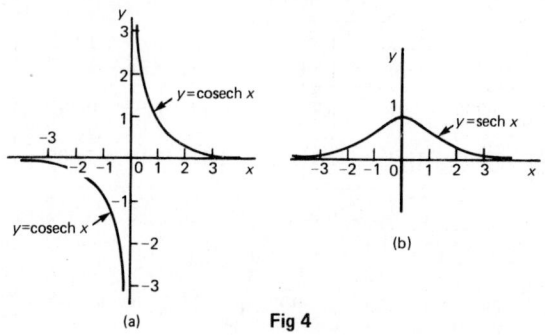

(a)    **Fig 4**    (b)

(b) A graph of $y = \text{sech } x$ is shown in *Fig 4(b)*. The graph is symmetrical about the $y$-axis and is thus an **even function**.

*Problem 7* Prove the hyperbolic identities (a) $\text{ch}^2 x - \text{sh}^2 x = 1$, (b) $1 - \text{th}^2 x = \text{sech}^2 x$, (c) $\text{coth}^2 x - 1 = \text{cosech}^2 x$.

(a) $\text{ch } x + \text{sh } x = \left(\dfrac{e^x + e^{-x}}{2}\right) + \left(\dfrac{e^x - e^{-x}}{2}\right) = e^x$

$\text{ch } x - \text{sh } x = \left(\dfrac{e^x + e^{-x}}{2}\right) - \left(\dfrac{e^x - e^{-x}}{2}\right) = e^{-x}$

$(\text{ch } x + \text{sh } x)(\text{ch } x - \text{sh } x) = (e^x)(e^{-x}) = e^0 = 1$
i.e., $\text{ch}^2 x - \text{sh}^2 x = 1$        (1)

(b) Dividing each term in equation (1) by $\text{ch}^2 x$ gives:

$\dfrac{\text{ch}^2 x}{\text{ch}^2 x} - \dfrac{\text{sh}^2 x}{\text{ch}^2 x} = \dfrac{1}{\text{ch}^2 x}$ , i.e. $1 - \text{th}^2 x = \text{sech}^2 x$.

(c) Dividing each term in equation (1) by $\text{sh}^2 x$ gives:

$\dfrac{\text{ch}^2 x}{\text{sh}^2 x} - \dfrac{\text{sh}^2 x}{\text{sh}^2 x} = \dfrac{1}{\text{sh}^2 x}$ , i.e. $\text{coth}^2 x - 1 = \text{cosech}^2 x$.

*Problem 8* Prove, using Osborne's rule (a) ch $2A = \text{ch}^2 A + \text{sh}^2 A$;
(b) $1 - \text{th}^2 x = \text{sech}^2 x$.

(a) From trigonometric ratios, $\cos 2A = \cos^2 A - \sin^2 A$     (1)

Osborne's rule states that trigonometric ratios may be replaced by their corresponding hyperbolic functions but the sign of any product of two sines has to be changed. In this case, $\sin^2 A = (\sin A)(\sin A)$, i.e. a product of two sines, thus the sign of the corresponding hyperbolic function, $\text{sh}^2 A$, is changed from $+$ to $-$. Hence, from (1), $\text{ch} 2A = \text{ch}^2 A + \text{sh}^2 A$.

(b) From trigonometric ratios, $1 + \tan^2 x = \sec^2 x$     (2)

$$\tan^2 x = \frac{\sin^2 x}{\cos^2 x} = \frac{(\sin x)(\sin x)}{\cos^2 x} \text{ , i.e., a product of two sines.}$$

Hence, in equation (2), the trigonometric ratios are changed to their equivalent hyperbolic function and the sign of $\text{th}^2 x$ changed from $+$ to $-$.
i.e., $1 - \text{th}^2 x = \text{sech}^2 x$.

*Problem 9* Prove that $1 + 2 \text{ sh}^2 x = \text{ch } 2x$.

Left hand side (L.H.S.) $= 1 + 2 \text{ sh}^2 x = 1 + 2\left(\dfrac{e^x - e^{-x}}{2}\right)^2 = 1 + 2\left(\dfrac{e^{2x} - 2e^x e^{-x} + e^{-2x}}{4}\right)$

$= 1 + \dfrac{e^{2x} - 2 + e^{-2x}}{2} = 1 + \left(\dfrac{e^{2x} + e^{-2x}}{2}\right) - \dfrac{2}{2} = \dfrac{e^{2x} + e^{-2x}}{2} = \text{ch } 2x = \text{R.H.S.}$

*Problem 10* Show that $\text{th}^2 x + \text{sech}^2 x = 1$.

L.H.S. $= \text{th}^2 x + \text{sech}^2 x = \dfrac{\text{sh}^2 x}{\text{ch}^2 x} + \dfrac{1}{\text{ch}^2 x} = \dfrac{\text{sh}^2 x + 1}{\text{ch}^2 x}$

Since $\text{ch}^2 x - \text{sh}^2 x = 1$ then $\text{sh}^2 x + 1 = \text{ch}^2 x$

Thus $\dfrac{\text{sh}^2 x + 1}{\text{ch}^2 x} = \dfrac{\text{ch}^2 x}{\text{ch}^2 x} = 1 = \text{R.H.S.}$

*Problem 11* Prove that $\text{ch } A \text{ ch } B + \text{sh } A \text{ sh } B = \text{ch}(A + B)$

L.H.S. $= \text{ch } A \text{ ch } B + \text{sh } A \text{ sh } B = \left(\dfrac{e^A + e^{-A}}{2}\right)\left(\dfrac{e^B + e^{-B}}{2}\right) + \left(\dfrac{e^A - e^{-A}}{2}\right)\left(\dfrac{e^B - e^{-B}}{2}\right)$

$= \dfrac{1}{4}(e^A e^B + e^A e^{-B} + e^{-A} e^B + e^{-A} e^{-B}) + \dfrac{1}{4}(e^A e^B - e^A e^{-B} - e^{-A} e^B + e^{-A} e^{-B})$

$= \dfrac{1}{4}(e^{A+B} + e^{A-B} + e^{-A+B} + e^{-A-B} + e^{A+B} - e^{A-B} - e^{-A+B} + e^{-A-B})$

$= \dfrac{1}{4}(2e^{A+B} + 2e^{-A-B}) = \dfrac{1}{2}(e^{(A+B)} + e^{-(A+B)}) = \text{ch }(A+B) = \text{R.H.S.}$

*Problem 12* Given $Ae^x + Be^{-x} \equiv 4 \text{ ch } x - 5 \text{ sh } x$, determine the values of $A$ and $B$.

$Ae^x + Be^{-x} \equiv 4 \text{ ch } x - 5 \text{ sh } x = 4\left(\dfrac{e^x + e^{-x}}{2}\right) - 5\left(\dfrac{e^x - e^{-x}}{2}\right)$

$= 2e^x + 2e^{-x} - \dfrac{5}{2}e^x + \dfrac{5}{2}e^{-x} = -\dfrac{1}{2}e^x + \dfrac{9}{2}e^{-x}$

Equating coefficients gives: $A = -\dfrac{1}{2}$ and $B = 4\dfrac{1}{2}$.

**Problem 13** If $4e^x - 3e^{-x} \equiv P \text{ sh } x + Q \text{ ch } x$ determine the values of $P$ and $Q$.

$$4e^x - 3e^{-x} \equiv P \text{ sh } x + Q \text{ ch } x = P\left(\frac{e^x - e^{-x}}{2}\right) + Q\left(\frac{e^x + e^{-x}}{2}\right)$$

$$= \frac{P}{2}e^x - \frac{P}{2}e^{-x} + \frac{Q}{2}e^x + \frac{Q}{2}e^{-x} = \left(\frac{P+Q}{2}\right)e^x + \left(\frac{Q-P}{2}\right)e^{-x}$$

Equating coefficients gives: $4 = \dfrac{P+Q}{2}$ and $-3 = \dfrac{Q-P}{2}$

i.e.  $P + Q = 8$        (1)

     $-P + Q = -3$      (2)

Adding equations (1) and (2) gives:  $2Q = 5$, i.e., $Q = 2\frac{1}{2}$

Substituting in equation (1) gives:  $P = 5\frac{1}{2}$.

---

**Problem 14** Determine the differential coefficient of (a) $y = \text{th } x$ and (b) $y = \text{sech } x$.

(a) $\dfrac{d}{dx}(\text{th } x) = \dfrac{d}{dx}\left(\dfrac{\text{sh } x}{\text{ch } x}\right) = \dfrac{(\text{ch } x)(\text{ch } x) - (\text{sh } x)(\text{sh } x)}{\text{ch}^2 x}$ , (using the quotient rule)

$$= \frac{\text{ch}^2 x - \text{sh}^2 x}{\text{ch}^2 x} = \frac{1}{\text{ch}^2 x} = \text{sech}^2 x$$

(b) $\dfrac{d}{dx}(\text{sech } x) = \dfrac{d}{dx}\left(\dfrac{1}{\text{ch } x}\right) = \dfrac{(\text{ch } x)(0) - (1)(\text{sh } x)}{\text{ch}^2 x}$

$$= \frac{-\text{sh } x}{\text{ch}^2 x} = -\left(\frac{1}{\text{ch } x}\right)\left(\frac{\text{sh } x}{\text{ch } x}\right) = -\text{sech } x \text{ th } x$$

---

**Problem 15** Determine $\dfrac{dy}{d\theta}$ given (a) $y = \text{cosech } \theta$; (b) $y = \text{coth } \theta$.

(a) $\dfrac{d}{d\theta}(\text{cosech } \theta) = \dfrac{d}{d\theta}\left(\dfrac{1}{\text{sh } \theta}\right) = \dfrac{(\text{sh } \theta)(0) - (1)(\text{ch } \theta)}{\text{sh}^2 \theta}$

$$= \frac{-\text{ch } \theta}{\text{sh}^2 \theta} = -\left(\frac{1}{\text{sh } \theta}\right)\left(\frac{\text{ch } \theta}{\text{sh } \theta}\right) = -\text{cosech } \theta \text{ coth } \theta.$$

(b) $\dfrac{d}{d\theta}(\text{coth } \theta) = \dfrac{d}{d\theta}\left(\dfrac{\text{ch } \theta}{\text{sh } \theta}\right) = \dfrac{(\text{sh } \theta)(\text{sh } \theta) - (\text{ch } \theta)(\text{ch } \theta)}{\text{sh}^2 \theta} = \dfrac{\text{sh}^2 \theta - \text{ch}^2 \theta}{\text{sh}^2 \theta}$

$$= \frac{-(\text{ch}^2 \theta - \text{sh}^2 \theta)}{\text{sh}^2 \theta} = \frac{-1}{\text{sh}^2 \theta} = -\text{cosech}^2 \theta$$

---

**Problem 16** Differentiate the following with respect to $x$:

(a) $y = 4 \text{ sh } 2x - \dfrac{3}{7} \text{ch } 3x$; (b) $y = 5 \text{ th } \dfrac{x}{2} - 2 \text{ coth } 4x$

(a) $y = 4 \text{ sh } 2x - \dfrac{3}{7} \text{ch } 3x$

$\dfrac{dy}{dx} = 4 (2 \cosh 2x) - \dfrac{3}{7}(3 \text{ sh } 3x) = 8 \cosh 2x - \dfrac{9}{7} \text{sh } 3x$

(b) $y = 5 \text{ th } \dfrac{x}{2} - 2 \text{ coth } 4x$

$\dfrac{dy}{dx} = 5\left(\dfrac{1}{2} \text{ sech}^2 \dfrac{x}{2}\right) - 2(-4 \text{ cosech}^2 4x) = \dfrac{5}{2} \text{sech}^2 \dfrac{x}{2} + 8 \text{ cosech}^2 4x.$

**Problem 17** Given $f(\theta) = \frac{1}{4}(\operatorname{sech} 5\theta - 2 \operatorname{cosech} 4\theta)$, find $f'(\theta)$.

$f(\theta) = \frac{1}{4}(\operatorname{sech} 5\theta - 2 \operatorname{cosech} 4\theta)$

$f'(\theta) = \frac{1}{4}[-5 \operatorname{sech} 5\theta \operatorname{th} 5\theta - 2(-4 \operatorname{cosech} 4\theta \operatorname{coth} 4\theta)]$

$\quad\ = \frac{1}{4}[8 \operatorname{cosech} 4\theta \operatorname{coth} 4\theta - 5 \operatorname{sech} 5\theta \operatorname{th} 5\theta]$

**Problem 18** Differentiate the following with respect to the variable:
(a) $y = 4 \sin 3t \operatorname{ch} 4t$; (b) $y = \ln(\operatorname{sh} 3\theta) - 2 \operatorname{ch}^2 3\theta$.

(a) $y = 4 \sin 3t \operatorname{ch} 4t$ (i.e. a product)

$\quad \dfrac{dy}{dt} = (4 \sin 3t)(4 \operatorname{sh} 4t) + (\operatorname{ch} 4t)(4)(3 \cos 3t)$

$\qquad\ = 16 \sin 3t \operatorname{sh} 4t + 12 \operatorname{ch} 4t \cos 3t = 4(4 \sin 3t \operatorname{sh} 4t + 3 \cos 3t \operatorname{ch} 4t)$

(b) $y = \ln(\operatorname{sh} 3\theta) - 4 \operatorname{ch}^2 3\theta$ (i.e. a function of a function)

$\quad \dfrac{dy}{d\theta} = \left(\dfrac{1}{\operatorname{sh} 3\theta}\right)(3 \operatorname{ch} 3\theta) - (4)(2 \operatorname{ch} 3\theta)(3 \operatorname{sh} 3\theta)$

$\qquad\ = 3 \coth 3\theta - 24 \operatorname{ch} 3\theta \operatorname{sh} 3\theta = 3(\coth 3\theta - 8 \operatorname{ch} 3\theta \operatorname{sh} 3\theta).$

**Problem 19** Show that the differential coefficient of $y = \dfrac{3x^2}{\operatorname{ch} 4x}$ is $6x \operatorname{sech} 4x (1 - 2x \operatorname{th} 4x)$.

$y = \dfrac{3x^2}{\operatorname{ch} 4x}$ (i.e., a quotient)

$\dfrac{dy}{dx} = \dfrac{(\operatorname{ch} 4x)(6x) - (3x^2)(4 \operatorname{sh} 4x)}{(\operatorname{ch} 4x)^2} = \dfrac{6x(\operatorname{ch} 4x - 2x \operatorname{sh} 4x)}{\operatorname{ch}^2 4x}$

$\quad\ = 6x\left[\dfrac{\operatorname{ch} 4x}{\operatorname{ch}^2 4x} - \dfrac{2x \operatorname{sh} 4x}{\operatorname{ch}^2 4x}\right] = 6x\left[\dfrac{1}{\operatorname{ch} 4x} - 2x\left(\dfrac{\operatorname{sh} 4x}{\operatorname{ch} 4x}\right)\left(\dfrac{1}{\operatorname{ch} 4x}\right)\right]$

$\quad\ = 6x[\operatorname{sech} 4x - 2x \operatorname{th} 4x \operatorname{sech} 4x] = 6x \operatorname{sech} 4x (1 - 2x \operatorname{th} 4x)$

**Problem 20** Solve the equation $\operatorname{sh} x = 3$, correct to 4 significant figures.

Following the procedure of para. 8:

(i) $\operatorname{sh} x = \left(\dfrac{e^x - e^{-x}}{2}\right) = 3$

(ii) $e^x - e^{-x} = 6$, i.e., $e^x - e^{-x} - 6 = 0$

(iii) $(e^x)^2 - (e^{-x})(e^x) - 6e^x = 0$, i.e., $(e^x)^2 - 6e^x - 1 = 0$

(iv) $e^x = \dfrac{-(-6) \pm \sqrt{[(-6)^2 - 4(1)(-1)]}}{2(1)} = \dfrac{6 \pm \sqrt{40}}{2} = \dfrac{6 \pm 6.3246}{2}$

Hence $e^x = 6.1623$ or $-0.1623$

(v) $x = \ln 6.1623$ or $x = \ln(-0.1623)$ which has no solution since it is not possible in real terms to find the logarithm of a negative number.
Hence $x = \ln 6.1623 = \mathbf{1.818}$, correct to 4 significant figures.

Following the procedure of para. 8:

(i) $4 \operatorname{ch} \theta - 5 = 0$, i.e., $4 \left( \dfrac{e^\theta + e^{-\theta}}{2} \right) - 5 = 0$; (ii) $2e^\theta + 2e^{-\theta} - 5 = 0$;

(iii) $2(e^\theta)^2 + 2 - 5e^\theta = 0$, i.e., $2(e^\theta)^2 - 5e^\theta + 2 = 0$;

(iv) $(2e^\theta - 1)(e^\theta - 2) = 0$, i.e., $e^\theta = \dfrac{1}{2}$ or $e^\theta = 2$; (v) $\theta = \ln\left(\dfrac{1}{2}\right)$ or $\theta = \ln 2$.

Hence $\theta = -0.6931$ **or** $0.6931$, correct to 4 significant figures.

*Problem 22* Solve the equation $2.6 \operatorname{ch} x + 5.1 \operatorname{sh} x = 8.73$, correct to 4 decimal places.

Following the procedure of para. 8:
(i) $2.6 \operatorname{ch} x + 5.1 \operatorname{sh} x = 8.73$

i.e. $2.6 \left( \dfrac{e^x + e^{-x}}{2} \right) + 5.1 \left( \dfrac{e^x - e^{-x}}{2} \right) = 8.73$

(ii) $1.3e^x + 1.3e^{-x} + 2.55e^x - 2.55e^{-x} = 8.73$
i.e., $3.85e^x - 1.25e^{-x} - 8.73 = 0$

(iii) $3.85(e^x)^2 - 8.73e^x - 1.25 = 0$

(iv) $e^x = \dfrac{-(-8.73) \pm \sqrt{[(-8.73)^2 - 4(3.85)(-1.25)]}}{2(3.85)} = \dfrac{8.73 \pm \sqrt{95.463}}{7.70}$

$$= \dfrac{8.73 \pm 9.7705}{7.70}$$

Hence $e^x = 2.4027$ or $e^x = -0.1351$
(v) $x = \ln 2.4027$ or $x = \ln(-0.1351)$ which has no real solution.
Hence $x = 0.8766$, correct to 4 decimal places.

*Problem 23* A chain hangs in the form given by $y = 40 \operatorname{ch} \dfrac{x}{40}$. Determine, correct to 4 significant figures, (a) the value of $y$ when $x$ is 25; and (b) the value of $x$ when $y$ is 54.30.

(a) $y = 40 \operatorname{ch} \dfrac{x}{40}$. When $x = 25$, $y = 40 \operatorname{ch} \dfrac{25}{40} = 40 \operatorname{ch} 0.625$

$= 40 \left( \dfrac{e^{0.625} + e^{-0.625}}{2} \right) = 20(1.8682 + 0.5353)$. Hence $y = 48.07$

(b) When $y = 54.30$, $54.30 = 40 \operatorname{ch} \dfrac{x}{40}$, from which $\operatorname{ch} \dfrac{x}{40} = \dfrac{54.30}{40} = 1.3575$

Following the procedure of para. 8:

(i) $\dfrac{e^{\frac{x}{40}} + e^{-\frac{x}{40}}}{2} = 1.3575$

(ii) $e^{\frac{x}{40}} + e^{-\frac{x}{40}} = 2.715$, i.e., $e^{\frac{x}{40}} + e^{-\frac{x}{40}} - 2.715 = 0$

(iii) $(e^{\frac{x}{40}})^2 + 1 - 2.715e^{\frac{x}{40}} = 0$, i.e., $(e^{\frac{x}{40}})^2 - 2.715e^{\frac{x}{40}} + 1 = 0$

(iv) $e^{\frac{x}{40}} = \dfrac{-(-2.715) \pm \sqrt{[(-2.715)^2 - 4(1)(1)]}}{2(1)} = \dfrac{2.715 \pm \sqrt{(3.3712)}}{2}$

$$= \dfrac{2.715 \pm 1.8361}{2}$$

Hence $e^{\frac{x}{40}} = 2.2756$ or $0.43945$

(v) $\dfrac{x}{40} = \ln 2.2756$ or $\dfrac{x}{40} = \ln (0.43945)$. Hence $\dfrac{x}{40} = 0.8222$

or $\dfrac{x}{40} = -0.8222$. Hence $x = 40(0.8222)$ or $x = 40(-0.8222)$.

i.e., $x = \pm\, 32.89$, correct to 4 significant figures.

*Problem 24* Using the series expansion for ch $x$ evaluate ch 1 correct to 4 decimal places.

ch $x = 1 + \dfrac{x^2}{2!} + \dfrac{x^4}{4!} + \ldots\ldots$ from para. 9

Let $x = 1$, then ch $1 = 1 + \dfrac{1^2}{2 \times 1} + \dfrac{1^4}{4 \times 3 \times 2 \times 1} + \dfrac{1^6}{6 \times 5 \times 4 \times 3 \times 2 \times 1} + \ldots$

$$= 1 + 0.5 + 0.04167 + 0.001389 + \ldots\ldots$$

i.e., ch $1 = 1.5431$, correct to 4 decimal places, which may be checked from tables or by using a calculator.

*Problem 25* Determine, correct to 3 decimal places, the value of sh 3 using the series expansion for sh $x$.

sh $x = x + \dfrac{x^3}{3!} + \dfrac{x^5}{5!} + \ldots\ldots$ from para. 9.

Let $x = 3$, then sh $3 = 3 + \dfrac{3^3}{3!} + \dfrac{3^5}{5!} + \dfrac{3^7}{7!} + \dfrac{3^9}{9!} + \dfrac{3^{11}}{11!} + \ldots\ldots$

$$= 3 + 4.5 + 2.025 + 0.43393 + 0.05424 + 0.00444 + \ldots\ldots$$

i.e. sh $3 = 10.018$, correct to 3 decimal places.

*Problem 26* Determine the power series for $2\,\text{ch}(\theta/2) - \text{sh}\,2\theta$ as far as the term in $\theta^5$.

In the series expansion for ch $x$, let $x = \dfrac{\theta}{2}$, then:

$$2\,\text{ch}\left(\dfrac{\theta}{2}\right) = 2\left[1 + \dfrac{\left(\frac{\theta}{2}\right)^2}{2!} + \dfrac{\left(\frac{\theta}{2}\right)^4}{4!} + \ldots\ldots\right] = 2 + \dfrac{\theta^2}{4} + \dfrac{\theta^4}{192} + \ldots\ldots$$

In the series expansion for sh $x$, let $x = 2\theta$, then:

sh $2\theta = 2\theta + \dfrac{(2\theta)^3}{3!} + \dfrac{(2\theta)^5}{5!} + \ldots\ldots = 2\theta + \dfrac{4}{3}\theta^3 + \dfrac{4}{15}\theta^5 + \ldots\ldots$

Hence $2\,\text{ch}\,\dfrac{\theta}{2} - \text{sh}\,2\theta = \left(2 + \dfrac{\theta^2}{4} + \dfrac{\theta^4}{192} + \ldots\right) - (2\theta + \dfrac{4}{3}\theta^3 + \dfrac{4}{15}\theta^5 + \ldots\ldots)$

$$= 2 - 2\theta + \dfrac{\theta^2}{4} - \dfrac{4}{3}\theta^3 + \dfrac{\theta^4}{192} - \dfrac{4}{15}\theta^5 + \ldots\ldots \text{ as far as the term in } \theta^5.$$

## C. FURTHER PROBLEMS ON HYPERBOLIC FUNCTIONS

In *Problems 1 to 5*, evaluate correct to 4 significant figures.

1   (a) sh 0.64; (b) sh 2.182; (c) sh 4.92.          [(a) 0.6846; (b) 4.376; (c) 68.50]
2   (a) sh 7.7; (b) sh 3.4681; (c) ch 0.72          [(a) 1104; (b) 16.02; (c) 1.271]
3   (a) ch 1.92; (b) ch 6.7; (c) ch 2.4625          [(a) 3.484; (b) 406.2; (c) 5.910]
4   (a) ch 8.2; (b) th 0.65; (c) th 1.81          [(a) 1820; (b) 0.5717; (c) 0.9478]
5   (a) cosech 0.543; (b) cosech 3.12; (c) sech 0.39

[(a) 1.754; (b) 0.08849; (c) 0.9285]

6   (a) sech 2.367; (b) coth 0.444; (c) coth 1.843  [(a) 0.1859; (b) 2.398; (c) 1.051]
7   A telegraph wire hangs so that its shape is described by $y = 50 \, \text{ch} \, \dfrac{x}{50}$.

Evaluate, correct to 4 significant figures, the value of $y$ when $x = 25$.          [56.38]

8   The length $l$ of a heavy cable hanging under gravity is given by $l = 2c \, \text{sh}(L/2c)$. Find the value of $l$ when $c = 40$ and $L = 30$.          [30.71]

9   $V^2 = 0.55L \tanh(6.3d/L)$ is a formula for velocity $V$ of waves over the bottom of shallow water, where $d$ is the depth and $L$ is the wavelength. If $d = 8.0$ and $L = 96$, calculate the value of $V$.          [5.042]

In *Problems 10 to 14*, prove the given identities.

10  (a) $\text{ch}(P-Q) \equiv \text{ch} \, P \, \text{ch} \, Q - \text{sh} \, P \, \text{sh} \, Q$; (b) $\text{ch} \, 2x \equiv \text{ch}^2 x + \text{sh}^2 x$.

11  (a) $\coth x \equiv 2 \, \text{cosech} \, 2x + \text{th} \, x$; (b) $\text{ch} \, 2\theta - 1 \equiv 2 \, \text{sh}^2 \, \theta$.

12  (a) $\text{sech} \, t - \text{cosech} \, t \equiv (\text{th} \, t - \coth t)(\text{ch} \, t - \text{sh} \, t)$; (b) $\text{th} \, 2x \equiv \dfrac{2 \, \text{th} \, x}{1 + \text{th}^2 x}$.

13  (a) $\text{th} \, (A-B) \equiv \dfrac{\text{th} \, A - \text{th} \, B}{1 - \text{th} \, A \, \text{th} \, B}$; (b) $\text{sh} \, 2A \equiv 2 \, \text{sh} \, A \, \text{ch} \, A$.

14  (a) $\text{sh} \, (A+B) \equiv \text{sh} \, A \, \text{ch} \, B + \text{ch} \, A \, \text{sh} \, B$; (b) $\dfrac{\text{sh}^2 \, x + \text{ch}^2 \, x - 1}{2 \text{ch}^2 \, x \coth^2 x} \equiv \tanh^4 x$.

15  Given $Pe^x - Qe^{-x} \equiv 6 \, \text{ch} \, x - 2 \, \text{sh} \, x$, find $P$ and $Q$.          $[P = 2, Q = -4]$
16  If $5e^x - 4e^{-x} \equiv A \, \text{sh} \, x + B \, \text{ch} \, x$, find $A$ and $B$.          $[A = 9, B = 1]$

In *Problems 17 to 22* differentiate the given functions with respect to the variable:

17  (a) $3 \, \text{sh} \, 2x$; (b) $2 \, \text{ch} \, 5\theta$; (c) $4 \, \text{th} \, 9t$     [(a) $6 \, \text{ch} \, 2x$; (b) $10 \, \text{sh} \, 5\theta$; (c) $36 \, \text{sech}^2 \, 9t$]

18  (a) $\dfrac{2}{3} \text{sech} \, 5x$; (b) $\dfrac{5}{8} \text{cosech} \, \dfrac{t}{2}$; (c) $2 \coth 7\theta$

$$\left[ \begin{array}{l} \text{(a)} - \dfrac{10}{3} \text{sech} \, 5x \, \text{th} \, 5x; \text{(b)} - \dfrac{5}{16} \text{cosech} \, \dfrac{t}{2} \coth \dfrac{t}{2}; \\ \text{(c)} -14 \, \text{cosech}^2 \, 7\theta \end{array} \right]$$

19  (a) $2 \ln (\text{sh} \, x)$; (b) $\dfrac{3}{4} \ln \left( \text{th} \, \left(\dfrac{\theta}{2}\right) \right)$          [(a) $2 \coth x$; (b) $\dfrac{3}{8} \text{sech} \, \dfrac{\theta}{2} \text{cosech} \, \dfrac{\theta}{2}$]

20  (a) $\dfrac{1}{2} \text{sech}^4 \, 2t$; (b) $4 \coth^3 2t$

[(a) $-4 \, \text{sech}^4 \, 2t \, \text{th} \, 2t$; (b) $-24 \coth^2 2t \, \text{cosech}^2 \, 2t$]

21  (a) $\text{sh} \, 2x \, \text{ch} \, 2x$; (b) $3e^{2x} \, \text{th} \, 2x$   [(a) $\text{sh}^2 \, 2x + \text{ch}^2 \, 2x$; (b) $6e^{2x} (\text{sech}^2 \, 2x + \text{th} \, 2x)$]

22  (a) $\dfrac{3 \, \text{sh} \, 4x}{2x^3}$; (b) $\dfrac{\text{ch} \, 2t}{\cos 2t}$

$$\left[ \text{(a)} \, \dfrac{12x \, \text{ch} \, 4x - 9 \, \text{sh} \, 4x}{2x^4}; \text{(b)} \, \dfrac{2(\cos 2t \, \text{sh} \, 2t + \text{ch} \, 2t \sin 2t)}{\cos^2 2t} \right]$$

In *Problems 23 to 28* solve the given equations correct to 4 decimal places.

23  $\text{sh} \, x = 1$          [0.8814]
24  $2 \, \text{ch} \, x = 3$          [±0.9624]
25  $3.5 \, \text{sh} \, x + 2.5 \, \text{ch} \, x = 0$          [−0.8959]

20

26  $2 \operatorname{sh} x + 3 \operatorname{ch} x = 5$   [0.6389 or −2.2484]

27  $4 \operatorname{th} x - 1 = 0$   [0.2554]

28  $2.76 \operatorname{ch} x + 4.32 \operatorname{sh} x = 5.44$   [0.5121]

29  A chain hangs so that its shape is of the form $y = 56 \operatorname{ch}(x/56)$. Determine, correct to 4 significant figures, (a) the value of $y$ when $x$ is 35, and (b) the value of $x$ when $y$ is 62.35.   [(a) 67.30; (b) 26.42]

30  Use the series expansion for $\operatorname{ch} x$ to evaluate, correct to 4 decimal places:
(a) ch 1.5; (b) ch 0.8; (c) ch 3.   [(a) 2.3524; (b) 1.3374; (c) 10.0677]

31  Use the series expansion for $\operatorname{sh} x$ to evaluate, correct to 4 decimal places:
(a) sh 0.5; (b) sh 2; (c) sh 2.5.   [(a) 0.5211; (b) 3.6269; (c) 6.0502]

32  Expand the following as a power series as far as the term in $x^5$ : (a) sh $3x$;

(b) ch $2x$   $\left[ \text{(a) } 3x + \dfrac{9}{2}x^3 + \dfrac{81}{40}x^5 ; \text{(b) } 1 + 2x^2 + \dfrac{2}{3}x^4 \right]$

In *Problems 33 to 35*, prove the given identities, the series being taken as far as the term in $\theta^5$ only.

33  $\operatorname{sh} 2\theta - \operatorname{sh} \theta \equiv \theta + \dfrac{7}{6}\theta^3 + \dfrac{31}{120}\theta^5$

34  $\operatorname{ch} 3\theta - \operatorname{ch} 2\theta \equiv \dfrac{\theta^2}{2}\left(5 + \dfrac{65}{12}\theta^2\right)$

35  $2 \operatorname{sh} \dfrac{\theta}{2} - \operatorname{ch} \dfrac{\theta}{2} \equiv -1 + \theta - \dfrac{\theta^2}{8} + \dfrac{\theta^3}{24} - \dfrac{\theta^4}{384} + \dfrac{\theta^5}{1920}$

# 3 De Moivre's theorem

## A. MAIN POINTS CONCERNED WITH DE MOIVRE'S THEOREM

1 Certain mathematical functions may be expressed as power series, three examples being:

(i) $e^x = 1 + x + \dfrac{x^2}{2!} + \dfrac{x^3}{3!} + \dfrac{x^4}{4!} + \dfrac{x^5}{5!} + \ldots\ldots\ldots$ (1)

(ii) $\sin x = x - \dfrac{x^3}{3!} + \dfrac{x^5}{5!} - \dfrac{x^7}{7!} + \ldots\ldots\ldots$ (2)

(iii) $\cos x = 1 - \dfrac{x^2}{2!} + \dfrac{x^4}{4!} - \dfrac{x^6}{6!} + \ldots\ldots\ldots$ (3)

2 Replacing $x$ in equation (1) by the imaginary number $j\theta$, gives:

$$e^{j\theta} = 1 + j\theta + \frac{(j\theta)^2}{2!} + \frac{(j\theta)^3}{3!} + \frac{(j\theta)^4}{4!} + \frac{(j\theta)^5}{5!} + \ldots\ldots$$
$$= 1 + j\theta + \frac{j^2\theta^2}{2!} + \frac{j^3\theta^3}{3!} + \frac{j^4\theta^4}{4!} + \frac{j^5\theta^5}{5!} + \ldots\ldots$$

By definition, $j = \sqrt{(-1)}$, hence $j^2 = -1, j^3 = -j, j^4 = 1, j^5 = j$, and so on.

Thus, $e^{j\theta} = 1 + j\theta - \dfrac{\theta^2}{2!} - j\dfrac{\theta^3}{3!} + \dfrac{\theta^4}{4!} + j\dfrac{\theta^5}{5!} - \ldots\ldots$

Grouping real and imaginary terms gives:

$$e^{j\theta} = (1 - \frac{\theta^2}{2!} + \frac{\theta^4}{4!} - \ldots\ldots) + j\,(\theta - \frac{\theta^3}{3!} + \frac{\theta^5}{5!} - \ldots\ldots)$$

However, from para. 1, equations (2) and (3):

$$(1 - \frac{\theta^2}{2!} + \frac{\theta^4}{4!} - \ldots\ldots) = \cos\theta \text{ and } (\theta - \frac{\theta^3}{3!} + \frac{\theta^5}{5!} - \ldots\ldots) = \sin\theta$$

Thus $e^{j\theta} = \cos\theta + j\sin\theta$ (4)

Writing $-\theta$ for $\theta$ in equation (4), gives:
$e^{j(-\theta)} = \cos(-\theta) + j\sin(-\theta)$
However, $\cos(-\theta) = \cos\theta$ and $\sin(-\theta) = -\sin\theta$
Thus $e^{-j\theta} = \cos\theta - j\sin\theta$ (5)

3 The polar form of a complex number $z$ is: $z = r(\cos\theta + j\sin\theta)$. But, from equation (4), $\cos\theta + j\sin\theta = e^{j\theta}$. Therefore, $z = re^{j\theta}$. When a complex number is written in this way, it is said to be expressed in **exponential form**.

22

4 **De Moivre's** theorem can be derived as follows:

$$z^2 = z \times z = re^{j\theta} \times re^{j\theta} = r^2 e^{(j\theta + j\theta)} = r^2 e^{j2\theta}$$

by applying the laws of indices. Similarly,

$$z^3 = z \times z \times z = re^{j\theta} \times re^{j\theta} \times re^{j\theta} = r^3 e^{(j\theta + j\theta + j\theta)} = r^3 e^{j3\theta}.$$

It may be deduced from these results that:

$$z^n = z \times z \times z \times \ldots \text{to } n \text{ terms} \quad = re^{j\theta} \times re^{j\theta} \times re^{j\theta} \ldots \text{to } n \text{ terms}$$
$$= r^n e^{jn\theta}.$$

From equation (4), $e^{j\theta} = \cos\theta + j\sin\theta$. Writing $n\theta$ for $\theta$ gives:

$e^{jn\theta} = \cos n\theta + j\sin n\theta$, i.e. $r^n e^{jn\theta} = r^n (\cos n\theta + j\sin n\theta) = r^n \angle n\theta$

This expression is known as de Moivres theorem and is true for all values of $n$ (positive, negative and fractional numbers). It is used for finding the values of **powers and roots of complex numbers**. (See *Problems 1 to 4*)

5 When finding the $n^{th}$ **root of a complex number**, there are $n$ solutions. For example, there are three solutions to a cube root, five solutions to a fifth root, and so on. In the solutions to the roots of a complex number, the modulus, $r$, is always the same, but the arguments, $\theta$, are different. It is shown in *Problem 3* that arguments are symmetrically spaced on an Argand diagram and are $(360/n)°$ apart, where $n$ is the number of the root required. Thus if one of the solutions to the cube root of a complex number is, say, $5\angle 20°$, the other two roots are symmetrically spaced $(360/3)°$, i.e. $120°$ from this root and the three roots are $5\angle 20°$, $5\angle 140°$ and $5\angle 260°$.

6 The complex number $z = r(\cos\theta + j\sin\theta)$, when raised to power $n$, becomes $z^n = r^n (\cos n\theta + j\sin n\theta)$ by de Moivres theorem.

Also $z^n = [r(\cos\theta + j\sin\theta)]^n = r^n (\cos\theta + j\sin\theta)^n$

Thus, $\cos n\theta + j\sin n\theta = (\cos\theta + j\sin\theta)^n$         (6)

The right hand side of this equation can be expanded by applying the binomial theorem, i.e.

$$(\cos\theta + j\sin\theta)^n = \cos^n\theta + n\cos^{n-1}\theta j\sin\theta + \frac{n(n-1)}{2!}\cos^{n-2}\theta j^2 \sin^2\theta + \ldots$$

But $j^2 = -1, j^3 = -j, j^4 = 1$, and so on. Hence

$$(\cos\theta + j\sin\theta)^n = \cos^n\theta + jn\cos^{n-1}\theta \sin\theta - \frac{n(n-1)}{2!}\cos^{n-2}\theta \sin^2\theta$$
$$- j\frac{n(n-1)(n-2)}{3!}\cos^{n-3}\theta \sin^3\theta + \ldots$$

Grouping real and imaginary terms gives:

$$(\cos\theta + j\sin\theta)^n = \cos^n\theta - \frac{n(n-1)}{2!}\cos^{n-2}\theta \sin^2\theta$$
$$+ \frac{n(n-1)(n-2)(n-3)}{4!}\cos^{n-4}\theta \sin^4\theta - \ldots$$
$$+ j(n\cos^{n-1}\theta \sin\theta - \frac{n(n-1)(n-2)}{3!}\cos^{n-3}\theta \sin^3\theta + \ldots)$$

and from equation (6):

$$(\cos n\theta + j\sin n\theta) = \cos^n\theta - \frac{n(n-1)}{2!}\cos^{n-2}\theta \sin^2\theta$$
$$+ \frac{n(n-1)(n-2)(n-3)}{4!}\cos^{n-4}\theta \sin^4\theta - \ldots$$
$$+ j(n\cos^{n-1}\theta \sin\theta - \frac{n(n-1)(n-2)}{3!}\cos^{n-3}\theta \sin^3\theta + \ldots)$$

Equating the real parts:

$$\cos n\theta = \cos^n \theta - \frac{n(n-1)}{2!} \cos^{n-2}\theta \sin^2 \theta + \tag{7}$$

Equating the imaginary parts:

$$\sin n\theta = n \cos^{n-1}\theta \sin \theta - \frac{n(n-1)(n-2)}{3!} \cos^{n-3}\theta \sin^3 \theta + \tag{8}$$

Equations (7) and (8) are used to express $\cos n\theta$ and $\sin n\theta$ in terms of powers of $\cos \theta$ and $\sin \theta$. (See *Problems 5 and 6.*)

7  From para. 2, equations (4) and (5):

$$z = e^{j\theta} = \cos \theta + j \sin \theta \tag{9}$$

$$\frac{1}{z} = \frac{1}{e^{j\theta}} = e^{-j\theta} = \cos \theta - j \sin \theta \tag{10}$$

Adding equations (9) and (10) gives: $z + \frac{1}{z} = 2 \cos \theta$ and it follows that

$$(z + \frac{1}{z})^n = 2^n \cos^n \theta \tag{11}$$

Subtracting equation (10) from equation (9) gives:

$$z - \frac{1}{z} = j\, 2 \sin \theta$$

It follows that $(z - \frac{1}{z})^n = j^n 2^n \sin^n \theta$ $\tag{12}$

If $z = \cos \theta + j \sin \theta$ then from de Moivres theorem, $z^n = \cos n\theta + j \sin n\theta$

Also, if $\frac{1}{z} = \cos \theta - j \sin \theta$, then from de Moivre's theorem, $\frac{1}{z^n} = \cos n\theta - j \sin n\theta$.

Adding gives $(z^n + \frac{1}{z^n}) = 2 \cos n\theta$ $\tag{13}$

Subtracting gives $(z^n - \frac{1}{z^n}) = 2j \sin n\theta$ $\tag{14}$

Equations (11) to (14) are used for expressing powers of $\cos \theta$ and $\sin \theta$ in terms of cosines and sines of multiples of $\theta$. (See *Problems 7 to 9*)

## B.  WORKED PROBLEMS ON DE MOIVRE'S THEOREM

*Problem 1* Determine the value of $(2+j3)^3$, expressing the result in both polar and rectangular forms.

The relationship between the various forms of a complex number are
$a+jb = r(\cos \theta + j \sin \theta) = r\angle\theta$, where $r = \sqrt{(a^2+b^2)}$ and $\theta = \arctan \frac{b}{a}$.

Thus: $(2+j3) = \sqrt{(2^2+3^2)}\angle\arctan \frac{3}{2} = 3.606\angle56° 19'$.

By de Moivre's theorem:

$(3.606\angle56° 19')^3 = 3.606^3\angle3 \times 56° 19' = 46.89\angle168° 57'$

Since $r\angle\theta = r \cos \theta + jr \sin \theta$

$46.89\angle168° 57' = 46.89 \cos 168° 57' + j46.89 \sin 168° 57'$

$\qquad\qquad\qquad = -46.02 + j8.987$

i.e. $(2+j3)^3 \quad = \mathbf{46.89\angle168° 57'}$ in polar form and

$\qquad\qquad\qquad \mathbf{-46.02 + j8.987}$ in rectangular form.

24

*Problem 2* Determine the value of $(-7+j5)^4$, expressing the result in rectangular form.

Since $a+jb = r\angle\theta$, where $r = \sqrt{(a^2+b^2)}$ and $\theta = \arctan b/a$

$(-7+j5) = \sqrt{[(-7)^2+5^2]} \angle\arctan \dfrac{5}{-7}\ = 8.602\angle144° \ 28'$

(Note, by considering the Argand diagram, $-7+j5$ must represent an angle in the second quadrant and **not** in the fourth quadrant.)

Applying de Moivre's theorem:

$(-7+j5)^4 = [8.602\angle144° \ 28']^4 = 8.602^4 \angle4 \times 144° \ 28'$
$\qquad\qquad = 5475\angle577° \ 52' = 5475\angle217° \ 52'$

Since $r\angle\theta = r\cos\theta+jr\sin\theta$,

$5475\angle217° \ 52' = 5475 \cos 217° \ 52'+j5475 \sin 217° \ 52'$
$\qquad\qquad\qquad\ = -4322-j3361$

i.e. $(-7+j5)^4 \ \ = -4322-j3361$ in rectangular form

*Problem 3* Find the roots of $(5+j3)^{\frac{1}{2}}$ in rectangular form, correct to 4 significant figures.

$(5+j3) = 5.831\angle30° \ 58'$ (see *Problems 1 and 2*).

Applying de Moivre's theorem:

$(5+j3)^{\frac{1}{2}} = 5.831^{\frac{1}{2}}\Big\angle \dfrac{1}{2} \times (30° \ 58') = 2.415\angle15° \ 29'.$

The second root may be obtained as shown in para. 5, i.e., having the same modulus but displaced $(360/2)°$ from the first root.

Thus, $(5+j3)^{\frac{1}{2}} = 2.415\angle(15° \ 29'+180°) = 2.415\angle195° \ 29'.$

In rectangular form:

$2.415\angle15° \ 29' = 2.415 \cos (15° \ 29')+j2.415 \sin (15° \ 29') = 2.327+j0.6447$
and $2.415\angle195° \ 29' = 2.415 \cos (195° \ 29')+j2.415 \sin (195° \ 29')$
$\qquad\qquad\qquad\qquad = -2.327-j0.6447$

[An alternative method of obtaining the second root is to express $(5+j3)$ as $5.831\angle30° \ 58'$ and also as $5.831\angle(360°+30° \ 58')$, i.e. $5.831\angle390° \ 58'$. It is necessary to do this since when applying de Moivre's theorem, the second angle of $390° \ 58'$, when multiplied by $\dfrac{1}{2}$, lies between 0 and 360°.

Thus $(5+j3)^{\frac{1}{2}}$ is also equal to $5.831^{\frac{1}{2}}\Big\angle (\dfrac{1}{2} \times 390° \ 58') = 2.415\angle195° \ 29'$ as obtained previously.]

*Problem 4* Express the roots of $(-14+j3)^{-2/5}$ in polar form.

$(-14+j3) \quad = 14.32\angle167° \ 54'$

$(-14+j3)^{-2/5} = 14.32^{-2/5}\Big\angle \Big[(-\dfrac{2}{5}) \times 167° \ 54'\Big]$
$\qquad\qquad\quad = 0.3448\angle-67° \ 10'$

There are five roots to this complex number, $\left(x^{-2/5} = \dfrac{1}{x^{2/5}} = \dfrac{1}{\sqrt[5]{x^2}}\right).$

The roots are symmetrically displaced from one another $(360/5)°$, i.e. $72°$ round an Argand diagram.

Thus the required roots are $0.3448\angle{-67°}\ 10'$, $0.3448\angle 4°\ 50'$, $0.3448\angle 76°\ 50'$, $0.3448\angle 148°\ 50'$ and $0.3448\angle 220°\ 50'$.

**Problem 5** Express $\sin 3A$ in terms of powers of $\sin A$.

From para. 6, equation (8),

$$\sin n\theta = n\cos^{n-1}\theta\ \sin\theta - \frac{n(n-1)(n-2)}{3!}\cos^{n-3}\theta\ \sin^3\theta + \ldots\ldots$$

Substituting 3 for $n$ and $A$ for $\theta$ gives:

$$\sin 3A = 3\cos^2 A\ \sin A - \frac{3(2)(1)}{3(2)(1)}\cos^0 A\ \sin^3 A.$$

(The next term is $+\dfrac{n(n-1)(n-2)(n-3)(n-4)}{5!}\cos^{n-5}A\ \sin^5 A$

$$= +\frac{(3)(2)(1)(0)(-1)}{5(4)(3)(2)(1)}\ \ldots\ldots$$

which is equal to 0 due to the 0 in the numerator. Thus all subsequent terms are zero.)

Thus $\sin 3A = 3\sin A\ \cos^2 A - \sin^3 A = \sin A\ (3\cos^2 A - \sin^2 A)$
Using the identity $\sin^2 A + \cos^2 A = 1$, $\cos^2 A = 1 - \sin^2 A$, thus
$\sin 3A = \sin A[3(1-\sin^2 A) - \sin^2 A] = \sin A\ (3-4\sin^2 A) = \mathbf{3\sin A - 4\sin^3 A}$

**Problem 6** Express $\cos 4B$ in terms of powers of $\cos B$.

From para. 6, equation (7), $\cos n\theta = \cos^n\theta - \dfrac{n(n-1)}{2!}\cos^{n-2}\theta\ \sin^2\theta + \ldots\ldots$

Substituting 4 for $n$ and $B$ for $\theta$ gives:

$$\cos 4B = \cos^4 B - \frac{(4)(3)}{(2)(1)}\cos^2 B\ \sin^2 B + \frac{(4)(3)(2)(1)}{(4)(3)(2)(1)}\cos^0 B\ \sin^4 B$$

$$= \cos^4 B - 6\cos^2 B\ \sin^2 B + \sin^4 B$$

Using the identity $\sin^2 B + \cos^2 B = 1$, $\sin^2 B = 1 - \cos^2 B$, thus
$\cos 4B = \cos^4 B - 6\cos^2 B(1-\cos^2 B) + (1-\cos^2 B)^2$
$\quad\quad\ = \cos^4 B - 6\cos^2 B + 6\cos^4 B + 1 - 2\cos^2 B + \cos^4 B$
$\quad\quad\ = \mathbf{8\cos^4 B - 8\cos^2 B + 1.}$

**Problem 7** Express $\sin^2 C$ in terms of cosines of multiples of $C$.

From para. 7, equation (12), $j^n 2^n\ \sin^n\theta = \left(z - \dfrac{1}{z}\right)^n$

When $n = 2$, $j^2 2^2\ \sin^2 C = \left(z - \dfrac{1}{z}\right)^2 = z^2 - 2 + \dfrac{1}{z^2} = \left(z^2 + \dfrac{1}{z^2}\right) - 2$

From equation (13), when $n = 2$, $\left(z^2 + \dfrac{1}{z^2}\right) = 2\cos 2C$

Hence $j^2 2^2\ \sin^2 C = 2\cos 2C - 2$, and since $j^2 = -1$,
$\quad\quad -4\sin^2 C\ = 2\cos 2C - 2$

$$\sin^2 C\ = \frac{1}{4}(2 - 2\cos 2C), \text{ i.e. } \sin^2 C = \frac{1}{2}(1 - \cos 2C)$$

*Problem 8* Express $\cos^5 \theta$ in terms of cosines of multiples of $\theta$.

From para. 7, equation (11), $2^n \cos^n \theta = \left(z + \dfrac{1}{z}\right)^n$

Expanding $\left(z + \dfrac{1}{z}\right)^5$ by the binomial theorem gives:

$$\left(z + \frac{1}{z}\right)^5 = z^5 + 5z^4 \cdot \frac{1}{z} + 10z^3 \cdot \frac{1}{z^2} + 10z^2 \cdot \frac{1}{z^3} + 5z \cdot \frac{1}{z^4} + \frac{1}{z^5}$$

$$= z^5 + 5z^3 + 10z + \frac{10}{z} + \frac{5}{z^3} + \frac{1}{z^5}$$

$$= \left(z^5 + \frac{1}{z^5}\right) + 5\left(z^3 + \frac{1}{z^3}\right) + 10\left(z + \frac{1}{z}\right)$$

But from para. 7, equation (13), $z^n + \dfrac{1}{z^n} = 2 \cos n\theta$

Hence $\left(z + \dfrac{1}{z}\right)^5 = 2 \cos 5\theta + 5(2 \cos 3\theta) + 10(2 \cos \theta)$

Thus $2^5 \cos^5 \theta = 2 \cos 5\theta + 10 \cos 3\theta + 20 \cos \theta$

i.e. $\cos^5 \theta = \dfrac{1}{2^4} (\cos 5\theta + 5 \cos 3\theta + 10 \cos \theta)$

*Problem 9* Express $(\sin^2 \theta \cos^5 \theta)$ in terms of cosines of multiples of $\theta$.

From the results of *Problems 7 and 8*:

$$\sin^2 \theta \cos^5 \theta = \left[\frac{1}{2}(1 - \cos 2\theta)\right]\left[\frac{1}{16}(\cos 5\theta + 5 \cos 3\theta + 10 \cos \theta)\right]$$

$$= \frac{1}{32}[\cos 5\theta + 5 \cos 3\theta + 10 \cos \theta - \cos 5\theta \cos 2\theta$$
$$- 5 \cos 3\theta \cos 2\theta - 10 \cos 2\theta \cos \theta]$$

However, from compound angle formulae, the product of two cosines may be expressed as a sum of two cosines,

i.e., $\cos A \cos B = \dfrac{1}{2}[\cos (A+B) + \cos(A-B)]$

Hence $(\cos 5\theta \cos 2\theta) = \dfrac{1}{2}(\cos 7\theta + \cos 3\theta)$,

$\cos 3\theta \cos 2\theta = \dfrac{1}{2}(\cos 5\theta + \cos \theta)$ and

$\cos 2\theta \cos \theta = \dfrac{1}{2}(\cos 3\theta + \cos \theta)$.

Thus $\sin^2 \theta \cos^5 \theta = \dfrac{1}{32}[\cos 5\theta + 5 \cos 3\theta + 10 \cos \theta - \dfrac{1}{2} \cos 7\theta - \dfrac{1}{2} \cos 3\theta$

$- \dfrac{5}{2} \cos 5\theta - \dfrac{5}{2} \cos \theta - 5 \cos 3\theta - 5 \cos \theta]$

$= \dfrac{1}{32}\left[-\dfrac{1}{2} \cos 7\theta - \dfrac{3}{2} \cos 5\theta - \dfrac{1}{2} \cos 3\theta + \dfrac{5}{2} \cos \theta\right]$

$= -\dfrac{1}{64}[\cos 7\theta + 3 \cos 5\theta + \cos 3\theta - 5 \cos \theta]$

## C. FURTHER PROBLEMS ON DE MOIVRE'S THEOREM

In *Problems 1 to 5*, express in both polar and rectangular forms.

1 $(6+j5)^3$          [476.4∠119° 25′, −234+j415]

2 $(3-j8)^5$       [45 530∠12° 47′, 44 400+j10 070]

3 $(-2+j7)^4$       [2809∠63° 47′, 1241+j2520]

4 $(-16-j9)^6$    [(38.27 × 10⁶)∠176° 9′, 10⁶ (−38.18+j2.570)]

5 $(-2+j2)^{10}$       [32 770∠270°, −j32 770]

In *Problems 6 to 10*, determine the modulii and arguments of the complex roots.

6 $(3+j4)^{1/3}$   [Modulii 1.710, arguments 17° 43′, 137° 43′ and 257° 43′]

7 $(-2+j)^{1/4}$   [Modulii 1.223, arguments 38° 22′, 128° 22′, 218° 22′ and 308° 22′]

8 $(-6-j5)^{1/2}$      [Modulii 2.795, arguments 109° 54′, 289° 54′]

9 $(4-j3)^{-2/3}$   [Modulii 0.3420, arguments 24° 35′, 144° 35′ and 264° 35′]

10 $(-2-j2)^{-3/4}$

[Modulii 0.4585, arguments 11° 15′, 101° 15′, 191° 15′ and 281° 15′]

11 Express $\cos 5\theta$ in terms of powers of $\cos\theta$ and $\sin\theta$.

$$[\cos^5\theta - 10\cos^3\theta\,\sin^2\theta + 5\cos\theta\,\sin^4\theta]$$

12 Express $\cos 8\theta$ in terms of powers of $\sin\theta$ and $\cos\theta$.

$$\begin{bmatrix}\cos^8\theta - 28\cos^6\theta\,\sin^2\theta + 70\cos^4\theta\,\sin^4\theta \\ -28\cos^2\theta\,\sin^6\theta + \sin^8\theta\end{bmatrix}$$

13 Express $\sin 9\theta$ in terms of powers of $\cos\theta$ and $\sin\theta$.

$$\begin{bmatrix}9\cos^8\theta\,\sin\theta - 84\cos^6\theta\,\sin^3\theta + 126\cos^4\theta\,\sin^5\theta \\ -36\cos^2\theta\,\sin^7\theta + \sin^9\theta\end{bmatrix}$$

14 Express $\sin 7\theta$ in terms of powers of $\sin\theta$.

$$[7\sin\theta - 56\sin^3\theta + 112\sin^5\theta - 64\sin^7\theta]$$

15 Express $\dfrac{\sin 6\theta}{\sin\theta}$ in terms of powers of $\cos\theta$.    $[32\cos^5\theta - 32\cos^3\theta + 6\cos\theta]$

16 Express $\sin^3\theta$ in terms of sines of multiples of $\theta$.     $\left[\dfrac{3}{4}\sin\theta - \dfrac{1}{4}\sin 3\theta\right]$

17 Express $\cos^3\theta$ in terms of cosines of multiples of $\theta$.     $\left[\dfrac{1}{4}\cos 3\theta + \dfrac{3}{4}\cos\theta\right]$

18 Express $\cos^6\theta$ in terms of cosines of multiples of $\theta$.

$$\left[\frac{1}{2^5}\,(\cos 6\theta + 6\cos 4\theta + 15\cos 2\theta + 10)\right]$$

19 Express $\sin^8\theta$ in terms of cosines of multiples of $\theta$.

$$\left[\frac{1}{2^7}\,(\cos 8\theta - 8\cos 6\theta + 28\cos 4\theta - 56\cos 2\theta + 35)\right]$$

20 Express $(\sin^4\theta)(\cos^2\theta)$ in terms of cosines of multiples of $\theta$.

$$\left[\frac{1}{2^4}(\frac{1}{2}\cos 6\theta - \cos 4\theta - \frac{1}{2}\cos 2\theta + 1)\right]$$

# 4 The relationship between trigonometric and hyperbolic functions

## A. MAIN POINTS CONCERNED WITH THE RELATIONSHIP BETWEEN TRIGONOMETRIC AND HYPERBOLIC FUNCTIONS

1   In chapter 3, it is shown that

$$\cos\theta + j\sin\theta = e^{j\theta} \tag{1}$$

and $\quad \cos\theta - j\sin\theta = e^{-j\theta} \tag{2}$

Adding equations (1) and (2) gives:

$$\cos\theta = \frac{1}{2}(e^{j\theta} + e^{-j\theta}) \tag{3}$$

Subtracting equation (2) from equation (1) gives:

$$\sin\theta = \frac{1}{2j}(e^{j\theta} - e^{-j\theta}) \tag{4}$$

2   Substituting $j\theta$ for $\theta$ in equations (3) and (4) gives:

$\cos j\theta = \frac{1}{2}(e^{j(j\theta)} + e^{-j(j\theta)})$ and

$\sin j\theta = \frac{1}{2j}(e^{j(j\theta)} - e^{-j(j\theta)})$.

Since $j^2 = -1$, $\cos j\theta = \frac{1}{2}(e^{-\theta} + e^{\theta}) = \frac{1}{2}(e^{\theta} + e^{-\theta})$

Hence from chapter 2, $\cos j\theta = \cosh\theta \tag{5}$

Similarly, $\sin j\theta = \frac{1}{2j}(e^{-\theta} - e^{\theta}) = -\frac{1}{2j}(e^{\theta} - e^{-\theta})$

$\qquad\qquad = \frac{-1}{j}\left[\frac{1}{2}(e^{\theta} - e^{-\theta})\right] = -\frac{1}{j}\sinh\theta$, (see chapter 2).

But $-\frac{1}{j} = -\frac{1}{j} \times \frac{j}{j} = -\frac{j}{j^2} = j$, hence $\sin j\theta = j\sinh\theta \tag{6}$

Equations (5) and (6) may be used to verify that in all standard trigonometric identities, $j\theta$ may be written for $\theta$ and the identity still remains true. (See *Problems 1 and 2*)

3   From chapter 2, $\cosh\theta = \frac{1}{2}(e^{\theta} + e^{-\theta})$

Substituting $j\theta$ for $\theta$ gives:

$\cosh j\theta = \frac{1}{2}(e^{j\theta} + e^{-j\theta}) = \cos\theta$, from equation (3), i.e., $\cosh j\theta = \cos \theta$            (7)

Similarly, from chapter 2, $\sinh \theta = \frac{1}{2}(e^\theta - e^{-\theta})$

Substituting $j\theta$ for $\theta$ gives:

$\sinh j\theta = \frac{1}{2}(e^{j\theta} - e^{-j\theta}) = j \sin \theta$, from equation (4).

Hence $\sinh j\theta = j \sin \theta$          (8)

4    $\text{Tan } j\theta = \dfrac{\sin j\theta}{\cos j\theta}$

From equations (5) and (6), $\dfrac{\sin j\theta}{\cos j\theta} = \dfrac{j \sinh \theta}{\cosh \theta} = j \tanh \theta$.

Hence $\tan j\theta = j \tanh \theta$         (9)

Similarly, $\tanh j\theta = \dfrac{\sinh j\theta}{\cosh j\theta}$.

From equations (7) and (8), $\dfrac{\sinh j\theta}{\cosh j\theta} = \dfrac{j \sin \theta}{\cos \theta} = j \tan \theta$

Hence $\tanh j\theta = j \tan \theta$         (10)

5    Two methods are commonly used to verify hyperbolic identities. These are (a) by substituting $j\theta$ (and $j\phi$) in the corresponding trigonometric identity and using the relationships given in equations (5) to (10), (see *Problems 3 to 5*), and (b) by applying Osborne's rule given in chapter 2, page 10.

## B. WORKED PROBLEMS ON THE RELATIONSHIP BETWEEN TRIGONOMETRIC AND HYPERBOLIC FUNCTIONS

*Problem 1*   Verify that $\cos^2 j\theta + \sin^2 j\theta = 1$.

From para. 2, equation (5), $\cos j\theta = \cosh \theta$, and
from equation (6), $\sin j\theta = j \sinh \theta$.
Thus, $\cos^2 j\theta + \sin^2 j\theta = \cosh^2 \theta + j^2 \sinh^2 \theta$, and since $j^2 = -1$,
$\cos^2 j\theta + \sin^2 j\theta = \cosh^2 \theta - \sinh^2 \theta$.
But from chapter 2, *Problem 7*, page 14,
$\cosh^2 \theta - \sinh^2 \theta = 1$, hence, $\cos^2 j\theta + \sin^2 j\theta = 1$.

*Problem 2*   Verify that $\sin j\, 2A = 2 \sin jA \cos jA$.

From para. 2, equation (6), writing $2A$ for $\theta$, $\sin j\, 2A = j \sinh 2A$, and from chapter 2, *Table 1*, page 10, $\sinh 2A = 2 \sinh A \cosh A$.
Hence, $\sin j\, 2A = j (2 \sinh A \cosh A)$

But, $\sinh A = \frac{1}{2}(e^A - e^{-A})$ and $\cosh A = \frac{1}{2}(e^A + e^{-A})$.

Hence, $\sin j\, 2A = j^2 \left(\dfrac{e^A - e^{-A}}{2}\right)\left(\dfrac{e^A + e^{-A}}{2}\right) = -\dfrac{2}{j}\left(\dfrac{e^A - e^{-A}}{2}\right)\left(\dfrac{e^A + e^{-A}}{2}\right)$

       $= 2 \sin j A \cos j A$, from para. 2, i.e. $\sin j\, 2A = 2 \sin j A \cos j A$

*Problem 3* By writing $j\,A$ for $\theta$ in $\cot^2\theta+1=\operatorname{cosec}^2\theta$, determine the corresponding hyperbolic identity.

Substituting $j\,A$ for $\theta$ gives:

$\cot^2 j\,A+1=\operatorname{cosec}^2 j\,A$, i.e. $\dfrac{\cos^2 j\,A}{\sin^2 j\,A}+1=\dfrac{1}{\sin^2 j\,A}$

But from equation (5), $\cos j\,A=\cosh A$ and from equation (6), $\sin j\,A=j\sinh A$.

Hence $\dfrac{\cosh^2 A}{j^2\sinh^2 A}+1=\dfrac{1}{j^2\sinh^2 A}$

and since $j^2=-1$, $\quad -\dfrac{\cosh^2 A}{\sinh^2 A}+1=-\dfrac{1}{\sinh^2 A}$

Multiplying throughout by $-1$, gives:

$\dfrac{\cosh^2 A}{\sinh^2 A}-1=\dfrac{1}{\sinh^2 A}$, i.e. $\coth^2 A-1=\operatorname{cosech}^2 A$.

*Problem 4* By substituting $j\,A$ and $j\,B$ for $\theta$ and $\phi$ respectively in the trigonometric identity for $\cos\theta-\cos\phi$, show that
$$\cosh A-\cosh B=2\sinh\left(\frac{A+B}{2}\right)\sinh\left(\frac{A-B}{2}\right)$$

$\cos\theta-\cos\phi=-2\sin\left(\dfrac{\theta+\phi}{2}\right)\sin\left(\dfrac{\theta-\phi}{2}\right)$, thus

$\cos j\,A-\cos j\,B=-2\sin j\left(\dfrac{A+B}{2}\right)\sin j\left(\dfrac{A-B}{2}\right)$.

But from equation (5), $\cos j\,A=\cosh A$ and from equation (6), $\sin j\,A=j\sinh A$.

Hence, $\cosh A-\cosh B=-2j\sinh\left(\dfrac{A+B}{2}\right)j\sinh\left(\dfrac{A-B}{2}\right)$

$\qquad\qquad\qquad = -2j^2\sinh\left(\dfrac{A+B}{2}\right)\sinh\left(\dfrac{A-B}{2}\right)$

But $j^2=-1$, hence $\cosh A-\cosh B=2\sinh\left(\dfrac{A+B}{2}\right)\sinh\left(\dfrac{A-B}{2}\right)$

*Problem 5* Develop the hyperbolic identity corresponding to $\sin 3\theta=3\sin\theta-4\sin^3\theta$, by writing $j\,A$ for $\theta$.

Substituting $j\,A$ for $\theta$ gives: $\sin 3j\,A=3\sin j\,A-4\sin^3 j\,A$
and since from equation (6), $\sin j\,A=j\sinh A$, $j\sinh 3A=3j\sinh A-4j^3\sinh^3 A$
Dividing throughout by $j$ gives: $\sinh 3A=3\sinh A-j^2\,4\sinh^3 A$
But $j^2=-1$, hence $\sinh 3A=3\sinh A+4\sinh^3 A$.

[An examination of *Problems 3 to 5* shows that whenever the trigonometric identity contains a term which is the product of two sines, or the implied product of two sines (e.g., $\tan^2\theta=\sin^2\theta/\cos^2\theta$, thus $\tan^2\theta$ is the implied product of two sines), the sign of the corresponding term in the hyperbolic function changes. This relationship between trigonometric and hyperbolic functions is known as Osborne's rule, as discussed in chapter 2, page 10.]

## C. FURTHER PROBLEMS ON THE RELATIONSHIP BETWEEN TRIGONOMETRIC AND HYPERBOLIC FUNCTIONS

In *Problems 1 to 5*, verify the identities given by expressing in exponential form.

1  $\sin j (A+B) = \sin jA \cos jB + \cos jA \sin jB$.

2  $\cos j (A-B) = \cos jA \cos jB - \sin jA \sin jB$.

3  $\cos j 2A = 1 - 2 \sin^2 jA$.

4  $\sin jA \cos jB = \frac{1}{2} [\sin j (A+B) + \sin j (A-B)]$.

5  $\sin jA - \sin jB = 2 \cos j \left(\frac{A+B}{2}\right) \sin j \left(\frac{A-B}{2}\right)$.

In *Problems 6 to 15*, use the substitution $A = j\theta$ (and $B = j\phi$) to obtain the hyperbolic identities corresponding to the trigonometric identities given.

6  $1 + \tan^2 A = \sec^2 A$.　　　　　　　　　　　$[1 - \tanh^2 \theta = \text{sech}^2 \theta]$

7  $\cos (A+B) = \cos A \cos B - \sin A \sin B$.

$[\cosh (\theta+\phi) = \cosh \theta \cosh \phi + \sinh \theta \sinh \phi]$

8  $\sin (A-B) = \sin A \cos B - \cos A \sin B$.

$[\sinh (\theta-\phi) = \sinh \theta \cosh \phi - \cosh \theta \sinh \phi]$

9  $\tan 2A = \frac{2 \tan A}{1 - \tan^2 A}$.　　　　　　　　$\left[\tanh 2\theta = \frac{2 \tanh \theta}{1 + \tanh^2 \theta}\right]$

10  $\cos A \sin B = \frac{1}{2} [\sin (A+B) - \sin (A-B)]$.

$\left[\cosh \theta \cosh \phi = \frac{1}{2} [\sinh (\theta+\phi) - \sinh (\theta-\phi)]\right]$

11  $\cos A - \cos B = -2 \sin \left(\frac{A+B}{2}\right) \sin \left(\frac{A-B}{2}\right)$.

$\left[\cosh \theta - \cosh \phi = 2 \sinh \left(\frac{\theta+\phi}{2}\right) \sinh \left(\frac{\theta-\phi}{2}\right)\right]$

12  $\sin^3 A = \frac{3}{4} \sin A - \frac{1}{4} \sin 3A$.　　　$\left[\sinh^3 \theta = \frac{1}{4} \sinh 3\theta - \frac{3}{4} \sinh \theta\right]$

13  $(\sin A + \cos A)^2 - 1 = 2 \sin A \cos A$.　　$[(\sinh \theta + \cosh \theta)^2 - 1 = 2 \sinh \theta \cosh \theta]$

14  $\cot^2 A (\sec^2 A - 1) = 1$.　　　　　　　　$[\coth^2 \theta (1 - \text{sech}^2 \theta) = 1]$

15  $\sec A (\cot A - 1) + \text{cosec } A (1 - \tan A) = 2 (\text{cosec } A - \sec A)$

$[\text{sech } \theta (\coth \theta - 1) + \text{cosech } \theta (1 - \tanh \theta) = 2 (\text{cosech } \theta - \text{sech } \theta)]$

# 5 Differentiation of implicit functions

## A. MAIN POINTS CONCERNED WITH THE DIFFERENTIATION OF IMPLICIT FUNCTIONS

1  When an equation can be written in the form $y = f(x)$ it is said to be an **explicit function** of $x$. Examples of explicit functions include $y = 2x^3 - 3x + 4$, $y = 2x \ln x$ and $y = \dfrac{3e^x}{\cos x}$. In these examples $y$ may be differentiated with respect to $x$ by using standard derivatives, the product rule and the quotient rule of differentiation respectively.

2  Sometimes with equations involving, say, $y$ and $x$, it is impossible to make $y$ the subject of the formula. The equation is then called an **implicit function** and examples of such functions include $y^3 + 2x^2 = y^2 - x$ and $\sin y = x^2 + 2xy$.

3  It is possible to **differentiate an implicit function** by using the **function of a function rule** (often called the chain rule), which may be stated as
$$\frac{du}{dx} = \frac{du}{dy} \times \frac{dy}{dx} \ .$$
Thus, to differentiate $y^3$ with respect to $x$, the substitution $u = y^3$ is made, from which, $\dfrac{du}{dy} = 3y^2$. Hence, $\dfrac{d}{dx}(y^3) = (3y^2) \times \dfrac{dy}{dx}$, by the function of a function rule.

4  A simple rule for differentiating an implicit function is summarised as:
$$\frac{d}{dx}[f(y)] = \frac{d}{dy}[f(y)] \times \frac{dy}{dx} \tag{1}$$
(See *Problems 1 and 2*)

5  The product and quotient rules of differentiation must be applied when differentiating functions containing products and quotients of two variables.

For example, $\dfrac{d}{dx}(x^2 y) = (x^2)\dfrac{d}{dx}(y) + (y)\dfrac{d}{dx}(x^2)$, by the product rule

$\qquad\qquad\qquad = (x^2)\left(1\dfrac{dy}{dx}\right) + y(2x)$, by using equation (1)

$\qquad\qquad\qquad = x^2\dfrac{dy}{dx} + 2xy.$

(See *Problems 3 to 6*)

33

6     An implicit function such as $3x^2 + y^2 - 5x + y = 2$, may be differentiated term by term with respect to $x$. This gives:

$$\frac{d}{dx}(3x^2) + \frac{d}{dx}(y^2) - \frac{d}{dx}(5x) + \frac{d}{dx}(y) = \frac{d}{dx}(2)$$

i.e., $6x + 2y\frac{dy}{dx} - 5 + 1\frac{dy}{dx} = 0$, using equation (1) and standard derivatives.

An expression for the derivative $\frac{dy}{dx}$ in terms of $x$ and $y$ may be obtained by rearranging this latter equation. Thus: $(2y+1)\frac{dy}{dx} = 5 - 6x$

from which, $\dfrac{dy}{dx} = \dfrac{5-6x}{2y+1}$ .

(See *Problems 7 to 13*)

## B.   WORKED PROBLEMS ON DIFFERENTIATING IMPLICIT FUNCTIONS

*Problem 1*   Differentiate the following functions with respect to $x$: (a) $2y^4$; (b) $\sin 3t$.

(a) Let $u = 2y^4$, then, by the function of a function rule:

$$\frac{du}{dx} = \frac{du}{dy} \times \frac{dy}{dx} = \frac{d}{dy}(2y^4) \times \frac{dy}{dx} = 8y^3\frac{dy}{dx}$$

(b) Let $u = 3\sin 3t$, then, by the function of a function rule:

$$\frac{du}{dx} = \frac{du}{dt} \times \frac{dt}{dx} = \frac{d}{dt}(\sin 3t) \times \frac{dt}{dx} = 3\cos 3t\frac{dt}{dx}.$$

*Problem 2*   Differentiate the following functions with respect to $x$: (a) $4\ln 5y$; (b) $\frac{1}{5}e^{3\theta-2}$

(a) Let $u = 4\ln 5y$, then, by the function of a function rule:

$$\frac{du}{dx} = \frac{du}{dy} \times \frac{dy}{dx} = \frac{d}{dy}(4\ln 5y) \times \frac{dy}{dx} = \frac{4}{y}\frac{dy}{dx}$$

(b) Let $u = \frac{1}{5}e^{3\theta-2}$, then, by the function of a function rule:

$$\frac{du}{dx} = \frac{du}{d\theta} \times \frac{d\theta}{dx} = \frac{d}{d\theta}\left(\frac{1}{5}e^{3\theta-2}\right) \times \frac{d\theta}{dx} = \frac{3}{5}e^{3\theta-2}\frac{d\theta}{dx}$$

*Problem 3*   Determine $\frac{d}{dx}(2x^3y^2)$.

In the product rule of differentiation let $u = 2x^3$ and $v = y^2$.

Thus $\frac{d}{dx}(2x^3y^2) = (2x^3)\frac{d}{dx}(y^2) + (y^2)\frac{d}{dx}(2x^3)$

$$= (2x^3)(2y\frac{dy}{dx}) + (y^2)(6x^2)$$

$$= 4x^3y\frac{dy}{dx} + 6x^2y^2 = 2x^2y\left(2x\frac{dy}{dx} + 3y\right)$$

*Problem 4* Find $\dfrac{d}{dx}\left(\dfrac{3y}{2x}\right)$.

In the quotient rule of differentiation let $u = 3y$ and $v = 2x$.

Thus $\dfrac{d}{dx}\left(\dfrac{3y}{2x}\right) = \dfrac{(2x)\dfrac{d}{dx}(3y)-(3y)\dfrac{d}{dx}(2x)}{(2x)^2}$

$= \dfrac{(2x)(3\dfrac{dy}{dx})-(3y)(2)}{4x^2} = \dfrac{6x\dfrac{dy}{dx}-6y}{4x^2} = \dfrac{3}{2x^2}(x\dfrac{dy}{dx}-y)$

*Problem 5* Given $z = 3y^2 \sin 2x$ find $\dfrac{dz}{dx}$.

Using the product rule of differentiation, $\dfrac{dz}{dx} = (3y^2)(2\cos 2x)+(\sin 2x)(6y\dfrac{dy}{dx})$

$= 6y^2 \cos 2x+6y \sin 2x \dfrac{dy}{dx} = \mathbf{6y\ (y\ cos\ 2x+sin\ 2x\dfrac{dy}{dx})}$

*Problem 6* Differentiate $z = x^2+3x \cos 3y$ with respect to $y$.

$\dfrac{dz}{dy} = \dfrac{d}{dy}(x^2)+\dfrac{d}{dy}(3x \cos 3y) = 2x\dfrac{dx}{dy}+[(3x)(-3 \sin 3y)+(\cos 3y)(3\dfrac{dx}{dy})]$

$= \mathbf{2x\dfrac{dx}{dy}-9x\ sin\ 3y+3\ cos\ 3y\dfrac{dx}{dy}}$.

*Problem 7* Given $2y^2-5x^4-2-7y^3 = 0$, determine $\dfrac{dy}{dx}$.

Each term in turn is differentiated with respect to $x$:

Hence $\dfrac{d}{dx}(2y^2)-\dfrac{d}{dx}(5x^4)-\dfrac{d}{dx}(2)-\dfrac{d}{dx}(7y^3) = \dfrac{d}{dx}(0)$

i.e. $4y\dfrac{dy}{dx}-20x^3-0-21y^2\dfrac{dy}{dx} = 0$

Rearranging gives: $(4y-21y^2)\dfrac{dy}{dx} = 20x^3$, i.e. $\mathbf{\dfrac{dy}{dx} = \dfrac{20x^3}{(4y-21y^2)}}$

*Problem 8* Find $\dfrac{dy}{dx}$ given $\sin 2y^2 = 4x^3$.

Differentiating each term in turn with respect to $x$ gives:

$\dfrac{d}{dx}(\sin 2y^2) = \dfrac{d}{dx}(4x^3)$, i.e., $(4y \cos 2y^2)\dfrac{dy}{dx} = 12x^2$.

Hence $\dfrac{dy}{dx} = \dfrac{12x^2}{4y \cos 2y^2} = \mathbf{\dfrac{3x^2}{y\ cos\ 2y^2}}$

*Problem 9* Determine the values of $\dfrac{dy}{dx}$ when $x = 4$ given that $x^2+y^2 = 25$.

Differentiating each term in turn with respect to $x$ gives:

$\frac{d}{dx}(x^2)+\frac{d}{dx}(y^2) = \frac{d}{dx}(25)$ i.e. $2x+2y\frac{dy}{dx} = 0$

Hence $\frac{dy}{dx} = -\frac{2x}{2y} = -\frac{x}{y}$

Since $x^2+y^2 = 25$, when $x = 4$, $y = \sqrt{(25-4^2)} = \pm3$.

Thus when $x = 4$ and $y = \pm3$, $\frac{dy}{dx} = \frac{-4}{\pm3} = \pm\frac{4}{3}$.

Problem 10 (a) Find $\frac{dy}{dx}$ in terms of $x$ and $y$ given $4x^2+2xy^3-5y^2 = 0$.

(b) Evaluate $\frac{dy}{dx}$ when $x = 1$ and $y = 2$.

(a) Differentiating each term in turn with respect to $x$ gives:

$\frac{d}{dx}(4x^2) +\frac{d}{dx}(2xy^3)-\frac{d}{dx}(5y^2) = \frac{d}{dx}(0)$

i.e. $8x+\left[(2x)(3y^2\frac{dy}{dx})+(y^3)(2)\right]-10y\frac{dy}{dx} = 0$

i.e. $8x+6xy^2\frac{dy}{dx}+2y^3-10y\frac{dy}{dx} = 0$

Rearranging gives: $8x+2y^3 = (10y-6xy^2)\frac{dy}{dx}$

and $\frac{dy}{dx} = \frac{8x+2y^3}{10y-6xy^2} = \frac{4x+y^3}{y(5-3xy)}$

(b) When $x = 1$ and $y = 2$, $\frac{dy}{dx} = \frac{4(1)+(2)^3}{2[5-(3)(1)(2)]} = \frac{12}{-2} = -6$

Problem 11 Given $2x^2(\sin t-4x)= \cos t$, find the value of $\frac{dx}{dt}$ when $t = \pi$ radians.

$2x^2(\sin t-4x) = \cos t$. Hence $2x^2\sin t-8x^3 = \cos t$.
Differentiating each term in turn with respect to $t$ gives:

$\frac{d}{dt}(2x^2\sin t) - \frac{d}{dt}(8x^3) = \frac{d}{dt}(\cos t)$

i.e. $\left[(2x^2)(\cos t)+(\sin t)(4x\frac{dx}{dt})\right]-24x^2\frac{dx}{dt} = -\sin t$

Hence $2x^2\cos t+4x\sin t\frac{dx}{dt}-24x^2\frac{dx}{dt} = -\sin t$.

Rearranging gives: $2x^2\cos t+\sin t = \frac{dx}{dt}(24x^2-4x\sin t)$

i.e. $\frac{dx}{dt} = \frac{2x^2\cos t+\sin t}{24x^2-4x\sin t}$

To evaluate $\frac{dx}{dt}$, the value of $x$ when $t = \pi$ radians is required. Substituting $t = \pi$

in the original equation gives: $2x^2(\sin \pi-4x) = \cos \pi$.
$\sin \pi = 0$ and $\cos \pi = -1$, hence $2x^2(-4x) = -1$
from which $-8x^3 = -1$

$x^3 = \frac{1}{8}$ and $x = \frac{1}{2}$

Substituting $x = \frac{1}{2}$ and $t = \pi$ into the expression for $\frac{dx}{dt}$ gives:

$$\frac{dx}{dt} = \frac{2(\frac{1}{2})^2 \cos \pi + \sin \pi}{24(\frac{1}{2})^2 - 4(\frac{1}{2}) \sin \pi} = \frac{-\frac{1}{2}}{6} = -\frac{1}{12}$$

*Problem 12* Find the gradients of the tangents drawn to the circle $x^2 + y^2 - 2x - 2y = 3$ at $x = 2$.

The gradient of the tangent is given by $\frac{dy}{dx}$.

Differentiating each term in turn with respect to $x$ gives:

$$\frac{d}{dx}(x^2) + \frac{d}{dx}(y^2) - \frac{d}{dx}(2x) - \frac{d}{dx}(2y) = \frac{d}{dx}(3)$$

i.e. $2x + 2y \frac{dy}{dx} - 2 - 2\frac{dy}{dx} = 0$

Hence $(2y-2)\frac{dy}{dx} = 2 - 2x$, from which $\frac{dy}{dx} = \frac{2-2x}{2y-2} = \frac{1-x}{y-1}$

The value of $y$ when $x = 2$ is determined from the original equation.
Hence $(2)^2 + y^2 - 2(2) - 2y = 3$
i.e. $4 + y^2 - 4 - 2y = 3$ or $y^2 - 2y - 3 = 0$.
Factorising gives: $(y+1)(y-3) = 0$, from which $y = -1$ or $y = 3$

When $x = 2$ and $y = -1$, $\frac{dy}{dx} = \frac{1-x}{y-1} = \frac{1-2}{-1-1} = \frac{-1}{-2} = \frac{1}{2}$

When $x = 2$ and $y = 3$, $\frac{dy}{dx} = \frac{1-2}{3-1} = -\frac{1}{2}$

**Hence the gradients of the tangents are $\pm \frac{1}{2}$.**

*Problem 13* Pressure $p$ and volume $v$ of a gas are related by the law $pv^\gamma = k$, where $\gamma$ and $k$ are constants. Show that the rate of change of pressure $\frac{dp}{dt} = -\gamma \frac{p}{v} \frac{dv}{dt}$.

Since $pv^\gamma = k$, then $p = \frac{k}{v^\gamma} = kv^{-\gamma}$.

$\frac{dp}{dt} = \frac{dp}{dv} \times \frac{dv}{dt}$ by the function of a function rule.

$\frac{dp}{dv} = \frac{d}{dv}(kv^{-\gamma}) = -\gamma kv^{-\gamma-1} = \frac{-\gamma k}{v^{\gamma+1}}$

$\frac{dp}{dt} = \frac{-\gamma k}{v^{\gamma+1}} \times \frac{dv}{dt}$

Since $k = pv^\gamma$, then $\frac{dp}{dt} = \frac{-\gamma(pv^\gamma)}{v^{\gamma+1}} \frac{dv}{dt} = \frac{-\gamma pv^\gamma}{v^\gamma v^1} \frac{dv}{dt}$

i.e. $\frac{dp}{dt} = -\gamma \frac{p}{v} \frac{dv}{dt}$

## C. FURTHER PROBLEMS ON DIFFERENTIATING IMPLICIT FUNCTIONS

In *Problems 1 and 2* differentiate the given functions with respect to $x$.

1 (a) $3y^5$; (b) $2\cos 4\theta$; (c) $\sqrt{k}$.

$$\left[\text{(a) } 15y^4 \frac{dy}{dx}; \text{(b) } -8\sin 4\theta \frac{d\theta}{dx}; \text{(c) } \frac{1}{2\sqrt{k}}\frac{dk}{dx}\right]$$

2 (a) $\frac{5}{2}\ln 3t$; (b) $\frac{3}{4}e^{2y+1}$; (c) $2\tan 3y$

$$\left[\text{(a) } \frac{5}{2t}\frac{dt}{dx}; \text{(b) } \frac{3}{2}e^{2y+1}\frac{dy}{dx}; \text{(c) } 6\sec^2 3y \frac{dy}{dx}\right]$$

3 Differentiate the following with respect to $y$:

(a) $3\sin 2\theta$; (b) $4\sqrt{x^3}$; (c) $\frac{2}{e^t}$

$$\left[\text{(a) } 6\cos 2\theta \frac{d\theta}{dy}; \text{(b) } 6\sqrt{x}\frac{dx}{dy}; \text{(c) } \frac{-2}{e^t}\frac{dt}{dy}\right]$$

4 Differentiate the following with respect to $u$:

(a) $\frac{2}{(3x+1)}$; (b) $3\sec 2\theta$; (c) $\frac{2}{\sqrt{y}}$

$$\left[\text{(a) } \frac{-6}{(3x+1)^2}\frac{dx}{du}; \text{(b) } 6\sec 2\theta \tan 2\theta \frac{d\theta}{du}; \text{(c) } \frac{-1}{\sqrt{y^3}}\frac{dy}{du}\right]$$

In *Problems 5 and 6* determine $\frac{dy}{dx}$.

5 $x^2+y^2+4x-3y+1 = 0$

$$\left[\frac{2x+4}{3-2y}\right]$$

6 $2y^3-y+3x-2 = 0$

$$\left[\frac{3}{1-6y^2}\right]$$

7 Determine $\frac{d}{dx}(3x^2 y^3)$

$$\left[3xy^2\left(3x\frac{dy}{dx}+2y\right)\right]$$

8 Show that $\frac{d}{dy}5\sqrt{(xy^3)} = \frac{5}{2}\sqrt{\left(\frac{y}{x}\right)}(3x+y)\frac{dx}{dy}$.

9 Find $\frac{d}{dx}\left(\frac{2y}{5x}\right)$

$$\left[\frac{2}{5x^2}\left(x\frac{dy}{dx}-y\right)\right]$$

10 Determine $\frac{d}{du}\left(\frac{3u}{4v}\right)$

$$\left[\frac{3}{4v^2}\left(v-u\frac{dv}{du}\right)\right]$$

11 Given $z = 3\sqrt{y}\cos 3x$ find $\frac{dz}{dx}$

$$\left[3\left(\frac{\cos 3x}{2\sqrt{y}}\right)\frac{dy}{dx}-3\sqrt{y}\sin 3x\right]$$

12 Determine $\frac{dz}{dy}$ given $z = 2x^3 \ln y$.

$$\left[2x^2\left(\frac{x}{y}+3\ln y \frac{dx}{dy}\right)\right]$$

13 Differentiate with respect to $x$: $2\cos 3y = 3x^2$.

$$\left[-\sin 3y \frac{dy}{dx} = x\right]$$

14 Given $x^2+y^2 = 9$, evaluate $\frac{dy}{dx}$ when $x =\sqrt{5}$ and $y = 2$.

$$\left[-\frac{\sqrt{5}}{2}\right]$$

In *Problems 15 to 18*, determine $\frac{dy}{dx}$.

15 $x^2+2x\sin 4y = 0$

$$\left[\frac{-(x+\sin 4y)}{4x\cos 4y}\right]$$

16 $3y^2+2xy-4x^2 = 0$

$$\left[\frac{4x-y}{3y+x}\right]$$

38

17    $2x^2 y + 3x^3 = \sin y$                     $\left[\dfrac{x(4y+9x)}{\cos y - 2x^2}\right]$

18    $3y + 2x \ln y = y^4 + x$             $\left[\dfrac{1 - 2 \ln y}{3 + \dfrac{2x}{y} - 4y^3}\right]$

19   Given $y \cos x = x \cos y$, show that $\dfrac{dy}{dx} = \dfrac{\cos y + y \sin x}{\cos x + x \sin y}$.

20   If $3x^2 + 2x^2 y^3 - \dfrac{5}{4}y^2 = 0$, evaluate $\dfrac{dy}{dx}$ when $x = \dfrac{1}{2}$ and $y = 1$.       [5]

21   Given $3x^3 \cos \theta - 2x^2 = \sin \theta$, find the values of $\dfrac{dx}{d\theta}$ when $\theta = \pi$ rads and $x = -\dfrac{2}{3}$.

                                                                          $\left[\dfrac{3}{4}\right]$

22   Determine the gradients of the tangents drawn to the circle $x^2 + y^2 = 16$ at the point where $x = 2$. Give the answer correct to 4 significant figures.     [±0.5774]

23   Find the gradients of the tangents drawn to the ellipse $\dfrac{x^2}{4} + \dfrac{y^2}{9} = 2$ at the point where $x = 2$.                                         [±1.5]

24   Determine the gradient of the curve $3xy + y^2 = -2$ at the point $(1, -2)$.     [−6]

# 6 Logarithmic differentiation

## A. MAIN POINTS CONCERNED WITH LOGARITHMIC DIFFERENTIATION

1   With certain functions containing more complicated products and quotients, differentiation is often made easier if the logarithm of the function is taken before differentiating. This technique, called 'logarithmic differentiation' is achieved with a knowledge of (i) the laws of logarithms, (ii) the differential coefficients of logarithmic functions, and (iii) the differentiation of implicit functions.

2   **Three laws of logarithms** may be expressed as:

(i) $\log (A \times B) = \log A + \log B$; (ii) $\log \left(\dfrac{A}{B}\right) = \log A - \log B$

(iii) $\log A^n = n \log A$.

3   In calculus, Naperian logarithms (i.e. logarithms to a base of 'e') are invariably used. Thus for two functions $f(x)$ and $g(x)$ the laws of logarithms may be expressed as:

(i) $\ln [f(x).g(x)] = \ln f(x) + \ln g(x)$; (ii) $\ln \left(\dfrac{f(x)}{g(x)}\right) = \ln f(x) - \ln g(x)$

(iii) $\ln [f(x)]^n = n \ln f(x)$

4   Taking Naperian logarithms of both sides of the equation $y = \dfrac{f(x).g(x)}{h(x)}$

gives: $\ln y = \ln \left(\dfrac{f(x).g(x)}{h(x)}\right)$

which may be simplified using the laws of logarithms given in para. 3, giving:
$\ln y = \ln f(x) + \ln g(x) - \ln h(x)$.

This latter form of the equation is often easier to differentiate.

5   (i) The differential coefficient of the logarithmic function $\ln x$ is given by:

$$\frac{d}{dx} (\ln x) = \frac{1}{x}$$

(ii) More generally, it may be shown that:

$$\frac{d}{dx} [\ln f(x)] = \frac{f'(x)}{f(x)} \tag{1}$$

For example, if $y = \ln (3x^2 + 2x - 1)$ then $\dfrac{dy}{dx} = \dfrac{6x+2}{3x^2+2x-1}$ .

Similarly, if $y = \ln (\sin 3x)$ then $\dfrac{dy}{dx} = \dfrac{3 \cos 3x}{\sin 3x} = 3 \cot 3x$.

6   As explained in chapter 5, by using the function of a function rule:

$$\frac{d}{dx}(\ln y) = \left(\frac{1}{y}\right)\frac{dy}{dx} \qquad (2)$$

7   Differentiation of an expression such as $y = \dfrac{(1+x)^2\sqrt{(x-1)}}{x\sqrt{(x+2)}}$ may be achieved by using the product and quotient rules of differentiation; however the working would be rather complicated. With logarithmic differentiation the following procedure is adopted:

(i) Take Naperian logarithms of both sides of the equation.

Thus   $\ln y = \ln\left\{\dfrac{(1+x)^2\sqrt{(x-1)}}{x\sqrt{(x+2)}}\right\} = \ln\left\{\dfrac{(1+x)^2(x-1)^{\frac{1}{2}}}{x(x+2)^{\frac{1}{2}}}\right\}$

(ii) Apply the laws of logarithms.

Thus   $\ln y = \ln(1+x)^2 + \ln(x-1)^{\frac{1}{2}} - \ln x - \ln(x+2)^{\frac{1}{2}}$,
         by laws (i) and (ii) of para. 3.

i.e.,   $\ln y = 2\ln(1+x) + \dfrac{1}{2}\ln(x-1) - \ln x - \dfrac{1}{2}\ln(x+2)$,
         by law (iii) of para. 3.

(iii) Differentiate each term in turn with respect to $x$ using equations (1) and (2).

Thus $\dfrac{1}{y}\dfrac{dy}{dx} = \dfrac{2}{(1+x)} + \dfrac{\frac{1}{2}}{(x-1)} - \dfrac{1}{x} - \dfrac{\frac{1}{2}}{(x+2)}$

(iv) Rearrange the equation to make $\dfrac{dy}{dx}$ the subject.

Thus   $\dfrac{dy}{dx} = y\left\{\dfrac{2}{(1+x)} + \dfrac{1}{2(x-1)} - \dfrac{1}{x} - \dfrac{1}{2(x+2)}\right\}$

(v) Substitute for $y$ in terms of $x$.

Thus   $\dfrac{dy}{dx} = \dfrac{(1+x)^2\sqrt{(x-1)}}{x\sqrt{(x+2)}}\left\{\dfrac{2}{(1+x)} + \dfrac{1}{2(x-1)} - \dfrac{1}{x} - \dfrac{1}{2(x+2)}\right\}$

(See *Problems 1 to 4*)

8   Whenever an expression to be differentiated contains a term raised to a power which is itself a function of the variable, then logarithmic differentiation must be used. For example, the differentiation of expressions such as $x^x$, $(x+2)^x$, $\sqrt[x]{(x-1)}$ and $x^{3x+2}$ can only be achieved using logarithmic differentiation. (See *Problems 5 to 8.*)

## B. WORKED PROBLEMS ON LOGARITHMIC DIFFERENTIATION

*Problem 1* Use logarithmic differentiation to differentiate $y = \dfrac{(x+1)(x-2)^3}{(x-3)}$

Following the procedure of para. 7:

(i) Since $y = \dfrac{(x+1)(x-2)^3}{(x-3)}$ then $\ln y = \ln\left\{\dfrac{(x+1)(x-2)^3}{(x-3)}\right\}$

(ii) $\ln y = \ln(x+1) + \ln(x-2)^3 - \ln(x-3)$, by laws (i) and (ii) of para. 3.
  i.e. $\ln y = \ln(x+1) + 3\ln(x-2) - \ln(x-3)$, by law (iii) of para. 3.

(iii) Differentiating with respect to $x$ gives:

$$\frac{1}{y}\frac{dy}{dx} = \frac{1}{(x+1)} + \frac{3}{(x-2)} - \frac{1}{(x-3)}$$ , by using equations (1) and (2).

(iv) Rearranging gives: $\dfrac{dy}{dx} = y\left\{\dfrac{1}{(x+1)} + \dfrac{3}{(x-2)} - \dfrac{1}{(x-3)}\right\}$

(v) Substituting for $y$ gives: $\dfrac{dy}{dx} = \dfrac{(x+1)(x-2)^3}{(x-3)}\left\{\dfrac{1}{(x+1)} + \dfrac{3}{(x-2)} - \dfrac{1}{(x-3)}\right\}$

---

***Problem 2*** Differentiate $y = \dfrac{\sqrt{(x-2)^3}}{(x+1)^2(2x-1)}$ with respect to $x$, and evaluate $\dfrac{dy}{dx}$ when $x = 3$.

---

Using logarithmic differentiation and following the procedure of para. 7:

(i) Since $y = \dfrac{\sqrt{(x-2)^3}}{(x+1)^2(2x-1)}$ then $\ln y = \ln\left\{\dfrac{\sqrt{(x-2)^3}}{(x+1)^2(2x-1)}\right\} = \ln\left\{\dfrac{(x-2)^{3/2}}{(x+1)^2(2x-1)}\right\}$

(ii) $\ln y = (x-2)^{3/2} - \ln(x+1)^2 - \ln(2x-1)$

i.e. $\ln y = \dfrac{3}{2}\ln(x-2) - 2\ln(x+1) - \ln(2x-1)$

(iii) $\dfrac{1}{y}\dfrac{dy}{dx} = \dfrac{3/2}{(x-2)} - \dfrac{2}{(x+1)} - \dfrac{2}{(2x-1)}$

(iv) $\dfrac{dy}{dx} = y\left\{\dfrac{3}{2(x-2)} - \dfrac{2}{(x+1)} - \dfrac{2}{(2x-1)}\right\}$

(v) $\dfrac{dy}{dx} = \dfrac{\sqrt{(x-2)^3}}{(x+1)^2(2x-1)}\left\{\dfrac{3}{2(x-2)} - \dfrac{2}{(x+1)} - \dfrac{2}{(2x-1)}\right\}$

When $x = 3$, $\dfrac{dy}{dx} = \dfrac{\sqrt{(1)^3}}{(4)^2(5)}\left(\dfrac{3}{2} - \dfrac{2}{4} - \dfrac{2}{5}\right) = \dfrac{1}{80}\left(\dfrac{3}{5}\right) = \dfrac{3}{400}$

---

***Problem 3*** Given $y = \dfrac{3e^{2\theta}\sec 2\theta}{\sqrt{(\theta-2)}}$ , determine $\dfrac{dy}{d\theta}$ .

---

Using logarithmic differentiation and following the procedure of para. 7 gives:

(i) Since $y = \dfrac{3e^{2\theta}\sec 2\theta}{\sqrt{(\theta-2)}}$ then $\ln y = \ln\left\{\dfrac{3e^{2\theta}\sec 2\theta}{\sqrt{(\theta-2)}}\right\} = \ln\left\{\dfrac{3e^{2\theta}\sec 2\theta}{(\theta-2)^{1/2}}\right\}$

(ii) $\ln y = \ln 3e^{2\theta} + \ln\sec 2\theta - \ln(\theta-2)^{1/2}$

i.e. $\ln y = \ln 3 + \ln e^{2\theta} + \ln\sec 2\theta - \dfrac{1}{2}\ln(\theta-2)$

i.e. $\ln y = \ln 3 + 2\theta + \ln\sec 2\theta - \dfrac{1}{2}\ln(\theta-2)$

(iii) Differentiating with respect to $\theta$ gives:

$$\frac{1}{y}\frac{dy}{d\theta} = 0 + 2 + \frac{2\sec 2\theta\tan 2\theta}{\sec 2\theta} - \frac{1/2}{(\theta-2)}$$ , from equations (1) and (2).

(iv) Rearranging gives: $\dfrac{dy}{d\theta} = y\left\{2 + 2\tan 2\theta - \dfrac{1}{2(\theta-2)}\right\}$

(v) Substituting for $y$ gives: $\dfrac{dy}{d\theta} = \dfrac{3e^{2\theta}\sec 2\theta}{\sqrt{(\theta-2)}}\left\{2 + 2\tan 2\theta - \dfrac{1}{2(\theta-2)}\right\}$

---

***Problem 4*** Differentiate $y = \dfrac{x^3\ln 2x}{e^x\sin x}$ with respect to $x$.

---

Using logarithmic differentiation and following the procedure of para. 7 gives:

(i) $\ln y = \ln \left\{ \dfrac{x^3 \ln 2x}{e^x \sin x} \right\}$

(ii) $\ln y = \ln x^3 + \ln (\ln 2x) - \ln (e^x) - \ln (\sin x)$

i.e. $\ln y = 3 \ln x + \ln (\ln 2x) - x - \ln (\sin x)$

(iii) $\dfrac{1}{y} \dfrac{dy}{dx} = \dfrac{3}{x} + \dfrac{\frac{1}{x}}{\ln 2x} - 1 - \dfrac{\cos x}{\sin x}$

(iv) $\dfrac{dy}{dx} = y \left\{ \dfrac{3}{x} + \dfrac{1}{x \ln 2x} - 1 - \cot x \right\}$

(v) $\dfrac{dy}{dx} = \dfrac{x^3 \ln 2x}{e^x \sin x} \left\{ \dfrac{3}{x} + \dfrac{1}{x \ln 2x} - 1 - \cot x \right\}$

**Problem 5** Determine $\dfrac{dy}{dx}$ given $y = x^x$.

Taking Naperian logarithms of both sides of $y = x^x$ gives:
$\ln y = \ln x^x = x \ln x$, by law (iii) of para. 3.
Differentiating both sides with respect to $x$ gives:

$\dfrac{1}{y} \dfrac{dy}{dx} = (x)(\dfrac{1}{x}) + (\ln x)(1)$, using the product rule.

i.e. $\dfrac{1}{y} \dfrac{dy}{dx} = 1 + \ln x$, from which, $\dfrac{dy}{dx} = y (1 + \ln x)$

i.e. $\dfrac{dy}{dx} = x^x (1 + \ln x)$

**Problem 6** Evaluate $\dfrac{dy}{dx}$ when $x = -1$ given $y = (x+2)^x$.

Taking Naperian logarithms of both sides of $y = (x+2)^x$ gives:
$\ln y = \ln (x+2)^x = x \ln (x+2)$, by law (iii) of para. 3.
Differentiating both sides with respect to $x$ gives:

$\dfrac{1}{y} \dfrac{dy}{dx} = (x) \left( \dfrac{1}{x+2} \right) + [\ln (x+2)] (1)$, by the product rule.

Hence $\dfrac{dy}{dx} = y \left( \dfrac{x}{(x+2)} + \ln (x+2) \right) = (x+2)^x \left\{ \dfrac{x}{(x+2)} + \ln (x+2) \right\}$

When $x = -1$, $\dfrac{dy}{dx} = (1)^{-1} \left( \dfrac{-1}{1} + \ln 1 \right) = (+1)(-1) = -1$

**Problem 7** Determine (a) the differential coefficient of $y = \sqrt[x]{(x-1)}$ and (b) evaluate $\dfrac{dy}{dx}$ when $x = 2$.

(a) $y = \sqrt[x]{(x-1)} = (x-1)^{\frac{1}{x}}$, since by the laws of indices $\sqrt[n]{a^m} = a^{\frac{m}{n}}$. Taking Naperian logarithms of both sides gives:

$\ln y = \ln (x-1)^{\frac{1}{x}} = \dfrac{1}{x} \ln (x-1)$, by law (iii) of para. 3.

Differentiating each side with respect to $x$ gives:

$\dfrac{1}{y} \dfrac{dy}{dx} = \left( \dfrac{1}{x} \right) \dfrac{1}{(x-1)} + [\ln (x-1)] \left( \dfrac{-1}{x^2} \right)$, by the product rule.

43

Hence $\dfrac{dy}{dx} = y\left\{\dfrac{1}{x(x-1)} - \dfrac{\ln(x-1)}{x^2}\right\}$

i.e. $\dfrac{dy}{dx} = \sqrt[x]{(x-1)}\left\{\dfrac{1}{x(x-1)} - \dfrac{\ln(x-1)}{x^2}\right\}$

(b) When $x = 2$, $\dfrac{dy}{dx} = \sqrt[2]{(1)}\left\{\dfrac{1}{2(1)} - \dfrac{\ln(1)}{4}\right\} = \pm1\left\{\dfrac{1}{2} - 0\right\} = \pm\dfrac{1}{2}$

---

*Problem 8* Differentiate $x^{3x+2}$ with respect to $x$.

Let $y = x^{3x+2}$
Taking Naperian logarithms of both sides gives: $\ln y = \ln x^{3x+2}$
i.e. $\ln y = (3x+2)\ln x$, by law (iii) of para. 3.
Differentiating each term with respect to $x$ gives:
$\dfrac{1}{y}\dfrac{dy}{dx} = (3x+2)(\dfrac{1}{x})+(\ln x)(3)$, by the product rule.

Hence $\dfrac{dy}{dx} = y\left\{\dfrac{3x+2}{x} + 3\ln x\right\}$

$= x^{3x+2}\left\{\dfrac{3x+2}{x} + 3\ln x\right\} = x^{3x+2}\left\{3 + \dfrac{2}{x} + 3\ln x\right\}$

---

## C. FURTHER PROBLEMS ON LOGARITHMIC DIFFERENTIATION

In *Problems 1 to 6*, use logarithmic differentiation to differentiate the given functions with respect to the variable.

1  $y = \dfrac{(x-2)(x+1)}{(x-1)(x+3)}$  $\left[\dfrac{(x-2)(x+1)}{(x-1)(x+3)}\left\{\dfrac{1}{(x-2)} + \dfrac{1}{(x+1)} - \dfrac{1}{(x-1)} - \dfrac{1}{(x+3)}\right\}\right]$

2  $y = \dfrac{(x+1)(2x+1)^3}{(x-3)^2(x+2)^4}$  $\left[\dfrac{(x+1)(2x+1)^3}{(x-3)^2(x+2)^4}\left\{\dfrac{1}{(x+1)} + \dfrac{6}{(2x+1)} - \dfrac{2}{(x-3)} - \dfrac{4}{(x+2)}\right\}\right]$

3  $y = \dfrac{(2x-1)\sqrt{(x+2)}}{(x-3)\sqrt{(x+1)^3}}$  $\left[\dfrac{(2x-1)\sqrt{(x+2)}}{(x-3)\sqrt{(x+1)^3}}\left\{\dfrac{2}{(2x-1)} + \dfrac{1}{2(x+2)} - \dfrac{1}{(x-3)} - \dfrac{3}{2(x+1)}\right\}\right]$

4  $y = \dfrac{e^{2x}\cos 3x}{\sqrt{(x-4)}}$  $\left[\dfrac{e^{2x}\cos 3x}{\sqrt{(x-4)}}\left\{2-3\tan 3x - \dfrac{1}{2(x-4)}\right\}\right]$

5  $y = 3\theta\sin\theta\cos\theta$  $\left[3\theta\sin\theta\cos\theta\left\{\dfrac{1}{\theta} + \cot\theta - \tan\theta\right\}\right]$

6  $y = \dfrac{2x^4\tan x}{e^{2x}\ln 2x}$  $\left[\dfrac{2x^4\tan x}{e^{2x}\ln 2x}\left\{\dfrac{4}{x} + \dfrac{1}{\sin x\cos x} - 2 - \dfrac{1}{x\ln 2x}\right\}\right]$

7  Evaluate $\dfrac{dy}{dx}$ when $x = 1$ given $y = \dfrac{(x+1)^2\sqrt{(2x-1)}}{\sqrt{(x+3)^3}}$. $\left[\dfrac{13}{16}\right]$

8  Evaluate $\dfrac{dy}{d\theta}$, correct to 3 significant figures, when $\theta = \dfrac{\pi}{4}$ given $y = \dfrac{2e^{\theta}\sin\theta}{\sqrt{\theta^5}}$.

$[-6.71]$

In *Problems 9 to 12*, differentiate with respect to $x$.

9  $y = x^{2x}$. $[2x^{2x}(1+\ln x)]$

10  $y = (2x-1)^x$. $\left[(2x-1)^x\left\{\dfrac{2x}{2x-1} + \ln(2x-1)\right\}\right]$

44

11  $y = \sqrt[x]{(x+3)}$                                $\left[\sqrt[x]{(x+3)}\left\{\dfrac{1}{x(x+3)} - \dfrac{\ln(x+3)}{x^2}\right\}\right]$

12  $y = 3x^{4x+1}$.                                     $\left[3x^{4x+1}\left\{4 + \dfrac{1}{x} + 4\ln x\right\}\right]$

13  Show that when $y = 2x^x$ and $x = 1$, $\dfrac{dy}{dx} = 2$.

14  Evaluate $\dfrac{d}{dx}\{\sqrt[x]{(x-2)}\}$ when $x = 3$.                       $\left[\dfrac{1}{3}\right]$

15  Show that if $y = \theta^\theta$ and $\theta = 2$, $\dfrac{dy}{d\theta} = 6.77$, correct to 3 significant figures.

# 7 Differentiation of inverse trigonometric and hyperbolic functions

## A. MAIN POINTS CONCERNED WITH DIFFERENTIATION OF INVERSE TRIGONOMETRIC AND HYPERBOLIC FUNCTIONS

1   If $y = 3x-2$, then by transposition, $x = \dfrac{y+2}{3}$. The function $x = \dfrac{y+2}{3}$ is called the **inverse function** of $y = 3x-2$.

2   **Inverse trigonometric functions** are denoted by prefixing the function with 'arc'. For example, if $y = \sin x$, then $x = \arcsin y$. Similarly, if $y = \cos x$, then $x = \arccos y$, and so on. A sketch of each of the inverse trigonometric functions is shown in *Fig 1*.

3   **Inverse hyperbolic functions** are denoted by prefixing the function with 'ar'. For example, if $y = \sinh x$, then $x = \operatorname{arsinh} y$. Similarly, if $y = \operatorname{sech} x$, then

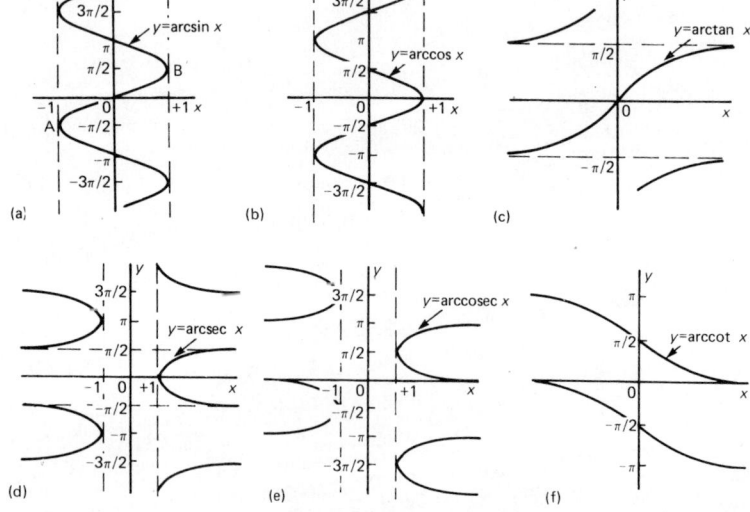

**Fig 1**

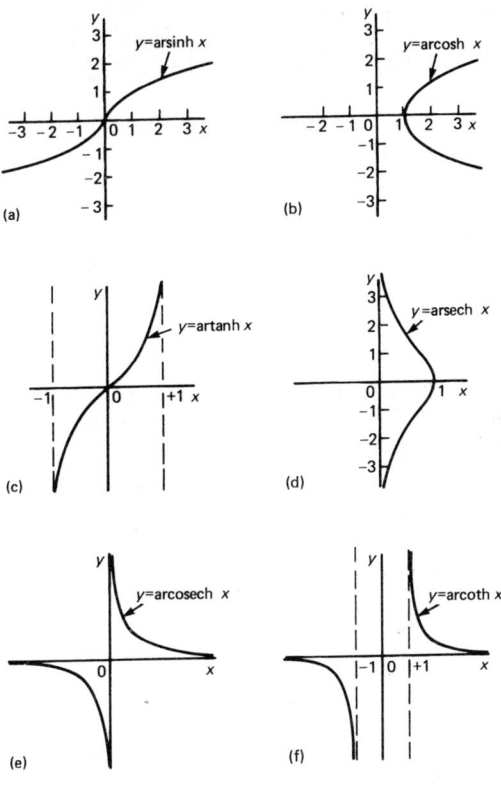

**Fig 2**

$x = \text{arsech } y$, and so on. A sketch of each of the inverse hyperbolic functions is shown in *Fig 2*.

4 **Differentiation of inverse trigonometric functions**

(i) If $y = \arcsin x$, then $x = \sin y$.

Differentiating both sides with respect to $y$ gives:

$$\frac{dx}{dy} = \cos y = \sqrt{(1 - \sin^2 y)}, \text{ since } \cos^2 y + \sin^2 y = 1$$

i.e. $\dfrac{dx}{dy} = \sqrt{(1 - x^2)}$, however $\dfrac{dy}{dx} = \dfrac{1}{\dfrac{dx}{dy}}$

**Hence when** $y = \arcsin x$ **then** $\dfrac{dy}{dx} = \dfrac{1}{\sqrt{(1 - x^2)}}$

(ii) A sketch of part of the curve of $y = \arcsin x$ is shown in *Fig 1(a)*. The principal value of $\arcsin x$ is defined as the value lying between $-(\pi/2)$ and $\pi/2$. The gradient of the curve between points A and B is positive for all values of $x$ and thus only the positive value is taken when evaluating $1/\sqrt{(1 - x^2)}$.

47

(iii) Given $y = \arcsin \dfrac{x}{a}$ then $\dfrac{x}{a} = \sin y$ and $x = a \sin y$.

Hence $\dfrac{dx}{dy} = a \cos y = a\sqrt{(1 - \sin^2 y)} = a\sqrt{\left[1 - \left(\dfrac{x}{a}\right)^2\right]} = a\sqrt{\left(\dfrac{a^2 - x^2}{a^2}\right)}$

$$= \dfrac{a\sqrt{(a^2 - x^2)}}{a} = \sqrt{(a^2 - x^2)}$$

Thus $\dfrac{dy}{dx} = \dfrac{1}{\dfrac{dx}{dy}} = \dfrac{1}{\sqrt{(a^2 - x^2)}}$.

Since integration is the reverse process of differentiation then:

$$\int \dfrac{1}{\sqrt{(a^2 - x^2)}}\, dx = \arcsin \dfrac{x}{a} + c.$$

TABLE 1. Differential coefficients of inverse trigonometric functions

| | $y$ or $f(x)$ | $\dfrac{dy}{dx}$ or $f'(x)$ |
|---|---|---|
| (i) | $\arcsin \dfrac{x}{a}$ | $\dfrac{1}{\sqrt{(a^2 - x^2)}}$ |
| | $\arcsin f(x)$ | $\dfrac{f'(x)}{\sqrt{\{1 - [f(x)]^2\}}}$ |
| (ii) | $\arccos \dfrac{x}{a}$ | $\dfrac{-1}{\sqrt{(a^2 - x^2)}}$ |
| | $\arccos f(x)$ | $\dfrac{-f'(x)}{\sqrt{\{1 - [f(x)]^2\}}}$ |
| (iii) | $\arctan \dfrac{x}{a}$ | $\dfrac{a}{a^2 + x^2}$ |
| | $\arctan f(x)$ | $\dfrac{f'(x)}{\{1 + [f(x)]^2\}}$ |
| (iv) | $\operatorname{arcsec} \dfrac{x}{a}$ | $\dfrac{a}{x\sqrt{(x^2 - a^2)}}$ |
| | $\operatorname{arcsec} f(x)$ | $\dfrac{f'(x)}{f(x)\sqrt{\{[f(x)]^2 - 1\}}}$ |
| (v) | $\operatorname{arccosec} \dfrac{x}{a}$ | $\dfrac{-a}{x\sqrt{(x^2 - a^2)}}$ |
| | $\operatorname{arccosec} f(x)$ | $\dfrac{-f'(x)}{f(x)\sqrt{\{[f(x)]^2 - 1\}}}$ |
| (vi) | $\operatorname{arccot} \dfrac{x}{a}$ | $\dfrac{-a}{a^2 + x^2}$ |
| | $\operatorname{arccot} f(x)$ | $\dfrac{-f'(x)}{\{1 + [f(x)]^2\}}$ |

(iv) Given $y = \arcsin f(x)$ the function of a function rule may be used to find $\dfrac{dy}{dx}$.

Let $u = f(x)$ then $y = \arcsin u$.

Then $\dfrac{du}{dx} = f'(x)$ and $\dfrac{dy}{du} = \dfrac{1}{\sqrt{(1-u^2)}}$  (see para. 4(i))

Thus $\dfrac{dy}{dx} = \dfrac{dy}{du} \times \dfrac{du}{dx} = \dfrac{1}{\sqrt{(1-u^2)}}\, f'(x) = \dfrac{f'(x)}{\sqrt{\{\,1 - [f(x)]^2\,\}}}$

(v) The differential coefficients of the remaining inverse trigonometric functions are obtained in a similar manner to that shown above and a summary of the results is shown in *Table 1*. (See *Problems 1 to 9*)

5  (i) Inverse hyperbolic functions may be evaluated most conveniently when expressed in a **logarithmic form**.

(ii) For example, if $y = \operatorname{arsinh}\dfrac{x}{a}$ then $\dfrac{x}{a} = \sinh y$.

From chapter 2, $e^y = \cosh y + \sinh y$ and $\cosh^2 y - \sinh^2 y = 1$, from which $\cosh y = \sqrt{(1+\sinh^2 y)}$ which is positive since $\cosh y$ is always positive (see *Fig 2*, chapter 2).

Hence $e^y = \sqrt{(1+\sinh^2 y)} + \sinh y$

$$= \sqrt{\left[1 + \left(\dfrac{x}{a}\right)^2\right]} + \dfrac{x}{a} = \sqrt{\left(\dfrac{a^2+x^2}{a^2}\right)} + \dfrac{x}{a}$$

$$= \dfrac{\sqrt{(a^2+x^2)}}{a} + \dfrac{x}{a} \text{ or } \dfrac{x+\sqrt{(a^2+x^2)}}{a}$$

Taking Naperian logarithms of both sides gives: $y = \ln\left\{\dfrac{x+\sqrt{(a^2+x^2)}}{a}\right\}$

Hence $\operatorname{arsinh}\dfrac{x}{a} = \ln\left\{\dfrac{x+\sqrt{(a^2+x^2)}}{a}\right\}$     (1)

Thus to evaluate $\operatorname{arsinh}\dfrac{3}{4}$, let $x = 3$ and $a = 4$ in equation (1).

Then $\operatorname{arsinh}\dfrac{3}{4} = \ln\left\{\dfrac{3+\sqrt{(4^2+3^2)}}{4}\right\} = \ln\left(\dfrac{3+5}{4}\right) = \ln 2 = 0.6931$

(iii) By similar reasoning to above it may be shown that:

$\operatorname{arcosh}\dfrac{x}{a} = \ln\left\{\dfrac{x+\sqrt{(x^2-a^2)}}{a}\right\}$

and $\operatorname{artanh}\dfrac{x}{a} = \dfrac{1}{a}\ln\left(\dfrac{a+x}{a-x}\right)$

(See *Problems 10 to 12*)

6  **Differentiation of inverse hyperbolic functions**

(i) If $y = \operatorname{arsinh}\dfrac{x}{a}$ then $\dfrac{x}{a} = \sinh y$ and $x = a \sinh y$

$\dfrac{dx}{dy} = a \cosh y$ (from chapter 2).

Also $\cosh^2 y - \sinh^2 y = 1$, from which $\cosh y = \sqrt{(1+\sinh^2 y)}$

$$= \sqrt{\left[1 + \left(\dfrac{x}{a}\right)^2\right]} = \dfrac{\sqrt{(a^2+x^2)}}{a}$$

Hence $\dfrac{dx}{dy} = a \cosh y = \dfrac{a\sqrt{(a^2+x^2)}}{a} = \sqrt{(a^2+x^2)}$

Then $\dfrac{dy}{dx} = \dfrac{1}{\dfrac{dx}{dy}} = \dfrac{1}{\sqrt{(a^2+x^2)}}$.

(An alternative method of differentiating $\text{arsinh}\,\dfrac{x}{a}$ is to differentiate the logarithmic form $\ln\left\{\dfrac{x+\sqrt{(a^2+x^2)}}{a}\right\}$ with respect to $x$.) From the sketch of $y = \text{arsinh}\,x$ shown in *Fig 2(a)* it is seen that the gradient (i.e. $dy/dx$) is always positive.

(ii) It follows from (i) that $\displaystyle\int \dfrac{1}{\sqrt{(x^2+a^2)}}\,dx = \text{arsinh}\,\dfrac{x}{a} + c$ or $\ln\left\{\dfrac{x+\sqrt{(x^2+a^2)}}{a}\right\}+c$.

(iii) It may be shown that $\dfrac{d}{dx}(\text{arsinh}\,x) = \dfrac{1}{\sqrt{(1+x^2)}}$

or more generally, $\dfrac{d}{dx}[\text{arsinh}\,f(x)] = \dfrac{f'(x)}{\sqrt{\{1+[f(x)]^2\}}}$,

by using the function of a function rule as in para. 4(iv).

TABLE 2. Differential coefficients of inverse hyperbolic functions

| | $y$ or $f(x)$ | $\dfrac{dy}{dx}$ or $f'(x)$ |
|---|---|---|
| (i) | $\text{arsinh}\,\dfrac{x}{a}$ | $\dfrac{1}{\sqrt{(x^2+a^2)}}$ |
| | $\text{arsinh}\,f(x)$ | $\dfrac{f'(x)}{\sqrt{\{[f(x)]^2+1\}}}$ |
| (ii) | $\text{arcosh}\,\dfrac{x}{a}$ | $\dfrac{1}{\sqrt{(x^2-a^2)}}$ |
| | $\text{arcosh}\,f(x)$ | $\dfrac{f'(x)}{\sqrt{\{[f(x)]^2-1\}}}$ |
| (iii) | $\text{artanh}\,\dfrac{x}{a}$ | $\dfrac{a}{a^2-x^2}$ |
| | $\text{artanh}\,f(x)$ | $\dfrac{f'(x)}{\{1-[f(x)]^2\}}$ |
| (iv) | $\text{arsech}\,\dfrac{x}{a}$ | $\dfrac{-a}{x\sqrt{(a^2-x^2)}}$ |
| | $\text{arsech}\,f(x)$ | $\dfrac{-f'(x)}{f(x)\sqrt{\{1-[f(x)]^2\}}}$ |
| (v) | $\text{arcosech}\,\dfrac{x}{a}$ | $\dfrac{-a}{x\sqrt{(x^2+a^2)}}$ |
| | $\text{arcosech}\,f(x)$ | $\dfrac{-f'(x)}{f(x)\sqrt{\{[f(x)]^2+1\}}}$ |
| (vi) | $\text{arcoth}\,\dfrac{x}{a}$ | $\dfrac{a}{a^2-x^2}$ |
| | $\text{arcoth}\,f(x)$ | $\dfrac{f'(x)}{\{1-[f(x)]^2\}}$ |

(iv) The remaining inverse hyperbolic functions are differentiated in a similar manner to that shown above and the results are summarised in *Table 2*. (See *Problems 13 to 20*)

## B. WORKED PROBLEMS ON DIFFERENTIATION OF INVERSE TRIGONOMETRIC AND HYPERBOLIC FUNCTIONS

*Problem 1* Find $\dfrac{dy}{dx}$ given $y = \arcsin 5x^2$.

From *Table 1(i)*, if $y = \arcsin f(x)$ then $\dfrac{dy}{dx} = \dfrac{f'(x)}{\sqrt{\{1-[f(x)]^2\}}}$

Hence, if $y = \arcsin 5x^2$ then $f(x) = 5x^2$ and $f'(x) = 10x$.

Thus $\dfrac{dy}{dx} = \dfrac{10x}{\sqrt{\{1-(5x^2)^2\}}} = \dfrac{10x}{\sqrt{\{1-25x^4\}}}$

*Problem 2* (a) Show that if $y = \arccos x$ then $\dfrac{dy}{dx} = -\dfrac{1}{\sqrt{(1-x^2)}}$.

(b) Hence obtain the differential coefficient of $y = \arccos(1-2x^2)$.

(a) If $y = \arccos x$ then $x = \cos y$.

Differentiating with respect to $y$ gives: $\dfrac{dx}{dy} = -\sin y = -\sqrt{(1-\cos^2 y)}$
$$= -\sqrt{(1-x^2)}.$$

Hence $\dfrac{dy}{dx} = \dfrac{1}{\dfrac{dx}{dy}} = -\dfrac{1}{\sqrt{(1-x^2)}}$.

The principal value of $y = \arccos x$ is defined as the angle lying between 0 and $\pi$, i.e., between points C and D shown in *Fig 1(b)*. The gradient of the curve is negative between C and D and thus the differential coefficient $dy/dx$ is negative as shown above.

(b) If $y = \arccos f(x)$ then by letting $u = f(x)$, $y = \arccos u$.

Then $\dfrac{dy}{du} = -\dfrac{1}{\sqrt{(1-u^2)}}$ (from part (a)) and $\dfrac{du}{dx} = f'(x)$.

From the function of a function rule, $\dfrac{dy}{dx} = \dfrac{dy}{du} \cdot \dfrac{du}{dx} = -\dfrac{1}{\sqrt{(1-u^2)}} f'(x)$
$$= \dfrac{-f'(x)}{\sqrt{\{1-[f(x)]^2\}}}$$

Hence when $y = \arccos(1-2x^2)$ then $\dfrac{dy}{dx} = \dfrac{-(-4x)}{\sqrt{\{1-(1-2x^2)^2\}}}$

$$= \dfrac{4x}{\sqrt{\{1-(1-4x^2+4x^4)\}}} = \dfrac{4x}{\sqrt{(4x^2-4x^4)}} = \dfrac{4x}{\sqrt{\{4x^2(1-x^2)\}}}$$

$$= \dfrac{4x}{2x\sqrt{(1-x^2)}} = \dfrac{2}{\sqrt{(1-x^2)}}$$

*Problem 3* Determine the differential coefficient of $y = \arctan x/a$ and show that the differential coefficient of $\arctan \dfrac{2x}{3}$ is $\dfrac{6}{9+4x^2}$.

If $y = \arctan\dfrac{x}{a}$ then $\dfrac{x}{a} = \tan y$ and $x = a\tan y$.

$\dfrac{dx}{dy} = a\sec^2 y = a\,(1+\tan^2 y)$, (since $\sec^2 y = 1+\tan^2 y$)

$$= a\left[1+\left(\dfrac{x}{a}\right)^2\right] = a\left(\dfrac{a^2+x^2}{a^2}\right) = \dfrac{a^2+x^2}{a}$$

Hence $\dfrac{dy}{dx} = \dfrac{1}{\dfrac{dx}{dy}} = \dfrac{a}{a^2+x^2}$ .

The principal value of $y = \arctan x$ is defined as the angle lying between $-\dfrac{\pi}{2}$ and $\dfrac{\pi}{2}$ and the gradient (i.e. $\dfrac{dy}{dx}$) between these two values is always positive (see *Fig 1(c)*).

Comparing $\arctan\dfrac{2x}{3}$ with $\arctan\dfrac{x}{a}$ shows that $a = \dfrac{3}{2}$.

Hence if $y = \arctan\dfrac{2x}{3}$ then $\dfrac{dy}{dx} = \dfrac{\dfrac{3}{2}}{\left(\dfrac{3}{2}\right)^2+x^2} = \dfrac{\dfrac{3}{2}}{\dfrac{9}{4}+x^2}$

$$= \dfrac{\dfrac{3}{2}}{\dfrac{9+4x^2}{4}} = \dfrac{\dfrac{3}{2}(4)}{9+4x^2} = \dfrac{6}{9+4x^2}$$

**Problem 4** Show that $\dfrac{d}{dx}\left(\operatorname{arcsec}\dfrac{x}{a}\right) = \dfrac{a}{x\sqrt{(x^2-a^2)}}$ and hence determine $\dfrac{d}{dx}(\operatorname{arcsec} 2x)$.

If $y = \operatorname{arcsec}\dfrac{x}{a}$ then $\dfrac{x}{a} = \sec y$ and $x = a\sec y$. Then $\dfrac{dx}{dy} = a\sec y\tan y$.

Since $1+\tan^2 y = \sec^2 y$ then $\tan y = \sqrt{(\sec^2 y-1)} = \sqrt{\left[\left(\dfrac{x}{a}\right)^2-1\right]}$

$$= \sqrt{\left(\dfrac{x^2-a^2}{a^2}\right)} = \dfrac{\sqrt{(x^2-a^2)}}{a}$$

Hence $\dfrac{dx}{dy} = a\sec y\,\dfrac{\sqrt{(x^2-a^2)}}{a} = a\left(\dfrac{x}{a}\right)\dfrac{\sqrt{(x^2-a^2)}}{a} = \dfrac{x}{a}\sqrt{(x^2-a^2)}$

Thus $\dfrac{dy}{dx} = \dfrac{1}{\dfrac{dx}{dy}} = \dfrac{a}{x\sqrt{(x^2-a^2)}}$ .

The principal value of $y = \operatorname{arcsec} x$ is defined as the angle lying between $0$ and $\pi$ and the gradient between these values is always positive (see *Fig 1(d)*).

Comparing $\operatorname{arcsec} 2x$ with $\operatorname{arcsec}\dfrac{x}{a}$ shows that $a = \dfrac{1}{2}$.

Hence $\dfrac{d}{dx}(\operatorname{arcsec} 2x) = \dfrac{\dfrac{1}{2}}{x\sqrt{[x^2-(\frac{1}{2})^2]}} = \dfrac{1}{2x\sqrt{\left(\dfrac{4x^2-1}{4}\right)}}$

$$= \dfrac{1}{2x\,\dfrac{\sqrt{(4x^2-1)}}{2}} = \dfrac{1}{x\sqrt{(4x^2-1)}}$$

*Problem 5* Find the differential coefficient of $y = \ln(\arccos 3x)$.

Let $u = \arccos 3x$ then $y = \ln u$.

By the function of a function rule, $\dfrac{dy}{dx} = \dfrac{dy}{du} \cdot \dfrac{du}{dx} = \dfrac{1}{u} \times \dfrac{d}{dx}(\arccos 3x)$

$$= \frac{1}{\arccos 3x} \left\{ \frac{-3}{\sqrt{[1-(3x)^2]}} \right\}$$

i.e. $\dfrac{d}{dx}(\ln \arccos 3x) = \dfrac{-3}{\sqrt{(1-9x^2)} \arccos 3x}$

*Problem 6* If $y = \arctan \dfrac{3}{t^2}$ find $\dfrac{dy}{dt}$.

Using the general form from *Table 1(iii)*, $f(t) = \dfrac{3}{t^2} = 3t^{-2}$, from which $f'(t) = \dfrac{-6}{t^3}$

Hence $\dfrac{d}{dt}\left(\arctan \dfrac{3}{t^2}\right) = \dfrac{f'(t)}{\{1+[f(t)]^2\}} = \dfrac{-\dfrac{6}{t^3}}{\{1+(\dfrac{3}{t^2})^2\}} = \dfrac{-\dfrac{6}{t^3}}{\dfrac{t^4+9}{t^4}} = \left(\dfrac{-6}{t^3}\right)\left(\dfrac{t^4}{t^4+9}\right)$

$$= \frac{-6t}{t^4+9}$$

*Problem 7* Differentiate $y = \dfrac{\text{arccot } 2x}{(1+4x^2)}$.

Using the quotient rule: $\dfrac{dy}{dx} = \dfrac{(1+4x^2)\left(\dfrac{-2}{1+(2x)^2}\right) - (\text{arccot } 2x)(8x)}{(1+4x^2)^2}$, from *Table 1(iv)*,

$$= \frac{-2(1+4x \text{ arccot } 2x)}{(1+4x^2)^2}$$

*Problem 8* Differentiate $y = x \text{ arccosec } x$.

Using the product rule: $\dfrac{dy}{dx} = (x)\left[\dfrac{-1}{x\sqrt{(x^2-1)}}\right] + (\text{arccosec } x)(1)$, from *Table 1(v)*

$$= \frac{-1}{\sqrt{(x^2-1)}} + \text{arccosec } x.$$

*Problem 9* Show that if $y = \arctan\left(\dfrac{\sin t}{\cos t - 1}\right)$ then $\dfrac{dy}{dt} = \dfrac{1}{2}$.

If $f(t) = \left(\dfrac{\sin t}{\cos t - 1}\right)$ then $f'(t) = \dfrac{(\cos t - 1)(\cos t) - (\sin t)(-\sin t)}{(\cos t - 1)^2}$

$= \dfrac{\cos^2 t - \cos t + \sin^2 t}{(\cos t - 1)^2} = \dfrac{1 - \cos t}{(\cos t - 1)^2}$ since $\sin^2 t + \cos^2 t = 1$

$= \dfrac{-(\cos t - 1)}{(\cos t - 1)^2} = \dfrac{-1}{(\cos t - 1)}$

Using *Table 1(iii)*, when $y = \arctan\left(\dfrac{\sin t}{\cos t - 1}\right)$

Then $\dfrac{dy}{dt} = \dfrac{\dfrac{-1}{\cos t - 1}}{1 + \left(\dfrac{\sin t}{\cos t - 1}\right)^2} = \dfrac{\left(\dfrac{-1}{\cos t - 1}\right)}{\dfrac{(\cos t - 1)^2 + (\sin t)^2}{(\cos t - 1)^2}}$

$\quad = \left(\dfrac{-1}{\cos t - 1}\right)\left(\dfrac{(\cos t - 1)^2}{\cos^2 t - 2\cos t + 1 + \sin^2 t}\right) = \dfrac{-(\cos t - 1)}{2 - 2\cos t} = \dfrac{1 - \cos t}{2(1 - \cos t)} = \dfrac{1}{2}$

*Problem 10* Evaluate, correct to 4 decimal places, arsinh 2.

From para. 5, $\text{arsinh}\,\dfrac{x}{a} = \ln\left\{\dfrac{x + \sqrt{(a^2 + x^2)}}{a}\right\}$

With $x = 2$ and $a = 1$, $\text{arsinh}\,2 = \ln\left\{\dfrac{2 + \sqrt{(1^2 + 2^2)}}{1}\right\} = \ln(2 + \sqrt 5) = \ln 4.2361$

$\qquad\qquad\qquad\qquad\qquad\qquad\qquad\qquad\quad = \textbf{1.4436, correct to 4 decimal}$
$\qquad\qquad\qquad\qquad\qquad\qquad\qquad\qquad\qquad\quad \textbf{places}$

*Problem 11* Show that $\text{artanh}\,\dfrac{x}{a} = \dfrac{1}{2}\ln\left(\dfrac{a+x}{a-x}\right)$ and evaluate, correct to 4 decimal places, $\text{artanh}\,\dfrac{3}{5}$.

If $y = \text{artanh}\,\dfrac{x}{a}$ then $\dfrac{x}{a} = \tanh y$.

From chapter 2,

$\tanh y = \dfrac{\sinh y}{\cosh y} = \dfrac{\dfrac{1}{2}(e^y - e^{-y})}{\dfrac{1}{2}(e^y + e^{-y})} = \dfrac{e^{2y} - 1}{e^{2y} + 1}$, by dividing each term by $e^{-y}$.

Thus, $\dfrac{x}{a} = \dfrac{e^{2y} - 1}{e^{2y} + 1}$, from which, $x(e^{2y} + 1) = a(e^{2y} - 1)$

Hence $x + a = ae^{2y} - xe^{2y} = e^{2y}(a - x)$ from which $e^{2y} = \left(\dfrac{a+x}{a-x}\right)$

Taking Naperian logarithms of both sides gives: $2y = \ln\left(\dfrac{a+x}{a-x}\right)$

$\qquad\qquad\qquad\qquad\qquad\qquad\qquad\text{and}\quad y = \dfrac{1}{2}\ln\left(\dfrac{a+x}{a-x}\right)$

Hence $\text{artanh}\,\dfrac{x}{a} = \dfrac{1}{2}\ln\left(\dfrac{a+x}{a-x}\right)$

Substituting $x = 3$ and $a = 5$ gives: $\text{artanh}\,\dfrac{3}{5} = \dfrac{1}{2}\ln\left(\dfrac{5+3}{5-3}\right) = \dfrac{1}{2}\ln 4$

$\qquad\qquad\qquad\qquad\qquad\qquad\qquad\qquad\qquad = \textbf{0.6931, correct to 4 decimal places.}$

*Problem 12* Prove that $\text{arcosh}\,\dfrac{x}{a} = \ln\left\{\dfrac{x + \sqrt{(x^2 - a^2)}}{a}\right\}$ and hence evaluate arcosh 1.4 correct to 4 decimal places.

If $y = \text{arcosh}\,\dfrac{x}{a}$ then $\dfrac{x}{a} = \cosh y$

$e^y = \cosh y + \sinh y = \cosh y \pm \sqrt{(\cosh^2 y - 1)}$

$\quad = \dfrac{x}{a} \pm \sqrt{\left[\left(\dfrac{x}{a}\right)^2 - 1\right]} = \dfrac{x}{a} \pm \dfrac{\sqrt{(x^2 - a^2)}}{a} = \dfrac{x \pm \sqrt{(x^2 - a^2)}}{a}$

Taking Naperian logarithms of both sides gives: $y = \ln \left\{ \dfrac{x \pm \sqrt{(x^2 - a^2)}}{a} \right\}$

Thus, assuming the principal value, $\text{arcosh} \dfrac{x}{a} = \ln \left\{ \dfrac{x + \sqrt{(x^2 - a^2)}}{a} \right\}$

$\text{arcosh } 1.4 = \text{arcosh } \dfrac{14}{10} = \text{arcosh } \dfrac{7}{5}$

In the equation for $\text{arcosh} \dfrac{x}{a}$, let $x = 7$ and $a = 5$.

Then $\text{arcosh } \dfrac{7}{5} = \ln \left\{ \dfrac{7 + \sqrt{(7^2 - 5^2)}}{5} \right\} = \ln 2.3798 = \mathbf{0.8670}$, correct to 4 decimal
places.

---

*Problem 13* Find the differential coefficient of $y = \text{arsinh } 2x$.

From *Table 2(i)*, $\dfrac{d}{dx} [\text{arsinh } f(x)] = \dfrac{f'(x)}{\sqrt{\{[f(x)]^2 + 1\}}}$

Hence $\dfrac{d}{dx} (\text{arsinh } 2x) = \dfrac{2}{\sqrt{\{(2x)^2 + 1\}}} = \dfrac{2}{\sqrt{\{4x^2 + 1\}}}$

---

*Problem 14* Determine the differential coefficient of $y = \text{arcosh } x/a$, and
hence find $\dfrac{d}{dx} [\text{arcosh } \sqrt{(x^2 + 1)}]$.

---

If $y = \text{arcosh} \dfrac{x}{a}$ then $\dfrac{x}{a} = \cosh y$ and $x = a \cosh y$.

$\dfrac{dx}{dy} = a \sinh y = a\sqrt{(\cosh^2 y - 1)}$, since $\cosh^2 y - \sinh^2 y = 1$

$\qquad = a\sqrt{\left[\left(\dfrac{x}{a}\right)^2 - 1\right]} = \dfrac{a\sqrt{(x^2 - a^2)}}{a} = \sqrt{(x^2 - a^2)}$

$\dfrac{dy}{dx} = \dfrac{1}{\dfrac{dx}{dy}} = \dfrac{1}{\sqrt{(x^2 - a^2)}}$

A sketch of part of $y = \text{arcosh } x$ is shown in *Fig 2(b)* where it can be seen that
when $x$ is greater than 1 there are two values of $y$ corresponding to a particular
value of $x$. The positive value of $y$ is defined as the principal value of $\text{arcosh } x/a$.
If $y = \text{arcosh } f(x)$, let $u = f(x)$.

Then $y = \text{arcosh } u$, $\dfrac{dy}{du} = \dfrac{1}{\sqrt{(u^2 - 1)}}$ and $\dfrac{du}{dx} = f'(x)$

From the function of a function rule, $\dfrac{dy}{dx} = \dfrac{dy}{du} \cdot \dfrac{du}{dx} = \dfrac{1}{\sqrt{(u^2 - 1)}} f'(x)$

i.e. $\dfrac{dy}{dx} = \dfrac{f'(x)}{\sqrt{\{[f(x)]^2 - 1\}}}$

If $y = \text{arcosh} \sqrt{(x^2 + 1)}$, then $f(x) = \sqrt{(x^2 + 1)}$ and $f'(x) = \dfrac{1}{2}(x^2 + 1)^{-\frac{1}{2}} (2x)$

$\qquad\qquad\qquad\qquad\qquad\qquad = \dfrac{x}{\sqrt{(x^2 + 1)}}$

55

Hence $\dfrac{d}{dx}[\text{arcosh}\sqrt{(x^2+1)}] = \dfrac{\dfrac{x}{\sqrt{(x^2+1)}}}{\sqrt{\{[\sqrt{(x^2+1)}]^2-1\}}} = \dfrac{\dfrac{x}{\sqrt{(x^2+1)}}}{\sqrt{(x^2+1-1)}}$.

$$= \dfrac{\dfrac{x}{\sqrt{(x^2+1)}}}{x} = \dfrac{1}{\sqrt{(x^2+1)}}$$

*Problem 15* Show that $\dfrac{d}{dx}[\text{artanh}\dfrac{x}{a}] = \dfrac{a}{a^2-x^2}$ and hence determine the differential coefficient of artanh $\dfrac{4x}{3}$.

If $y = \text{artanh}\dfrac{x}{a}$ then $\dfrac{x}{a} = \tanh y$ and $x = a\tanh y$.

$\dfrac{dx}{dy} = a\,\text{sech}^2\,y = a\,(1-\tanh^2\,y)$, since $1-\text{sech}^2\,y = \tanh^2\,y$

$$= a\left[1-\left(\dfrac{x}{a}\right)^2\right] = a\left(\dfrac{a^2-x^2}{a^2}\right) = \dfrac{a^2-x^2}{a}$$

Hence $\dfrac{dy}{dx} = \dfrac{1}{\dfrac{dx}{dy}} = \dfrac{a}{a^2-x^2}$

Comparing artanh $\dfrac{4x}{3}$ with artanh $\dfrac{x}{a}$ shows that $a = \dfrac{3}{4}$.

Hence $\dfrac{d}{dx}[\text{artanh}\dfrac{4x}{3}] = \dfrac{\dfrac{3}{4}}{\left(\dfrac{3}{4}\right)^2-x^2} = \dfrac{\dfrac{3}{4}}{\dfrac{9}{16}-x^2} = \dfrac{\dfrac{3}{4}}{\dfrac{9-16x^2}{16}} = \dfrac{3}{4}\cdot\dfrac{16}{(9-16x^2)}$

$$= \dfrac{12}{9-16x^2}$$

*Problem 16* Differentiate arcosech $(\sinh\theta)$.

From *Table 2(v)*, $\dfrac{d}{dx}[\text{arcosech}\,f(x)] = \dfrac{-f'(x)}{f(x)\sqrt{\{[f(x)]^2+1\}}}$

Hence $\dfrac{d}{d\theta}(\text{arcosech}\,(\sinh\theta)) = \dfrac{-\cosh\theta}{\sinh\theta\sqrt{\{\sinh^2\theta+1\}}}$

$$= \dfrac{-\cosh\theta}{\sinh\theta\sqrt{(\cosh^2\theta)}}\text{, since } \cosh^2\theta-\sinh^2\theta = 1$$

$$= \dfrac{-\cosh\theta}{\sinh\theta\cosh\theta} = \dfrac{-1}{\sinh\theta} = -\text{cosech}\,\theta.$$

*Problem 17* Find the differential coefficient of $y = \text{arsech}\,(2x-1)$.

From *Table 2(iv)*, $\dfrac{d}{dx}[\text{arsech}\,f(x)] = \dfrac{-f'(x)}{f(x)\sqrt{\{1-[f(x)]^2\}}}$

Hence $\dfrac{d}{dx}[\text{arsech}\,(2x-1)] = \dfrac{-2}{(2x-1)\sqrt{\{1-(2x-1)^2\}}} = \dfrac{-2}{(2x-1)\sqrt{\{1-(4x^2-4x+1)\}}}$

$$= \dfrac{-2}{(2x-1)\sqrt{(4x-4x^2)}} = \dfrac{-2}{(2x-1)\sqrt{[4x(1-x)]}}$$

$$= \dfrac{-2}{(2x-1)2\sqrt{[x(1-x)]}} = \dfrac{-1}{(2x-1)\sqrt{[x(1-x)]}}$$

**Problem 18** Show that $\frac{d}{dx}[\text{arcoth}(\sin x)] = \sec x$.

From *Table 2(vi)*, $\frac{d}{dx}[\text{arcoth}\,f(x)] = \frac{f'(x)}{\{1-[f(x)]^2\}}$

Hence $\frac{d}{dx}[\text{arcoth}(\sin x)] = \frac{\cos x}{\{1-(\sin x)^2\}} = \frac{\cos x}{\cos^2 x}$, since $\cos^2 x + \sin^2 x = 1$,

$$= \frac{1}{\cos x} = \sec x.$$

**Problem 19** Differentiate $y = (x^2-1)\,\text{artanh}\,x$.

Using the product rule, $\frac{dy}{dx} = (x^2-1)\left(\frac{1}{1-x^2}\right) + (\text{artanh}\,x)(2x)$

$$= \frac{-(1-x^2)}{(1-x^2)} + 2x\,\text{artanh}\,x$$

$$= 2x\,\text{artanh}\,x - 1.$$

**Problem 20** Differentiate $y = \frac{\text{arsinh}(\sinh x)}{x}$.

Using the quotient rule, $\frac{dy}{dx} = \dfrac{(x)\left(\dfrac{\cosh x}{\sqrt{(\sinh^2 x + 1)}}\right) - \{\text{arsinh}(\sinh x)\}\ (1)}{x^2}$

$$= \frac{(x)\dfrac{\cosh x}{\sqrt{(\cosh^2 x)}} - \text{arsinh}(\sinh x)}{x^2}$$

$$= \frac{x - \text{arsinh}(\sinh x)}{x^2} = \frac{x-x}{x^2} = 0$$

**Problem 21** Determine $\int \frac{dx}{\sqrt{(x^2+4)}}$.

Since $\frac{d}{dx}\left(\text{arsinh}\frac{x}{a}\right) = \frac{1}{\sqrt{(x^2+a^2)}}$ then $\int \frac{dx}{\sqrt{(x^2+a^2)}} = \text{arsinh}\frac{x}{a} + c$

Hence $\int \frac{1}{\sqrt{(x^2+4)}}\,dx = \int \frac{1}{\sqrt{(x^2+2^2)}}\,dx = \text{arsinh}\frac{x}{2} + c.$

**Problem 22** Determine $\int \frac{4}{\sqrt{(x^2-3)}}\,dx$

Since $\frac{d}{dx}\left(\text{arcosh}\frac{x}{a}\right) = \frac{1}{\sqrt{(x^2-a^2)}}$ then $\int \frac{1}{\sqrt{(x^2-a^2)}}\,dx = \text{arcosh}\frac{x}{a} + c$

Hence $\int \frac{4}{\sqrt{(x^2-3)}}\,dx = 4\int \frac{1}{\sqrt{[x^2-(\sqrt{3})^2]}}\,dx = 4\,\text{arcosh}\frac{x}{\sqrt{3}} + c.$

**Problem 23** Find $\int \frac{2}{(9-4x^2)}\,dx$.

Since $\operatorname{artanh} \dfrac{x}{a} = \dfrac{a}{a^2-x^2}$ then $\displaystyle\int \dfrac{a}{a^2-x^2}\,dx = \operatorname{artanh}\dfrac{x}{a}+c,$

i.e. $\displaystyle\int \dfrac{1}{(a^2-x^2)}\,dx = \dfrac{1}{a}\operatorname{artanh}\dfrac{x}{a}+c.$

Hence $\displaystyle\int\dfrac{2}{9-4x^2}\,dx = 2\int\dfrac{1}{4\left(\frac{9}{4}-x^2\right)}\,dx = \dfrac{1}{2}\int\dfrac{1}{[\left(\frac{3}{2}\right)^2-x^2]}\,dx$

$$= \dfrac{1}{2}\left[\dfrac{1}{\left(\frac{3}{2}\right)}\operatorname{artanh}\dfrac{x}{\frac{3}{2}}+c\right]$$

i.e. $\displaystyle\int\dfrac{2}{(9-4x^2)}\,dx = \dfrac{1}{3}\operatorname{artanh}\dfrac{2x}{3}+c.$

## C. FURTHER PROBLEMS ON DIFFERENTIATION OF INVERSE TRIGONOMETRIC AND HYPERBOLIC FUNCTIONS

In *Problems 1 to 6*, differentiate with respect to the variable.

1 (a) $\arcsin 4x$; (b) $\arcsin\dfrac{x}{2}$ $\qquad\left[\text{(a) }\dfrac{4}{\sqrt{(1-16x^2)}}; \text{(b) }\dfrac{1}{\sqrt{(4-x^2)}}\right]$

2 (a) $\arccos 3x$; (b) $\dfrac{2}{3}\arccos\dfrac{x}{3}$. $\qquad\left[\text{(a) }\dfrac{-3}{\sqrt{(1-9x^2)}}; \text{(b) }\dfrac{-2}{3\sqrt{(9-x^2)}}\right]$

3 (a) $3\arctan 2x$; (b) $\dfrac{1}{2}\arctan\sqrt{x}$. $\qquad\left[\text{(a) }\dfrac{6}{1+4x^2}; \text{(b) }\dfrac{1}{4\sqrt{x}(1+x)}\right]$

4 (a) $2\operatorname{arcsec} 2t$; (b) $\operatorname{arcsec}\dfrac{3}{4}x$. $\qquad\left[\text{(a) }\dfrac{2}{t\sqrt{(4t^2-1)}}; \text{(b) }\dfrac{4}{x\sqrt{(9x^2-16)}}\right]$

5 (a) $\dfrac{5}{2}\operatorname{arccosec}\dfrac{\theta}{2}$; (b) $\operatorname{arccosec} x^2$. $\qquad\left[\text{(a) }\dfrac{-5}{\theta\sqrt{(\theta^2-4)}}; \text{(b) }\dfrac{-2}{x\sqrt{(x^4-1)}}\right]$

6 (a) $3\operatorname{arccot} 2t$; (b) $\operatorname{arccot}\sqrt{(\theta^2-1)}$. $\qquad\left[\text{(a) }\dfrac{-6}{1+4t^2}; \text{(b) }\dfrac{-1}{\theta\sqrt{(\theta^2-1)}}\right]$

7 Show that $\dfrac{d}{dx}[\operatorname{arccot} f(x)] = \dfrac{-f'(x)}{\{1+[f(x)]^2\}}$ and hence determine

$\dfrac{d}{dx}\operatorname{arccot}(\sinh x)$. $\qquad\qquad\qquad [-\operatorname{sech} x]$

8 Differentiate (a) $\ln(\arccos x)$; (b) $2e^{\arctan x}$.

$\qquad\left[\text{(a) }\dfrac{-1}{\sqrt{(1-x^2)}\arccos x}; \text{(b) }\dfrac{2e^{\arctan x}}{1+x^2}\right]$

9 Show that the differential coefficient of $\arctan\dfrac{2x}{1-x^2}$ is $\dfrac{2}{1+x^2}$.

10 Show that the differential coefficient of $\operatorname{arccosec}\dfrac{x}{a}$ is $\dfrac{-a}{x\sqrt{(x^2-a^2)}}$

In *Problems 11 to 15* differentiate with respect to the variable.

11 (a) $2x\arcsin 3x$; (b) $t^2\operatorname{arcsec} 2t$

$\qquad\left[\begin{array}{l}\text{(a) }\dfrac{6x}{\sqrt{(1-9x^2)}}+2\arcsin 3x \\[2mm] \text{(b) }\dfrac{t}{\sqrt{(4t^2-1)}}+2t\operatorname{arcsec} 2t\end{array}\right]$

12 (a) $\theta^2 \arccos(\theta^2-1)$; (b) $(1-x^2)\arctan x$.

$$\begin{bmatrix} \text{(a) } 2\theta \arccos(\theta^2-1) - \dfrac{2\theta^2}{\sqrt{(2-\theta^2)}}; \\ \text{(b) } \left(\dfrac{1-x^2}{1+x^2}\right) - 2x \arctan x \end{bmatrix}$$

13 (a) $2\sqrt{t}\,\text{arccot}\,t$; (b) $x\,\text{arccosec}\sqrt{x}$.

$$\begin{bmatrix} \text{(a) } \dfrac{-2\sqrt{t}}{(1+t^2)} + \dfrac{1}{\sqrt{t}}\,\text{arccot}\,t; \\ \text{(b) } \text{arccosec}\sqrt{x} - \dfrac{1}{2\sqrt{(x+1)}} \end{bmatrix}$$

14 (a) $\dfrac{\arcsin 3x}{x^2}$; (b) $\dfrac{\arccos x}{\sqrt{(1-x^2)}}$.

$$\begin{bmatrix} \text{(a) } \dfrac{1}{x^3}\left\{\dfrac{3x}{\sqrt{(1-9x^2)}} - 2\arcsin 3x\right\}; \\ \text{(b) } \dfrac{-(1+\dfrac{x}{\sqrt{(1-x^2)}}\text{arccosec}\,x)}{(1-x^2)} \end{bmatrix}$$

15 (a) $\dfrac{\text{arccosec}\,x}{\sqrt{(x^2-1)}}$; (b) $\dfrac{\text{arccot}\,2\theta}{(1+4\theta^2)}$

$$\begin{bmatrix} \text{(a) } \dfrac{-\left(\dfrac{1}{x}+\dfrac{x}{\sqrt{(x-1)}}\text{arccosec}\,x\right)}{(x^2-1)}; \\ \text{(b) } \dfrac{-2(1+4\theta\,\text{arccot}\,2\theta)}{(1+4\theta^2)^2} \end{bmatrix}$$

In *Problems 16 to 18* use logarithmic equivalents of inverse hyperbolic functions to evaluate correct to 4 decimal places.

16 (a) $\text{arsinh}\,\dfrac{1}{2}$; (b) $\text{arsinh}\,4$; (c) $\text{arsinh}\,0.9$  [(a) 0.4812; (b) 2.0947; (c) 0.8089]

17 (a) $\text{arcosh}\,\dfrac{5}{4}$; (b) $\text{arcosh}\,3$; (c) $\text{arcosh}\,4.3$  [(a) 0.6931; (b) 1.7627; (c) 2.1380]

18 (a) $\text{artanh}\,\dfrac{1}{4}$; (b) $\text{artanh}\,\dfrac{5}{8}$; (c) $\text{artanh}\,0.7$  [(a) 0.1277; (b) 0.1833; (c) 0.1735]

In *Problems 19 to 31*, differentiate with respect to the variable.

19 (a) $\text{arsinh}\,\dfrac{x}{3}$; (b) $\text{arsinh}\,4x$.

$$\left[\text{(a) } \dfrac{1}{\sqrt{(x^2+9)}}; \text{(b) } \dfrac{4}{\sqrt{(16x^2+1)}}\right]$$

20 (a) $2\,\text{arcosh}\,\dfrac{t}{3}$; (b) $\dfrac{1}{2}\text{arcosh}\,2\theta$.

$$\left[\text{(a) } \dfrac{2}{\sqrt{(t^2-9)}}; \text{(b) } \dfrac{1}{\sqrt{(4\theta^2-1)}}\right]$$

21 (a) $\text{artanh}\,\dfrac{2x}{5}$; (b) $3\,\text{artanh}\,3x$.

$$\left[\text{(a) } \dfrac{10}{25-4x^2}; \text{(b) } \dfrac{9}{(1-9x^2)}\right]$$

22 (a) $\text{arsech}\,\dfrac{3x}{4}$; (b) $-\dfrac{1}{2}\text{arsech}\,2x$

$$\left[\text{(a) } \dfrac{-4}{x\sqrt{(16-9x^2)}}; \text{(b) } \dfrac{1}{2x\sqrt{(1-4x^2)}}\right]$$

23 (a) $\text{arcosech}\,\dfrac{x}{4}$; (b) $\dfrac{1}{2}\text{arcosech}\,4x$.

$$\left[\text{(a) } \dfrac{-4}{x\sqrt{(x^2+16)}}; \text{(b) } \dfrac{-1}{2x\sqrt{(16x^2+1)}}\right]$$

24 (a) $\text{arcoth}\,\dfrac{2x}{7}$; (b) $\dfrac{1}{4}\text{arcoth}\,3t$

$$\left[\text{(a) } \dfrac{14}{49-4x^2}; \text{(b) } \dfrac{3}{4(1-9t^2)}\right]$$

25 (a) $2\,\text{arsinh}\sqrt{(x^2-1)}$; (b) $\dfrac{1}{2}\text{arcosh}\sqrt{(x^2+1)}$.

$$\left[\text{(a) } \dfrac{2}{\sqrt{(x^2-1)}}; \text{(b) } \dfrac{1}{2\sqrt{(x^2+1)}}\right]$$

26 (a) $\text{arsech}\,(x-1)$; (b) $\text{artanh}\,(\tanh x)$.

$$\left[\text{(a) } \dfrac{-1}{(x-1)\sqrt{[x(2-x)]}}; \text{(b) } 1\right]$$

27 (a) $\text{arcosh}\left(\dfrac{t}{t-1}\right)$; (b) $\text{arcoth}\,(\cos x)$.

$$\left[\text{(a)} \dfrac{-1}{(t-1)\sqrt{(2t-1)}}; \text{(b) } -\text{cosec}\,x\right]$$

28  (a) $\theta$ arsinh $\theta$; (b) $\sqrt{x}$ arcosh $x$.

$$\left[\text{(a) } \frac{\theta}{\sqrt{(\theta^2+1)}} + \text{arsinh } \theta; \text{(b) } \frac{\sqrt{x}}{\sqrt{(x^2-1)}} + \frac{\text{arcosh } x}{2\sqrt{x}}\right]$$

29  (a) $x^2$ arcosech $\sqrt{x}$; (b) $\theta$ artanh $(1+\theta^2)$.

$$\left[\begin{array}{l} \text{(a) } 2x \text{ arcosech} \sqrt{x} - \dfrac{x}{2\sqrt{(x+1)}}; \\[2mm] \text{(b) artanh } (1+\theta^2) - \dfrac{2}{(2+\theta^2)} \end{array}\right]$$

30  (a) $\dfrac{\text{arsinh } (x-2)}{x}$; (b) $\dfrac{\text{arcoth } (3x-1)}{x^2}$

$$\left[\begin{array}{l} \text{(a) } \dfrac{1}{x^2}\left\{\dfrac{x}{\sqrt{(x^2-4x+5)}} - \text{arsinh } (x-2)\right\}; \\[3mm] \text{(b) } \dfrac{1}{x^3}\left\{\dfrac{1}{(2-3x)} -2 \text{ arcoth } (3x-1)\right\} \end{array}\right]$$

31  (a) $\dfrac{2 \text{ arsech} \sqrt{t}}{t^2}$; (b) $\dfrac{\text{artanh } x}{(1-x^2)}$

$$\left[\begin{array}{l} \text{(a) } \dfrac{-1}{t^3}\left\{\dfrac{1}{\sqrt{(1-t)}} + 4 \text{ arsech} \sqrt{t}\right\}; \\[3mm] \text{(b) } \dfrac{1+2x \text{ artanh } x}{(1-x^2)^2} \end{array}\right]$$

32  Show that $\dfrac{d}{dx}$ [$x$ arcosh $(\cosh x)$] $= 2x$.

In *Problems 33 to 35*, determine the given integrals.

33  (a) $\displaystyle\int \frac{1}{\sqrt{(x^2+9)}}\, dx$; (b) $\displaystyle\int \frac{3}{\sqrt{(4x^2+25)}}\, dx$

$$\left[\begin{array}{l} \text{(a) arsinh } \dfrac{x}{3}+ c; \\[3mm] \text{(b) } \dfrac{3}{2}\text{arsinh } \dfrac{2x}{5}+ c \end{array}\right]$$

34  (a) $\displaystyle\int \frac{1}{\sqrt{(x^2-16)}}\, dx$; (b) $\displaystyle\int \frac{2}{\sqrt{(t^2-5)}}\, dt$

$$\left[\begin{array}{l} \text{(a) arcosh } \dfrac{x}{4}+ c; \\[3mm] \text{(b) } 2 \text{ arcosh } \dfrac{x}{\sqrt{5}}+ c \end{array}\right]$$

35  (a) $\displaystyle\int \frac{d\theta}{(36+\theta^2)}$ ; (b) $\displaystyle\int \frac{3}{(16-2x^2)}\, dx$

$$\left[\begin{array}{l} \text{(a) } \dfrac{1}{6}\text{arctan } \dfrac{\theta}{6}+ c; \\[3mm] \text{(b) } \dfrac{3}{2\sqrt{8}} \text{ artanh } \dfrac{x}{\sqrt{8}} + c \end{array}\right]$$

# 8 Partial differentiation

## A. MAIN POINTS CONCERNED WITH PARTIAL DIFFERENTIATION

1  In engineering, it sometimes happens that the variation of one quantity depends on changes taking place in two, or more, other quantities. For example, the volume $V$ of a cylinder is given by $V = \pi r^2 h$. The volume will change if either radius $r$ or height $h$ is changed. The formula for volume may be stated mathematically as $V = f(r,h)$ which means '$v$ is some function of $r$ and $h$'. Some other practical examples include:

(i)   time of oscillation, $t = 2\pi\sqrt{\left(\dfrac{l}{g}\right)}$ , i.e. $t = f(l,g)$

(ii)  torque $T = I\alpha$, i.e. $T = f(I,\alpha)$

(iii) pressure of an ideal gas $p = \dfrac{mRT}{V}$ , i.e. $p = f(T,V)$

(iv)  resonant frequency $f_0 = \dfrac{1}{2\pi\sqrt{(LC)}}$ , i.e. $f_0 = f(L,C)$, and so on.

2  (i)   When differentiating a function having two variables, one variable is kept constant and the differential coefficient of the other variable is found with respect to that variable. The differential coefficient obtained is called a **partial derivative** of the function.

(ii)  A 'curly dee', $\partial$ is used to denote a differential coefficient in an expression containing more than one variable.

Hence if $V = \pi r^2 h$ then $\dfrac{\partial V}{\partial r}$ means 'the partial derivative of $V$ with respect to $r$, with $h$ remaining constant'.

Thus $\dfrac{\partial V}{\partial r} = (\pi h)\dfrac{d}{dr}(r^2) = (\pi h)(2r) = 2\pi rh.$

Similarly, $\dfrac{\partial V}{\partial h}$ means 'the partial derivative of $V$ with respect to $h$, with $r$ remaining constant'.

Thus $\dfrac{\partial V}{\partial h} = (\pi r^2)\dfrac{d}{dh}(h) = \pi r^2(1) = \pi r^2$

61

(iii) $\dfrac{\partial V}{\partial r}$ and $\dfrac{\partial V}{\partial h}$ are examples of **first order partial derivatives,**

since $n = 1$ when written in the form $\dfrac{\partial^n V}{\partial r^n}$

(See *Problems 1 to 6*)

3   As with ordinary differentiation, where a differential coefficient may be differentiated again, a partial derivative may be differentiated partially again to give higher order partial derivatives.

(i) Differentiating $\dfrac{\partial V}{\partial r}$ of para. 2 with respect to $r$, keeping $h$ constant, gives

$\dfrac{\partial}{\partial r}\left(\dfrac{\partial V}{\partial r}\right)$, which is written as $\dfrac{\partial^2 V}{\partial r^2}$ .

Thus $\dfrac{\partial^2 V}{\partial r^2} = \dfrac{\partial}{\partial r}(2\pi r h) = 2\pi h$

(ii) Differentiating $\dfrac{\partial V}{\partial h}$ with respect to $h$, keeping $r$ constant, gives $\dfrac{\partial}{\partial h}\left(\dfrac{\partial V}{\partial h}\right)$,

which is written as $\dfrac{\partial^2 V}{\partial h^2}$ .

Thus $\dfrac{\partial^2 V}{\partial h^2} = \dfrac{\partial}{\partial h}(\pi r^2) = 0$

(iii) Differentiating $\dfrac{\partial V}{\partial h}$ with respect to $r$, keeping $h$ constant, gives $\dfrac{\partial}{\partial r}\left(\dfrac{\partial V}{\partial h}\right)$,

which is written as $\dfrac{\partial^2 V}{\partial r \partial h}$ .

Thus $\dfrac{\partial^2 V}{\partial r \partial h} = \dfrac{\partial}{\partial r}\left(\dfrac{\partial V}{\partial h}\right) = \dfrac{\partial}{\partial r}(\pi r^2) = 2\pi r$

(iv) Differentiating $\dfrac{\partial V}{\partial r}$ with respect to $h$, keeping $r$ constant, gives $\dfrac{\partial}{\partial h}\left(\dfrac{\partial V}{\partial r}\right)$,

which is written as $\dfrac{\partial^2 V}{\partial h \partial r}$ .

Thus $\dfrac{\partial^2 V}{\partial h \partial r} = \dfrac{\partial}{\partial h}\left(\dfrac{\partial V}{\partial r}\right) = \dfrac{\partial}{\partial h}(2\pi r h) = 2\pi r$

(v) $\dfrac{\partial^2 V}{\partial r^2}$ , $\dfrac{\partial^2 V}{\partial h^2}$ , $\dfrac{\partial^2 V}{\partial r \partial h}$ and $\dfrac{\partial^2 V}{\partial h \partial r}$ are examples of **second order partial**

**derivatives.**

(vi) It is seen from (iii) and (iv) that $\dfrac{\partial^2 V}{\partial r \partial h} = \dfrac{\partial^2 V}{\partial h \partial r}$ and such a result is always

true for continuous functions (i.e. a graph of the function has no sudden jumps or breaks).

(See *Problems 5 to 9*)

4  (i) First order partial derivatives are used when finding the total differential, rates of change and errors for functions of two or more variables (see chapter 9).

(ii) Second order partial derivatives are used in the solution of partial differential equations, in waveguide theory, and in such areas of thermodynamics covering entropy and the continuity theorem.

# B. WORKED PROBLEMS ON PARTIAL DIFFERENTIATION

*Problem 1* If $Z = 5x^4 + 2x^3y^2 - 3y$ find (a) $\dfrac{\partial Z}{\partial x}$ and (b) $\dfrac{\partial Z}{\partial y}$.

(a) To find $\dfrac{\partial Z}{\partial x}$, $y$ is kept constant.

Since $Z = 5x^4 + (2y^2)x^3 - (3y)$

then $\dfrac{\partial Z}{\partial x} = \dfrac{d}{dx}(5x^4) + (2y^2)\dfrac{d}{dx}(x^3) - (3y)\dfrac{d}{dx}(1), = 20x^3 + (2y^2)(3x^2) - 0$

Hence $\dfrac{\partial Z}{\partial x} = 20x^3 + 6x^2y^2$

(b) To find $\dfrac{\partial Z}{\partial y}$, $x$ is kept constant.

Since $Z = (5x^4) + (2x^3)y^2 - 3y$

then $\dfrac{\partial Z}{\partial y} = (5x^4)\dfrac{d}{dy}(1) + (2x^3)\dfrac{d}{dy}(y^2) - 3\dfrac{d}{dy}(y) = 0 + (2x^3)(2y) - 3$

Hence $\dfrac{\partial Z}{\partial y} = 4x^3y - 3$

*Problem 2* Given $y = 4 \sin 3x \cos 2t$, find $\dfrac{\partial y}{\partial x}$ and $\dfrac{\partial y}{\partial t}$.

To find $\dfrac{\partial y}{\partial x}$, $t$ is kept constant.

Hence $\dfrac{\partial y}{\partial x} = (4 \cos 2t)\dfrac{d}{dx}(\sin 3x) = (4 \cos 2t)(3 \cos 3x)$

i.e. $\dfrac{\partial y}{\partial x} = 12 \cos 3x \cos 2t$.

To find $\dfrac{\partial y}{\partial t}$, $x$ is kept constant.

Hence $\dfrac{\partial y}{\partial t} = (4 \sin 3x)\dfrac{d}{dt}(\cos 2t) = (4 \sin 3x)(-2 \sin 2t)$

i.e. $\dfrac{\partial y}{\partial t} = -8 \sin 3x \sin 2t$.

*Problem 3* If $Z = \sin xy$ show that $\dfrac{1}{y}\dfrac{\partial Z}{\partial x} = \dfrac{1}{x}\dfrac{\partial Z}{\partial y}$.

$\dfrac{\partial Z}{\partial x} = y \cos xy$, since $y$ is kept constant.

$\dfrac{\partial Z}{\partial y} = x \cos xy$, since $x$ is kept constant.

$\dfrac{1}{y}\dfrac{\partial Z}{\partial x} = \left(\dfrac{1}{y}\right)(y \cos xy) = \cos xy;\quad \dfrac{1}{x}\dfrac{\partial Z}{\partial y} = \left(\dfrac{1}{x}\right)(x \cos xy) = \cos xy$

Hence $\dfrac{1}{y}\dfrac{\partial Z}{\partial x} = \dfrac{1}{x}\dfrac{\partial Z}{\partial y}$

*Problem 4* Determine $\dfrac{\partial Z}{\partial x}$ and $\dfrac{\partial Z}{\partial y}$ when $Z = \dfrac{1}{\sqrt{(x^2 + y^2)}}$.

$$Z = \frac{1}{\sqrt{(x^2+y^2)}} = (x^2+y^2)^{-\frac{1}{2}}$$

$$\frac{\partial Z}{\partial x} = -\frac{1}{2}(x^2+y^2)^{-\frac{3}{2}}(2x), \text{ by the function of a function rule (keeping } y \text{ constant)}$$

$$= \frac{-x}{(x^2+y^2)^{3/2}} = \frac{-x}{\sqrt{(x^2+y^2)^3}}$$

$$\frac{\partial Z}{\partial y} = -\frac{1}{2}(x^2+y^2)^{-\frac{3}{2}}(2y), \text{ keeping } x \text{ constant} = \frac{-y}{\sqrt{(x^2+y^2)^3}}$$

**Problem 5** Pressure $p$ of a mass of gas is given by $pV = mRT$, where $m$ and $R$ are constants, $V$ is the volume and $T$ the temperature. Find expressions for $\frac{\partial p}{\partial T}$ and $\frac{\partial p}{\partial V}$.

Since $pV = mRT$ then $p = \frac{mRT}{V}$

To find $\frac{\partial p}{\partial T}$, $V$ is kept constant

Hence $\frac{\partial p}{\partial T} = \left(\frac{mR}{V}\right) \frac{d}{dT}(T) = \frac{mR}{V}$

To find $\frac{\partial p}{\partial V}$, $T$ is kept constant

Hence $\frac{\partial p}{\partial V} = (mRT)\frac{d}{dV}\left(\frac{1}{V}\right) = (mRT)\frac{d}{dV}(V^{-1}) = (mRT)(-V^{-2}) = \frac{-mRT}{V^2}$

**Problem 6** The time of oscillation, $t$, of a pendulum is given by $t = 2\pi\sqrt{\left(\frac{l}{g}\right)}$, where $l$ is the length of the pendulum and $g$ the free fall acceleration due to gravity. Determine $\frac{\partial T}{\partial l}$ and $\frac{\partial T}{\partial g}$.

To find $\frac{\partial t}{\partial l}$, $g$ is kept constant.

$$t = 2\pi\sqrt{\left(\frac{l}{g}\right)} = \left(\frac{2\pi}{\sqrt{g}}\right)\sqrt{l} = \left(\frac{2\pi}{\sqrt{g}}\right)l^{\frac{1}{2}}$$

Hence $\frac{\partial t}{\partial l} = \left(\frac{2\pi}{\sqrt{g}}\right)\frac{d}{dl}(l^{\frac{1}{2}}) = \left(\frac{2\pi}{\sqrt{g}}\right)\left(\frac{1}{2}l^{-\frac{1}{2}}\right) = \left(\frac{2\pi}{\sqrt{g}}\right)\left(\frac{1}{2\sqrt{l}}\right) = \frac{\pi}{\sqrt{(lg)}}$

To find $\frac{\partial t}{\partial g}$, $l$ is kept constant.

$$t = 2\pi\sqrt{\left(\frac{l}{g}\right)} = (2\pi\sqrt{l})\left(\frac{1}{\sqrt{g}}\right) = (2\pi\sqrt{l})g^{-\frac{1}{2}}$$

Hence $\frac{\partial t}{\partial g} = (2\pi\sqrt{l})\frac{d}{dg}(g^{-\frac{1}{2}}) = (2\pi\sqrt{l})(-\frac{1}{2}g^{-\frac{3}{2}})$

$$= (2\pi\sqrt{l})\left(\frac{-1}{2\sqrt{g^3}}\right) = \frac{-\pi\sqrt{l}}{\sqrt{g^3}} = -\pi\sqrt{\left(\frac{l}{g^3}\right)}$$

**Problem 7** Given $Z = 4x^2y^3 - 2x^3 + 7y^2$ find (a) $\frac{\partial^2 Z}{\partial x^2}$ (b) $\frac{\partial^2 Z}{\partial y^2}$ (c) $\frac{\partial^2 Z}{\partial x \partial y}$ (d) $\frac{\partial^2 Z}{\partial y \partial x}$

(a) $\dfrac{\partial Z}{\partial x} = 8xy^3 - 6x^2$

$\dfrac{\partial^2 Z}{\partial x^2} = \dfrac{\partial}{\partial x}\left(\dfrac{\partial Z}{\partial x}\right) = \dfrac{\partial}{\partial x}(8xy^3 - 6x^2) = \mathbf{8y^3 - 12x}$

(b) $\dfrac{\partial Z}{\partial y} = 12x^2 y^2 + 14y$

$\dfrac{\partial^2 Z}{\partial y^2} = \dfrac{\partial}{\partial y}\left(\dfrac{\partial Z}{\partial y}\right) = \dfrac{\partial}{\partial y}(12x^2 y^2 + 14y) = \mathbf{24x^2 y + 14}.$

(c) $\dfrac{\partial^2 Z}{\partial x \partial y} = \dfrac{\partial}{\partial x}\left(\dfrac{\partial Z}{\partial y}\right) = \dfrac{\partial}{\partial x}(12x^2 y^2 + 14y) = \mathbf{24xy^2}$

(d) $\dfrac{\partial^2 Z}{\partial y \partial x} = \dfrac{\partial}{\partial y}\left(\dfrac{\partial Z}{\partial x}\right) = \dfrac{\partial}{\partial y}(8xy^3 - 6x^2) = \mathbf{24xy^2}$

(it is noted that $\dfrac{\partial^2 Z}{\partial x \partial y} = \dfrac{\partial^2 Z}{\partial y \partial x}$ )

*Problem 8* Show that when $Z = e^{-t}\sin\theta$, (a) $\dfrac{\partial^2 Z}{\partial t^2} = -\dfrac{\partial^2 Z}{\partial \theta^2}$ , and

(b) $\dfrac{\partial^2 Z}{\partial t \partial \theta} = \dfrac{\partial^2 Z}{\partial \theta \partial t}$

(a) $\dfrac{\partial Z}{\partial t} = -e^{-t}\sin\theta;\quad \dfrac{\partial^2 Z}{\partial t^2} = e^{-t}\sin\theta.$

$\dfrac{\partial Z}{\partial \theta} = e^{-t}\cos\theta;\quad \dfrac{\partial^2 Z}{\partial \theta^2} = -e^{-t}\sin\theta$

**Hence** $\dfrac{\partial^2 Z}{\partial t^2} = -\dfrac{\partial^2 Z}{\partial \theta^2}$ .

(b) $\dfrac{\partial^2 Z}{\partial t \partial \theta} = \dfrac{\partial}{\partial t}\left(\dfrac{\partial Z}{\partial \theta}\right) = \dfrac{\partial}{\partial t}(e^{-t}\cos\theta) = -e^{-t}\cos\theta$

$\dfrac{\partial^2 Z}{\partial \theta \partial t} = \dfrac{\partial}{\partial \theta}\left(\dfrac{\partial Z}{\partial t}\right) = \dfrac{\partial}{\partial \theta}(-e^{-t}\sin\theta) = -e^{-t}\cos\theta$

Hence $\dfrac{\partial^2 Z}{\partial t \partial \theta} = \dfrac{\partial^2 Z}{\partial \theta \partial t}$

*Problem 9* Show that if $Z = \dfrac{x}{y}\ln y$, then (a) $\dfrac{\partial Z}{\partial y} = x\dfrac{\partial^2 Z}{\partial y \partial x}$ , and

(b) evaluate $\dfrac{\partial^2 Z}{\partial y^2}$ when $x = -3$ and $y = 1$,

(a) To find $\dfrac{\partial Z}{\partial x}$, $y$ is kept constant.

Hence $\dfrac{\partial Z}{\partial x} = \left(\dfrac{1}{y}\ln y\right)\dfrac{d}{dx}(x) = \dfrac{1}{y}\ln y$

To find $\dfrac{\partial Z}{\partial y}$ , $x$ is kept constant.

Hence $\dfrac{\partial Z}{\partial y} = (x)\dfrac{d}{dy}\left(\dfrac{\ln y}{y}\right) = (x)\left\{\dfrac{(y)(\frac{1}{y}) - (\ln y)(1)}{y^2}\right\}$, using the quotient rule,

$= x\left(\dfrac{1 - \ln y}{y^2}\right) = \dfrac{x}{y^2}(1 - \ln y)$

$$\frac{\partial^2 Z}{\partial y \partial x} = \frac{\partial}{\partial y}\left(\frac{\partial Z}{\partial x}\right) = \frac{\partial}{\partial y}\left(\frac{\ln y}{y}\right) = \frac{(y)(\frac{1}{y}) - (\ln y)(1)}{y^2} \quad \text{, using the quotient rule,}$$

$$= \frac{1}{y^2}(1 - \ln y)$$

**Hence** $x\dfrac{\partial^2 Z}{\partial y \partial x} = \dfrac{x}{y^2}(1 - \ln y) = \dfrac{\partial Z}{\partial y}.$

(b) $\dfrac{\partial^2 Z}{\partial y^2} = \dfrac{\partial}{\partial y}\left(\dfrac{\partial Z}{\partial y}\right) = \dfrac{\partial}{\partial y}\left\{\dfrac{x}{y^2}(1 - \ln y)\right\} = (x)\dfrac{d}{dy}\left(\dfrac{1 - \ln y}{y^2}\right)$

$$= (x)\left\{\frac{(y^2)(-\frac{1}{y}) - (1 - \ln y)(2y)}{y^4}\right\}, \text{ using the quotient rule,}$$

$$= \frac{x}{y^4}\{-y - 2y + 2y \ln y\} = \frac{xy}{y^4}\{-3 + 2 \ln y\} = \frac{x}{y^3}(2 \ln y - 3)$$

When $x = -3$ and $y = 1$, $\dfrac{\partial^2 Z}{\partial y^2} = \dfrac{(-3)}{(1)^3}(2 \ln 1 - 3) = (-3)(-3) = \mathbf{9}.$

## C. FURTHER PROBLEMS ON PARTIAL DIFFERENTIATION

In *Problems 1 to 6*, find $\dfrac{\partial Z}{\partial x}$ and $\dfrac{\partial Z}{\partial y}$ .

1   $Z = 2xy.$          $\left[\dfrac{\partial Z}{\partial x} = 2y; \dfrac{\partial Z}{\partial y} = 2x\right]$

2   $Z = x^3 - 2xy + y^2$       $\left[\dfrac{\partial Z}{\partial x} = 3x^2 - 2y; \dfrac{\partial Z}{\partial y} = -2x + 2y\right]$

3   $Z = \dfrac{x}{y}$          $\left[\dfrac{\partial Z}{\partial x} = \dfrac{1}{y}; \dfrac{\partial Z}{\partial y} = \dfrac{-x}{y^2}\right]$

4   $Z = \sin(4x + 3y)$       $\left[\dfrac{\partial Z}{\partial x} = 4\cos(4x + 3y); \dfrac{\partial Z}{\partial y} = 3\cos(4x + 3y)\right]$

5   $Z = x^3 y^2 - \dfrac{y}{x^2} + \dfrac{1}{y}$    $\left[\dfrac{\partial Z}{\partial x} = 3x^2 y^2 + \dfrac{2y}{x^3}; \dfrac{\partial Z}{\partial y} = 2x^3 y - \dfrac{1}{x^2} - \dfrac{1}{y^2}\right]$

6   $Z = \cos 3x \sin 4y$     $\left[\dfrac{\partial Z}{\partial x} = -3\sin 3x \sin 4y; \dfrac{\partial Z}{\partial y} = 4\cos 3x \cos 4y\right]$

7   The volume of a cone of height $h$ and base radius $r$ is given by

$V = \dfrac{1}{3}\pi r^2 h$. Determine $\dfrac{\partial V}{\partial h}$ and $\dfrac{\partial V}{\partial r}$ .    $\left[\dfrac{\partial V}{\partial h} = \dfrac{1}{3}\pi r^2; \dfrac{\partial V}{\partial r} = \dfrac{2}{3}\pi r h\right]$

8   The resonant frequency $f_0$ in a series electrical circuit is given by

$f_0 = \dfrac{1}{2\pi\sqrt{LC}}$. Show that $\dfrac{\partial f_0}{\partial L} = \dfrac{-1}{4\pi\sqrt{(CL^3)}}$

In *Problems 9 to 12*, find (a) $\dfrac{\partial^2 Z}{\partial x^2}$; (b) $\dfrac{\partial^2 Z}{\partial y^2}$; (c) $\dfrac{\partial^2 Z}{\partial x \partial y}$; (d) $\dfrac{\partial^2 Z}{\partial y \partial x}$

9   $Z = (2x - 3y)^2$        [(a) 8; (b) 18; (c) −12; (d) −12]

10 $Z = 2 \ln xy$ $\qquad\qquad$ $\left[\text{(a) } \dfrac{-2}{x^2} \text{ ; (b) } \dfrac{-2}{y^2} \text{ ; (c) } 0; \text{ (d) } 0\right]$

11 $Z = \dfrac{(x-y)}{(x+y)}$ $\qquad\left[\begin{array}{l}\text{(a) } \dfrac{-4y}{(x+y)^3}\text{ ; (b) } \dfrac{4x}{(x+y)^3}\text{; (c) } \dfrac{2(x-y)}{(x+y)^3}\text{ ;} \\[2ex] \text{(d) } \dfrac{2(x-y)}{(x+y)^3}\end{array}\right]$

12 $Z = \sinh x \cosh 2y$ $\qquad\left[\begin{array}{l}\text{(a) } \sinh x \cosh 2y; \text{ (b) } 4 \sinh x \cosh 2y; \\ \text{(c) } 2 \cosh x \sinh 2y; \text{ (d) } 2 \cosh x \sinh 2y\end{array}\right]$

13 Given $Z = x^2 \sin (x-2y)$ find (a) $\dfrac{\partial^2 Z}{\partial x^2}$ and (b) $\dfrac{\partial^2 Z}{\partial y^2}$ .

Show also that $\dfrac{\partial^2 Z}{\partial x \partial y} = \dfrac{\partial^2 Z}{\partial y \partial x} = 2x^2 \sin (x-2y) - 4x \cos (x-2y)$.

$\qquad\qquad\left[\begin{array}{l}\text{(a) } (2-x^2) \sin (x-2y) + 4x \cos (x-2y) \\ \text{(b) } -4x^2 \sin (x-2y)\end{array}\right]$

14 Find $\dfrac{\partial^2 Z}{\partial x^2}$, $\dfrac{\partial^2 Z}{\partial y^2}$ and show that $\dfrac{\partial^2 Z}{\partial x \partial y} = \dfrac{\partial^2 Z}{\partial y \partial x}$ when $Z = \arccos \dfrac{x}{y}$ .

$\left[\begin{array}{l}\dfrac{\partial^2 Z}{\partial x^2} = \dfrac{-x}{\sqrt{(y^2-x^2)^3}}\text{; } \dfrac{\partial^2 Z}{\partial y^2} = \dfrac{-x}{\sqrt{(y^2-x^2)}}\left\{\dfrac{1}{y^2} + \dfrac{1}{(y^2-x^2)}\right\} \\[3ex] \dfrac{\partial^2 Z}{\partial x \partial y} = \dfrac{\partial^2 Z}{\partial y \partial x} = \dfrac{y}{\sqrt{(y^2-x^2)^3}}\end{array}\right]$

15 Given $Z = \sqrt{\left(\dfrac{3x}{y}\right)}$, show that $\dfrac{\partial^2 Z}{\partial x \partial y} = \dfrac{\partial^2 Z}{\partial y \partial x}$ , and evaluate $\dfrac{\partial^2 Z}{\partial x^2}$

when $x = \dfrac{1}{2}$ and $y = 3$. $\qquad\qquad\qquad\qquad\left[-\dfrac{1}{\sqrt{2}}\right]$

# 9 Total differential, rates of change and small changes

## A. MAIN POINTS CONCERNED WITH THE TOTAL DIFFERENTIAL, RATES OF CHANGE AND SMALL CHANGES

1 In chapter 8, partial differentiation is introduced for the case where only one variable changes at a time, the other variables being kept constant. In practice, variables may all be changing at the same time. If $Z = f(u, v, w, \ldots)$, i.e. the variables in the equation are $u, v, w, \ldots$, then the **total differential**, $dZ$, is given by the sum of the separate partial differentials of Z.

i.e.
$$dZ = \frac{\partial Z}{\partial u} \, du + \frac{\partial Z}{\partial v} \, dv + \frac{\partial Z}{\partial w} \, dw + \ldots \ldots \quad (1)$$

(See *Problems 1 to 3*)

2 Sometimes it is necessary to solve problems in which different quantities have different **rates of change**. From equation (1), the rate of change of Z, $\frac{dZ}{dt}$ is given by:

$$\frac{dZ}{dt} = \frac{\partial Z}{\partial u} \frac{du}{dt} + \frac{\partial Z}{\partial v} \frac{dv}{dt} + \frac{\partial Z}{\partial w} \frac{dw}{dt} + \ldots \ldots \quad (2)$$

(See *Problems 4 to 7*)

3 It is often useful to find an approximate value for the change (or error) of a quantity caused by small changes (or errors) in the variables associated with the quantity. If $Z = f(u, v, w, \ldots)$ and $\delta u, \delta v, \delta w, \ldots$ denote **small changes** in $u, v, w, \ldots$ respectively, then the corresponding approximate change $\delta Z$ in $Z$ is obtained from equation (1) by replacing the differentials by the small changes.

Thus
$$\delta Z \simeq \frac{\partial Z}{\partial u} \, \delta u + \frac{\partial Z}{\partial v} \, \delta v + \frac{\partial Z}{\partial w} \, \delta w + \ldots \ldots \quad (3)$$

(See *Problems 8 to 11*)

## B. WORKED PROBLEMS ON THE TOTAL DIFFERENTIAL, RATES OF CHANGE AND SMALL CHANGES

*Problem 1* If $Z = f(x, y)$ and $Z = x^2 y^3 + (2x/y) + 1$, determine the total differential, $dZ$.

The total differential is the sum of the partial differentials,

i.e. $dZ = \dfrac{\partial Z}{\partial x} dx + \dfrac{\partial Z}{\partial y} dy$

$\dfrac{\partial Z}{\partial x} = 2xy^3 + \dfrac{2}{y}$ (i.e., $y$ is kept constant)

$\dfrac{\partial Z}{\partial y} = 3x^2 y^2 - \dfrac{2x}{y^2}$ (i.e., $x$ is kept constant)

Hence $dZ = (2xy^3 + \dfrac{2}{y}) dx + (3x^2 y^2 - \dfrac{2x}{y^2}) dy$.

*Problem 2* If $Z = f(u, v, w)$ and $Z = 3u^2 - 2v + 4w^3 v^2$ find the total differential, $dZ$.

The total differential $dZ = \dfrac{\partial Z}{\partial u} du + \dfrac{\partial Z}{\partial v} dv + \dfrac{\partial Z}{\partial w} dw$.

$\dfrac{\partial Z}{\partial u} = 6u$ (i.e., $v$ and $w$ are kept constant)

$\dfrac{\partial Z}{\partial v} = -2 + 8w^3 v$ (i.e., $u$ and $w$ are kept constant)

$\dfrac{\partial Z}{\partial w} = 12v^2 w^2$ (i.e., $u$ and $v$ are kept constant).

Hence $dZ = 6u\,du + (8vw^3 - 2)dv + (12v^2 w^2)\,dw$.

*Problem 3* The pressure $p$, volume $V$ and temperature $T$ of a gas are related by $pV = kT$, where $k$ is a constant. Determine the total differentials (a) $dp$ and (b) $dT$ in terms of $p$, $V$ and $T$.

(a) Total differential $dp = \dfrac{\partial p}{\partial T} dT + \dfrac{\partial p}{\partial V} dV$

Since $pV = kT$ then $p = \dfrac{kT}{V}$

$\dfrac{\partial p}{\partial T} = \dfrac{k}{V}$ and $\dfrac{\partial p}{\partial V} = -\dfrac{kT}{V^2}$

Hence $dp = \dfrac{k}{V} dT - \dfrac{kT}{V^2} dV$

Since $pV = kT$, $k = \dfrac{pV}{T}$

Hence $dp = \dfrac{(\frac{pV}{T})}{V} dT - \dfrac{(\frac{pV}{T})T}{V^2} dV$

i.e. $dp = \dfrac{p}{T} dT - \dfrac{p}{V} dV$.

(b) Total differential $dT = \dfrac{\partial T}{\partial p} dp + \dfrac{\partial T}{\partial V} dV$

Since $pV = kT$, $T = \dfrac{pV}{k}$

$\dfrac{\partial T}{\partial p} = \dfrac{V}{k}$ and $\dfrac{\partial T}{\partial V} = \dfrac{p}{k}$

Hence $dT = \dfrac{V}{k} dp + \dfrac{p}{k} dV$

Substituting $k = \dfrac{pV}{T}$ gives: $dT = \dfrac{V}{\left(\dfrac{pV}{T}\right)} dp + \dfrac{p}{\left(\dfrac{pV}{T}\right)} dV$ i.e. $\boldsymbol{dT = \dfrac{T}{p} dp + \dfrac{T}{V} dV}$

---

*Problem 4* If $Z = f(x, y)$ and $Z = 2x^3 \sin 2y$ find the rate of change of $Z$, correct to 4 significant figures, when $x$ is 2 units and $y$ is $\pi/6$ radians and when $x$ is increasing at 4 units/s and $y$ is decreasing at 0.5 units/s.

Using equation (2), the rate of change of $Z$, $\dfrac{dZ}{dt} = \dfrac{\partial Z}{\partial x} \dfrac{dx}{dt} + \dfrac{\partial Z}{\partial y} \dfrac{dy}{dt}$

Since $Z = 2x^3 \sin 2y$, then $\dfrac{\partial Z}{\partial x} = 6x^2 \sin 2y$ and $\dfrac{\partial Z}{\partial y} = 4x^3 \cos 2y$.

Since $x$ is increasing at 4 units/s, $\dfrac{dx}{dt} = +4$

and since $y$ is decreasing at 0.5 units/s, $\dfrac{dy}{dt} = -0.5$

Hence $\dfrac{dZ}{dt} = (6x^2 \sin 2y)(+4) + (4x^3 \cos 2y)(-0.5)$

$= 24x^2 \sin 2y - 2x^3 \cos 2y$

When $x = 2$ units and $y = \dfrac{\pi}{6}$ radians, then:

$\dfrac{dZ}{dt} = 24(2)^2 \sin \left(2\dfrac{\pi}{6}\right) - 2(2)^3 \cos \left(2\dfrac{\pi}{6}\right) = 83.138 - 8.0$

Hence the rate of change of $Z$, $\dfrac{dZ}{dt} = \boldsymbol{75.14}$ **units/s**, correct to 4 significant figures.

---

*Problem 5* The height of a right circular cone is increasing at 33 mm/s and its radius is decreasing at 2 mm/s. Determine, correct to 3 significant figures, the rate at which the volume is changing (in cm/s) when the height is 3.2 cm and the radius is 1.5 cm.

Volume of a right circular cone, $V = \dfrac{1}{3} \pi r^2 h$

Using equation (2), the rate of change of volume, $\dfrac{dV}{dt} = \dfrac{\partial V}{\partial r} \dfrac{dr}{dt} + \dfrac{\partial V}{\partial h} \dfrac{dh}{dt}$

$\dfrac{\partial V}{\partial r} = \dfrac{2}{3} \pi r h$ and $\dfrac{\partial V}{\partial h} = \dfrac{1}{3} \pi r^2$

Since the height is increasing at 3 mm/s, i.e. 0.3 cm/s, then $\dfrac{dh}{dt} = +0.3$ and since the radius is decreasing at 2 mm/s, i.e. 0.2 cm/s, then $\dfrac{dr}{dt} = -0.2$

Hence $\dfrac{dV}{dt} = \left(\dfrac{2}{3} \pi r h\right)(-0.2) + \left(\dfrac{1}{3} \pi r^2\right)(+0.3) = \dfrac{-0.4}{3} \pi r h + 0.1 \pi r^2$

However, $h = 3.2$ cm and $r = 1.5$ cm.

Hence $\dfrac{dV}{dt} = \dfrac{-0.4}{3}\ \pi(1.5)(3.2)+(0.1)\pi(1.5)^2$

$= -2.011+0.707 = -1.304 \text{ cm}^3/\text{s}$

**Thus the rate of change of volume is 1.30 cm³/s decreasing.**

*Problem 6* The area $A$ of a triangle is given by $A = \frac{1}{2} ac \sin B$, where $B$ is the angle between sides $a$ and $c$. If $a$ is increasing at 0.4 units/s, $c$ is decreasing at 0.8 units/s and $B$ is increasing at 0.2 units/s, find the rate of change of the area of the triangle, correct to 3 significant figures, when $a$ is 3 units, $c$ is 4 units and $B$ is $\pi/6$ radians.

Using equation (2), the rate of change of area $\dfrac{dA}{dt} = \dfrac{\partial A}{\partial a}\dfrac{da}{dt} + \dfrac{\partial A}{\partial c}\dfrac{dc}{dt} + \dfrac{\partial A}{\partial B}\dfrac{dB}{dt}$.

Since $A = \frac{1}{2}ac \sin B$, $\dfrac{\partial A}{\partial a} = \frac{1}{2}c \sin B$, $\dfrac{\partial A}{\partial c} = \frac{1}{2}a \sin B$ and $\dfrac{\partial A}{\partial B} = \frac{1}{2}ac \cos B$.

$\dfrac{da}{dt} = 0.4$ units/s, $\dfrac{dc}{dt} = -0.8$ units/s and $\dfrac{dB}{dt} = 0.2$ units/s.

Hence $\dfrac{dA}{dt} = (\frac{1}{2}c \sin B)(0.4) + (\frac{1}{2}a \sin B)(-0.8)+(\frac{1}{2}ac \cos B)(0.2)$.

When $a = 3$, $c = 4$ and $B = \dfrac{\pi}{6}$ then:

$\dfrac{dA}{dt} = \left[\frac{1}{2}(4) \sin \frac{\pi}{6}\right](0.4) + \left[\frac{1}{2}(3) \sin \frac{\pi}{6}\right](-0.8) + \left[\frac{1}{2}(3)(4) \cos \frac{\pi}{6}\right](0.2)$

$= 0.4-0.6+1.039 = \mathbf{0.839}\ \textbf{units}^2/\textbf{s}$, correct to 3 significant figures.

*Problem 7* Determine the rate of increase of diagonal AC of the rectangular solid, shown in *Fig 1*, correct to 2 significant figures, if the sides $x$, $y$ and $z$ increase at 6 mm/s, 5 mm/s and 4 mm/s when these three sides are 5 cm, 4 cm and 3 cm respectively.

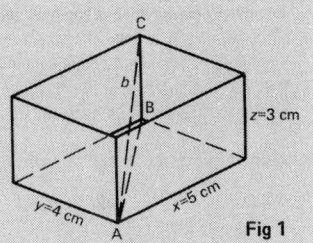

*Fig 1*

Diagonal AB $= \sqrt{(x^2+y^2)}$

Diagonal AC $= \sqrt{(BC^2+AB^2)} = \sqrt{\{ z^2 +[\sqrt{(x^2+y^2)}]^2 \}} = \sqrt{(z^2+x^2+y^2)}$

Let AC $= b$, then $b = \sqrt{(x^2+y^2+z^2)}$

Using equation (2), the rate of change of diagonal $b$ is given by:

$\dfrac{db}{dt} = \dfrac{\partial b}{\partial x}\dfrac{dx}{dt} + \dfrac{\partial b}{\partial y}\dfrac{dy}{dt} + \dfrac{\partial b}{\partial z}\dfrac{dz}{dt}$

Since $b = \sqrt{(x^2+y^2+z^2)}$, $\dfrac{\partial b}{\partial x} = \frac{1}{2}(x^2+y^2+z^2)^{-\frac{1}{2}}(2x) = \dfrac{x}{\sqrt{(x^2+y^2+z^2)}}$

Similarly, $\dfrac{\partial b}{\partial y} = \dfrac{y}{\sqrt{(x^2+y^2+z^2)}}$ and $\dfrac{\partial b}{\partial z} = \dfrac{z}{\sqrt{(x^2+y^2+z^2)}}$

$\dfrac{dx}{dt} = 6$ mm/s $= 0.6$ cm/s; $\dfrac{dy}{dt} = 5$ mm/s $= 0.5$ cm/s; $\dfrac{dz}{dt} = 4$ mm/s $= 0.4$ cm/s.

Hence $\dfrac{db}{dt} = \left[\dfrac{x}{\sqrt{(x^2+y^2+z^2)}}\right](0.6) + \left[\dfrac{y}{\sqrt{(x^2+y^2+z^2)}}\right](0.5) + \left[\dfrac{z}{\sqrt{(x^2+y^2+z^2)}}\right](0.4)$

When $x = 5$ cm, $y = 4$ cm and $z = 3$ cm, then:

$$\frac{db}{dt} = \frac{5}{\sqrt{(5^2+4^2+3^2)}}\,(0.6) + \frac{4}{\sqrt{(5^2+4^2+3^2)}}(0.5) + \frac{3}{\sqrt{(5^2+4^2+3^2)}}\,(0.4)$$

$$= 0.4243+0.2828+0.1697 = 0.8768 \text{ cm/s}.$$

**Hence the rate of increase of diagonal AC = 0.88 cm/s or 8.8 mm/s, correct to 2 significant figures.**

*Problem 8* Pressure $p$ and volume $V$ of a gas are connected by the equation $pV^{1.4} = k$. Determine the approximate percentage error in $k$ when the pressure is increased by 4% and the volume is decreased by 1.5%.

Using equation (3), the approximate error in $k$, $\delta k \simeq \dfrac{\partial k}{\partial p}\,\delta p + \dfrac{\partial k}{\partial V}\,\delta V$.

Let $p$, $V$ and $k$ refer to the initial values.

Since $k = pV^{1.4}$ then $\dfrac{\partial k}{\partial p} = V^{1.4}$ and $\dfrac{\partial k}{\partial V} = 1.4pV^{0.4}$

Since the pressure is increased by 4%, the change in pressure $\delta p = \dfrac{4}{100} \times p = 0.04p$

Since the volume is decreased by 1.5%, the change in volume $\delta V = \dfrac{-1.5}{100} \times V = -0.015V$

Hence the approximate error in $k$, $\delta k \simeq (V^{1.4})(0.04p) + (1.4pV^{0.4})(-0.015V)$

$$\simeq pV^{1.4}\,[0.04 - 1.4(0.015)]$$

$$\simeq \frac{1.9}{100}\,pV^{1.4} \simeq \frac{1.9}{100}\,k$$

i.e. **the approximate error in $k$ is a 1.9% increase.**

*Problem 9* Modulus of rigidity $G = (R^4\theta)/L$, where $R$ is the radius, $\theta$ the angle of twist and $L$ the length. Determine the approximate percentage error in $G$ when $R$ is increased by 2%, $\theta$ is reduced by 5% and $L$ is increased by 4%.

Using equation (3), the approximate error in $G$, $\delta G$ is given by:

$$\delta G \simeq \frac{\partial G}{\partial R}\,\delta R + \frac{\partial G}{\partial \theta}\,\delta \theta + \frac{\partial G}{\partial L}\,\delta L$$

Since $G = \dfrac{R^4\theta}{L}$, $\dfrac{\partial G}{\partial R} = \dfrac{4R^3\theta}{L}$; $\dfrac{\partial G}{\partial \theta} = \dfrac{R^4}{L}$; $\dfrac{\partial G}{\partial L} = \dfrac{-R^4\theta}{L^2}$

Since $R$ is increased by 2%, $\delta R = \dfrac{2}{100}\,R = 0.02R$.

Similarly, $\delta \theta = -0.05\theta$ and $\delta L = 0.04L$.

Hence $\delta G \simeq \left(\dfrac{4R^3\theta}{L}\right)(0.02R) + \left(\dfrac{R^4}{L}\right)(-0.05\theta) + \left(-\dfrac{R^4\theta}{L^2}\right)(0.04L)$

$$\simeq \frac{R^4\theta}{L}\,[0.08 - 0.05 - 0.04] \simeq -0.01\,\frac{R^4\theta}{L} \quad \text{i.e.} \quad \delta G \simeq -\frac{1}{100}\,G$$

**Hence the approximate percentage error in $G$ is a 1% decrease.**

*Problem 10* The second moment of area of a rectangle is given by $I = (bl^3)/3$. If $b$ and $l$ are measured as 40 mm and 90 mm respectively and the measurement errors are $-5$ mm in $b$ and $+8$ mm in $l$, find the approximate error in the calculated value of $I$.

Using equation (3), the approximate error in $I$, $\delta I \simeq \dfrac{\partial I}{\partial b}\,\delta b + \dfrac{\partial I}{\partial l}\,\delta l$

$\dfrac{\partial I}{\partial b} = \dfrac{l^3}{3}$ and $\dfrac{\partial I}{\partial l} = \dfrac{3bl^2}{3} = bl^2$.

$\delta b = -5$ mm and $\delta l = +8$ mm.

Hence $\delta I \simeq \left(\dfrac{l^3}{3}\right)(-5)+(bl^2)(+8)$

Since $b = 40$ mm and $l = 90$ mm then $\delta I \simeq \dfrac{90^3}{3}(-5)+40(90)^2(8)$

$\simeq -1215000 + 2592000 \simeq 1377000$ mm$^4 \simeq 137.7$ cm$^4$

**Hence the approximate error in the calculated value of $I$ is 137.7 cm$^4$ increase.**

*Problem 11* The time of oscillation $t$ of a pendulum is given by $t = 2\pi\sqrt{(l/g)}$. Determine the approximate percentage error in $t$ when $l$ has an error of 0.2% too large and $g$ 0.1% too small.

Using equation (3), the approximate change in $t$, $\delta t \simeq \dfrac{\partial t}{\partial l}\,\delta l + \dfrac{\partial t}{\partial g}\,\delta g$.

Since $t = 2\pi\sqrt{\dfrac{l}{g}}$ , $\dfrac{\partial t}{\partial l} = \dfrac{\pi}{\sqrt{(lg)}}$ and $\dfrac{\partial t}{\partial g} = -\pi\sqrt{\dfrac{l}{g^3}}$ (from *Problem 6*, chapter 8)

$\delta l = \dfrac{0.2}{100}\,l = 0.002l$ and $\delta g = -0.001g$

Hence $\delta t \simeq \dfrac{\pi}{\sqrt{(lg)}}(0.002l) + -\pi\sqrt{\dfrac{l}{g^3}}(-0.001g)$

$\simeq 0.002\pi\sqrt{\dfrac{l}{g}} + 0.001\pi\sqrt{\dfrac{l}{g}}$

$\simeq (.001)\left[2\pi\sqrt{\dfrac{l}{g}}\right] + (0.0005)\left[2\pi\sqrt{\dfrac{l}{g}}\right] \simeq 0.0015\left[2\pi\sqrt{\dfrac{l}{g}}\right]$

$\simeq 0.0015t \simeq \dfrac{0.15}{100}\,t.$

**Hence the approximate error in $t$ is a 0.15% increase.**

## C. FURTHER PROBLEMS ON THE TOTAL DIFFERENTIAL, RATES OF CHANGE AND SMALL CHANGES

In *Problems 1 to 5*, find the total differential $dz$.

1 $z = x^3 + y^2$ $\qquad\qquad\qquad\qquad\qquad$ $[3x^2\,dx + 2y\,dy]$

2 $z = 2xy - \cos x$ $\qquad\qquad\qquad\quad$ $[(2y + \sin x)dx + 2x\,dy]$

3 $z = \dfrac{x-y}{x+y}$ $\qquad\qquad\qquad\qquad$ $\dfrac{2y}{(x+y)^2}\,dx - \dfrac{2x}{(x+y)^2}\,dy$

4 $z = x \ln y$ $\qquad\qquad\qquad\qquad\qquad$ $[\ln y\,dx + \dfrac{x}{y}\,dy]$

5 $z = xy + \dfrac{\sqrt{x}}{y} - 4$ $\qquad\qquad$ $\left[\left(y + \dfrac{1}{2y\sqrt{x}}\right)dx + \left(x - \dfrac{\sqrt{x}}{y^2}\right)dy\right]$

6 If $z = f(a, b, c)$ and $z = 2ab - 3b^2c + abc$, find the total differential, $dz$.

$\qquad\qquad\qquad\qquad$ $[b(2+c)da + (2a - 6bc + ac)db + b(a - 3b)dc]$

7 Given $u = \ln\sin(xy)$ show that $du = \cot(xy)[y\,dx + x\,dy]$.

8 The radius of a right cylinder is increasing at a rate of 8 mm/s and the height is decreasing at a rate of 15 mm/s. Find the rate at which the volume is changing in $cm^3/s$ when the radius is 40 mm and the height is 150 mm. [+226.2 $cm^3$/s]

9 If $z = f(x, y)$ and $z = 3x^2 y^5$, find the rate of change of $z$ when $x$ is 3 units and $y$ is 2 units when $x$ is decreasing at 5 units/s and $y$ is increasing at 2.5 units/s.
[2 520 units/s]

10 Find the rate of change of $k$, correct to 4 significant figures, given the following data: $k = f(a, b, c); k = 2b \ln a + c^2 e^a$; $a$ is increasing at 2 cm/s; $b$ is decreasing at 3 cm/s; $c$ is decreasing at 1 cm/s; $a = 1.5$ cm, $b = 6$ cm and $c = 8$ cm. [515.5 cm/s]

11 A rectangular box has sides of length $x$ cm, $y$ cm and $z$ cm. Sides $x$ and $z$ are expanding at rates of 3 mm/s and 5 mm/s respectively and side $y$ is contracting at a rate of 2 mm/s. Determine the rate of change of volume when $x$ is 3 cm, $y$ is 1.5 cm and $z$ is 6 cm. [1.35 $cm^3$/s]

12 Find the rate of change of the total surface area of a right circular cone at the instant when the base radius is 5 cm and the height is 12 cm if the radius is increasing at 5 mm/s and the height is decreasing at 15 mm/s. [17.4 $cm^2$/s]

13 The power $P$ consumed in a resistor is given by $P = V^2/R$ watts. Determine the approximate change in power when $V$ increases by 5% and $R$ decreases by 0.5% if the original values of $V$ and $R$ are 50 volts and 12.5 ohms respectively. [+21 watts]

14 An equation for heat generated $H$ is $H = i^2 R t$. Determine the error in the calculated values of $H$ if the error in measuring current $i$ is +2%, the error in measuring resistance $R$ is −3% and the error in measuring time $t$ is +1%. [+2%]

15 $f_0 = \dfrac{1}{2\pi\sqrt{(LC)}}$ represents the resonant frequency of a series connected circuit containing inductance $L$ and capacitance $C$. Determine the approximate percentage change in $f_0$ when $L$ is decreased by 3% and $C$ is increased by 5%. [−1%]

16 The second moment of area of a rectangle about its centroid parallel to side $b$ is given by $I = bd^3/12$. If $b$ and $d$ are measured as 15 cm and 6 cm respectively and the measurement errors are +12 mm in $b$ and −1.5 mm in $d$, find the error in the calculated value of $I$. [+1.35 $cm^4$]

17 The side $b$ of a triangle is calculated using $b^2 = a^2 + c^2 - 2ac \cos B$. If $a$, $c$ and $B$ are measured as 3 cm, 4 cm and $\pi/4$ radians respectively and the measurement errors which occur are +0.8 mm, −0.5 mm and +$\pi/90$ radians respectively, determine the error in the calculated value of $b$. [+0.156 cm]

# 10 Integration using substitutions and partial fractions

## A. MAIN POINTS CONCERNED WITH INTEGRATION USING SUBSTITUTION AND PARTIAL FRACTIONS

1   Integration is introduced in chapter 17 of *Mathematics 3 Checkbook* and a list of **standard integrals** is summarised in *Table 1*.

2   Functions which require integrating are not always in the 'standard' form shown in *Table 1*. However, it is often possible to change a function into a form which can be integrated by using either:
   (i) an algebraic substitution;    (ii) a trigonometric or hyperbolic substitution,
   (iii) partial fractions;       or (iv) integration by parts (see chapter 11).

TABLE 1.  Standard integrals

|   | $y$ | $\int y \, dx$ |
|---|---|---|
| 1 | $ax^n$ | $\dfrac{ax^{n+1}}{n+1} + c$ (except when $n = -1$) |
| 2 | $\sin ax$ | $-\dfrac{1}{a} \cos ax + c$ |
| 3 | $\cos ax$ | $\dfrac{1}{a} \sin ax + c$ |
| 4 | $\sec^2 ax$ | $\dfrac{1}{a} \tan ax + c$ |
| 5 | $\operatorname{cosec}^2 ax$ | $-\dfrac{1}{a} \cot ax + c$ |
| 6 | $\operatorname{cosec} ax \cot ax$ | $-\dfrac{1}{a} \operatorname{cosec} ax + c$ |
| 7 | $\sec ax \tan ax$ | $\dfrac{1}{a} \sec ax + c$ |
| 8 | $e^{ax}$ | $\dfrac{1}{a} e^{ax} + c$ |
| 9 | $\dfrac{1}{x}$ | $\ln ax + c$ |

3   With **Algebraic substitutions**, the substitution usually made is to let $u$ be equal to $f(x)$ such that $f(u)du$ is a standard integral. It is found that integrals of the forms

$$k \int [f(x)]^n f'(x)dx \quad \text{and} \quad k \int \frac{f'(x)}{[f(x)]^n}\, dx$$

(where $k$ and $n$ are constants) can both be integrated by substituting $u$ for $f(x)$. (See *Problems 1 to 11*)

4   When evaluating definite integrals involving substitutions it is sometimes more convenient to **change the limits** of the integral, (see *Problems 10 and 11*).

5   *Table 2* gives a summary of the integrals that require the use of a **trigonometric or hyperbolic substitution**.

6   The process of expressing a fraction in terms of simpler fractions—called **partial**

TABLE 2.

| $f(x)$ | $\int f(x)\, dx$ | *Method* | *See problem* |
|---|---|---|---|
| 1   $\cos^2 x$ | $\frac{1}{2}\left(x + \frac{\sin 2x}{2}\right)+c$ | Use $\cos 2x = 2\cos^2 x - 1$ | 12 |
| 2   $\sin^2 x$ | $\frac{1}{2}\left(x - \frac{\sin 2x}{2}\right)+c$ | Use $\cos 2x = 1 - 2\sin^2 x$ | 13 |
| 3   $\tan^2 x$ | $\tan x - x + c$ | Use $1 + \tan^2 x = \sec^2 x$ | 14 |
| 4   $\cot^2 x$ | $-\cot x - x + c$ | Use $\cot^2 x + 1 = \operatorname{cosec}^2 x$ | 15 |
| 5   $\cos^m x \sin^n x$ | (a) If either $m$ or $n$ is odd (but not both), use $\cos^2 x + \sin^2 x = 1$ | | 16, 17 |
| | (b) If both $m$ and $n$ are even, use either $\cos 2x = 2\cos^2 x - 1$ or $\cos 2x = 1 - 2\sin^2 x$ | | 18, 19 |
| 6   $\sin A \cos B$ | | Use $\frac{1}{2}[\sin(A+B)+\sin(A-B)]$ | 20 |
| 7   $\cos A \sin B$ | | Use $\frac{1}{2}[\sin(A+B)-\sin(A-B)]$ | 21 |
| 8   $\cos A \cos B$ | | Use $\frac{1}{2}[\cos(A+B)+\cos(A-B)]$ | 22 |
| 9   $\sin A \sin B$ | | Use $-\frac{1}{2}[\cos(A+B)-\cos(A-B)]$ | 23 |
| 10   $\dfrac{1}{\sqrt{(a^2-x^2)}}$ | $\arcsin\dfrac{x}{a}+c$ | Use $x = a\sin\theta$ substitution | 24, 25 |
| 11   $\sqrt{(a^2-x^2)}$ | $\dfrac{a^2}{2}\arcsin\dfrac{x}{a}+\dfrac{x}{2}\sqrt{(a^2-x^2)}+c$ | | 26, 27 |
| 12   $\dfrac{1}{a^2+x^2}$ | $\dfrac{1}{a}\arctan\dfrac{x}{a}+c$ | Use $x = a\tan\theta$ substitution | 28–30 |
| 13   $\dfrac{1}{\sqrt{(x^2+a^2)}}$ | $\operatorname{arsinh}\dfrac{x}{a}+c$ or $\ln\left\{\dfrac{x+\sqrt{(x^2+a^2)}}{a}\right\}+c$ | Use $x = a\sinh\theta$ substitution | 31–33 |
| 14   $\sqrt{(x^2+a^2)}$ | $\dfrac{a^2}{2}\operatorname{arsinh}\dfrac{x}{a}+\dfrac{x}{2}\sqrt{(x^2+a^2)}+c$ | | 34 |
| 15   $\dfrac{1}{\sqrt{(x^2-a^2)}}$ | $\operatorname{arcosh}\dfrac{x}{a}+c$ or $\ln\left\{\dfrac{x+\sqrt{(x^2-a^2)}}{a}\right\}+c$ | Use $x = a\cosh\theta$ substitution | 35, 36 |
| 16   $\sqrt{(x^2-a^2)}$ | $\dfrac{x}{2}\sqrt{(x^2-a^2)}-\dfrac{a^2}{2}\operatorname{arcosh}\dfrac{x}{a}+c$ | | 37, 38 |

**TABLE 3.** Forms of partial fractions

| Denominator containing: | Expression | Form of partial fraction |
|---|---|---|
| Linear factors (see *Problems 39 to 42*) | $\dfrac{f(x)}{(x+a)(x-b)(x+c)}$ | $\dfrac{A}{(x+a)} + \dfrac{B}{(x-b)} + \dfrac{C}{(x+c)}$ |
| Repeated linear factors (see *Problems 43 to 45*) | $\dfrac{f(x)}{(x+a)^3}$ | $\dfrac{A}{(x+a)} + \dfrac{B}{(x+a)^2} + \dfrac{C}{(x+a)^3}$ |
| Quadratic factors (see *Problem 46*) | $\dfrac{f(x)}{(ax^2+bx+c)(x+d)}$ | $\dfrac{Ax+B}{(ax^2+bx+c)} + \dfrac{C}{(x+d)}$ |

**fractions**—is discussed in chapter 5 of *Mathematics 3 Checkbook*. The forms of partial fractions used are summarised in *Table 3*.
Certain functions have to be resolved into partial fractions before they can be integrated. (See *Problems 39 to 50*)

## B. WORKED PROBLEMS ON INTEGRATION USING SUBSTITUTIONS AND PARTIAL FRACTIONS

(a) ALGEBRAIC SUBSTITUTIONS

*Problem 1* Determine $\int \cos{(3x+7)}\, dx$.

$\int \cos{(3x+7)}\, dx$ is not a standard integral of the form shown in *Table 1*, thus an algebraic substitution is made.

Let $u = 3x+7$, then $\dfrac{du}{dx} = 3$ and rearranging gives $dx = \dfrac{du}{3}$.

Hence $\int \cos{(3x+7)}\, dx = \int (\cos u)\,\dfrac{du}{3} = \dfrac{1}{3}\int \cos u\, du$, which is a standard integral

$$= \dfrac{1}{3}\sin u + c$$

Rewriting $u$ as $(3x+7)$ gives: $\int \cos{(3x+7)}\, dx = \dfrac{1}{3}\sin{(3x+7)} + c$, which may be checked by differentiation.

*Problem 2* Find $\int (2x-5)^7\, dx$.

$(2x-5)$ may be multiplied by itself 7 times and then each term of the result integrated. However, this would be a lengthy process and thus an algebraic substitution is made.

Let $u = (2x-5)$ then $\dfrac{du}{dx} = 2$ and $dx = \dfrac{du}{2}$.

Hence $\int (2x-5)^7\, dx = \int u^7\,\dfrac{du}{2} = \dfrac{1}{2}\int u^7\, du = \dfrac{1}{2}\left(\dfrac{u^8}{8}\right) + c = \dfrac{1}{16}u^8 + c$.

Rewriting $u$ as $(2x-5)$ gives: $\int (2x-5)^7 dx = \frac{1}{16}(2x-5)^8 + c$.

*Problem 3* Find $\int \frac{4}{(5x-3)} dx$.

Let $u = (5x-3)$ then $\frac{du}{dx} = 5$ and $dx = \frac{du}{5}$.

Hence $\int \frac{4}{(5x-3)} dx = \int \frac{4}{u} \frac{du}{5} = \frac{4}{5} \int \frac{1}{u} du = \frac{4}{5} \ln u + c = \frac{4}{5} \ln (5x-3) + c$.

*Problem 4* Evaluate $\int_0^1 2e^{6x-1} dx$, correct to 4 significant figures.

Let $u = 6x-1$ then $\frac{du}{dx} = 6$ and $dx = \frac{du}{6}$.

Hence $\int 2e^{6x-1} dx = \int 2e^u \frac{du}{6} = \frac{1}{3} \int e^u du = \frac{1}{3} e^u + c = \frac{1}{3} e^{6x-1} + c$.

Thus $\int_0^1 2e^{6x-1} dx = \frac{1}{3} \left[ e^{6x-1} \right]_0^1 = \frac{1}{3} [e^5 - e^{-1}] = 49.35$, correct to 4 significant figures.

*Problem 5* Determine $\int 3x(4x^2+3)^5 dx$.

Let $u = (4x^2+3)$ then $\frac{du}{dx} = 8x$ and $dx = \frac{du}{8x}$.

Hence $\int 3x(4x^2+3)^5 dx = \int 3x(u)^5 \frac{du}{8x} = \frac{3}{8} \int u^5 du$, by cancelling.

The original variable '$x$' has been completely removed and the integral is now only in terms of $u$ and is a standard integral.

Hence $\frac{3}{8} \int u^5 du = \frac{3}{8} \left( \frac{u^6}{6} \right) + c = \frac{1}{16} u^6 + c = \frac{1}{16}(4x^2+3)^6 + c$.

*Problem 6* Evaluate $\int_0^{\frac{\pi}{6}} 24 \sin^5 \theta \cos \theta \, d\theta$.

Let $u = \sin \theta$ then $\frac{du}{d\theta} = \cos \theta$ and $d\theta = \frac{du}{\cos \theta}$.

Hence $\int 24 \sin^5 \theta \cos \theta \, d\theta = \int 24u^5 \cos \theta \frac{du}{\cos \theta} = 24 \int u^5 du$, by cancelling,

$$= 24 \frac{u^6}{6} + c = 4u^6 + c = 4(\sin \theta)^6 + c = 4 \sin^6 \theta + c$$

Thus $\int_0^{\frac{\pi}{6}} 24 \sin^5 \theta \cos \theta \, d\theta = [4 \sin^6 \theta]_0^{\frac{\pi}{6}} = 4 \left[ (\sin \frac{\pi}{6})^6 - (\sin 0)^6 \right]$

$$= 4 \left[ \left( \frac{1}{2} \right)^6 - 0 \right] = \frac{1}{16} \text{ or } 0.0625$$

*Problem 7* Find $\int \frac{x}{2+3x^2} dx$.

Let $u = 2+3x^2$ then $\frac{du}{dx} = 6x$ and $dx = \frac{du}{6x}$.

Hence $\int \dfrac{x}{2+3x^2}\,dx = \int \dfrac{x}{u}\,\dfrac{du}{6x} = \dfrac{1}{6}\int \dfrac{1}{u}\,du$, by cancelling,

$$= \dfrac{1}{6}\,\ln u + c = \dfrac{1}{6}\ln(2+3x^2)+c.$$

**Problem 8** Determine $\int \dfrac{2x}{\sqrt{(4x^2-1)}}\,dx$.

Let $u = 4x^2 - 1$ then $\dfrac{du}{dx} = 8x$ and $dx = \dfrac{du}{8x}$

Hence $\int \dfrac{2x}{\sqrt{(4x^2-1)}}\,dx = \int \dfrac{2x}{\sqrt{u}}\,\dfrac{du}{8x} = \dfrac{1}{4}\int \dfrac{1}{\sqrt{u}}\,du$, by cancelling,

$$= \dfrac{1}{4}\int u^{-\frac{1}{2}}\,du = \dfrac{1}{4}\left[\dfrac{u^{-\frac{1}{2}+1}}{-\frac{1}{2}+1}\right]+c = \dfrac{1}{4}\left[\dfrac{u^{\frac{1}{2}}}{\frac{1}{2}}\right]+c$$

$$= \dfrac{1}{2}\sqrt{u}+c = \dfrac{1}{2}\sqrt{(4x^2-1)}+c.$$

**Problem 9** Show that $\int \tan\theta\,d\theta = \ln(\sec\theta)+c.$

$\int \tan\theta\,d\theta = \int \dfrac{\sin\theta}{\cos\theta}\,d\theta$. Let $u = \cos\theta$ then $\dfrac{du}{d\theta} = -\sin\theta$ and $d\theta = \dfrac{-du}{\sin\theta}$

Hence $\int \dfrac{\sin\theta}{\cos\theta}\,d\theta = \int \dfrac{\sin\theta}{u}\left(\dfrac{-du}{\sin\theta}\right) = -\int \dfrac{1}{u}\,du = -\ln u + c$

$$= -\ln(\cos\theta)+c = \ln(\cos\theta)^{-1}+c,\text{ by the laws of logarithms.}$$

Hence $\int \tan\theta\,d\theta = \ln(\sec\theta)+c$, since $(\cos\theta)^{-1} = \sec\theta.$

**Problem 10** Evaluate $\int_1^3 5x\sqrt{(2x^2+7)}\,dx$, taking positive values of square roots only.

Let $u = 2x^2 + 7$, then $\dfrac{du}{dx} = 4x$ and $dx = \dfrac{du}{4x}$.

It is possible in this case to change the limits of integration. Thus when $x = 3$, $u = 2(3)^2 + 7 = 25$ and when $x = 1$, $u = 2(1)^2 + 7 = 9$.

Hence $\int_{x=1}^{x=3} 5x\sqrt{(2x^2+7)}\,dx = \int_{u=9}^{u=25} 5x\sqrt{u}\,\dfrac{du}{4x} = \dfrac{5}{4}\int_9^{25}\sqrt{u}\,du = \dfrac{5}{4}\int_9^{25} u^{\frac{1}{2}}\,du.$

Thus the limits have been changed, and it is unnecessary to change the integral back in terms of $x$.

Thus $\int_{x=1}^{x=3} 5x\sqrt{(2x^2+7)}\,dx = \dfrac{5}{4}\left[\dfrac{u^{3/2}}{3/2}\right]_9^{25} = \dfrac{5}{6}\left[\sqrt{u^3}\right]_9^{25}$

$$= \dfrac{5}{6}[\sqrt{25^3}-\sqrt{9^3}] = \dfrac{5}{6}(125-27) = 81\dfrac{2}{3}.$$

**Problem 11** Evaluate $\int_0^2 \dfrac{3x}{\sqrt{(2x^2+1)}}\,dx$, taking positive values of square roots only.

Let $u = 2x^2 + 1$ then $\dfrac{du}{dx} = 4x$ and $dx = \dfrac{du}{4x}$.

Hence $\displaystyle\int_0^2 \frac{3x}{\sqrt{(2x^2+1)}}\, dx = \int_{x=0}^{x=2} \frac{3x}{\sqrt{u}}\, \frac{du}{4x} = \frac{3}{4}\int_{x=0}^{x=2} u^{-\frac{1}{2}}\, du$

Since $u = 2x^2 + 1$, when $x = 2$, $u = 9$ and when $x = 0$, $u = 1$.

Thus $\dfrac{3}{4}\displaystyle\int_{x=0}^{x=2} u^{-\frac{1}{2}}\, du = \frac{3}{4}\int_{u=1}^{u=9} u^{-\frac{1}{2}}\, du$, i.e. the limits have been changed,

$$= \frac{3}{4}\left[\frac{u^{\frac{1}{2}}}{\frac{1}{2}}\right]_1^9 = \frac{3}{2}[\sqrt{9}-\sqrt{1}] = 3,\ \text{taking positive values of square roots only.}$$

## (b) TRIGONOMETRIC AND HYPERBOLIC SUBSTITUTIONS

*Problem 12* Evaluate $\displaystyle\int_0^{\frac{\pi}{4}} 2\cos^2 4t\, dt.$

Since $\cos 2t = 2\cos^2 t - 1$, then $\cos^2 t = \dfrac{1}{2}(1+\cos 2t)$ and $\cos^2 4t = \dfrac{1}{2}(1+\cos 8t)$

Hence $\displaystyle\int_0^{\frac{\pi}{4}} 2\cos^2 4t\, dt = 2\int_0^{\frac{\pi}{4}} \frac{1}{2}(1+\cos 8t)\, dt = \left[t + \frac{\sin 8t}{8}\right]_0^{\frac{\pi}{4}}$

$$= \left[\frac{\pi}{4} + \frac{\sin 8(\frac{\pi}{4})}{8}\right] - \left[0 + \frac{\sin 0}{8}\right] = \frac{\pi}{4}$$

*Problem 13* Determine $\displaystyle\int \sin^2 3x\, dx.$

Since $\cos 2x = 1 - 2\sin^2 x$, then $\sin^2 x = \dfrac{1}{2}(1-\cos 2x)$ and $\sin^2 3x = \dfrac{1}{2}(1-\cos 6x)$.

Hence $\displaystyle\int \sin^2 3x\, dx = \int \frac{1}{2}(1-\cos 6x)\, dx$

$$= \frac{1}{2}\left(x - \frac{\sin 6x}{6}\right) + c.$$

*Problem 14* Find $3\displaystyle\int \tan^2 4x\, dx.$

Since $1+\tan^2 x = \sec^2 x$, then $\tan^2 x = \sec^2 x - 1$ and $\tan^2 4x = \sec^2 4x - 1$.

Hence $3\displaystyle\int \tan^2 4x\, dx = 3\int(\sec^2 4x - 1)\, dx = 3\left(\frac{\tan 4x}{4} - x\right) + c$

*Problem 15* Evaluate $\displaystyle\int_{\frac{\pi}{6}}^{\frac{\pi}{3}} \frac{1}{2}\cot^2 2\theta\, d\theta.$

Since $\cot^2 \theta + 1 = \operatorname{cosec}^2 \theta$, then $\cot^2 \theta = \operatorname{cosec}^2 \theta - 1$ and $\cot^2 2\theta = \operatorname{cosec}^2 2\theta - 1$

Hence $\displaystyle\int_{\frac{\pi}{6}}^{\frac{\pi}{3}} \frac{1}{2} \cot^2 2\theta \, d\theta = \frac{1}{2} \int_{\frac{\pi}{6}}^{\frac{\pi}{3}} (\operatorname{cosec}^2 2\theta - 1) \, d\theta = \frac{1}{2} \left[ \frac{-\cot 2\theta}{2} - \theta \right]_{\frac{\pi}{6}}^{\frac{\pi}{3}}$

$$= \frac{1}{2} \left[ \left( \frac{-\cot 2(\frac{\pi}{3})}{2} - \frac{\pi}{3} \right) - \left( \frac{-\cot 2(\frac{\pi}{6})}{2} - \frac{\pi}{6} \right) \right]$$

$$= \frac{1}{2} [- -0.2887 - 1.0472) - (-0.2887 - 0.5236)] = \mathbf{0.0269}$$

*Problem 16* Determine $\displaystyle\int \sin^5 \theta \, d\theta$.

Since $\cos^2 \theta + \sin^2 \theta = 1$, then $\sin^2 \theta = (1 - \cos^2 \theta)$

Hence $\displaystyle\int \sin^5 \theta \, d\theta = \int \sin \theta \, (\sin^2 \theta)^2 \, d\theta = \int \sin \theta \, (1 - \cos^2 \theta)^2 \, d\theta$

$$= \int \sin \theta \, (1 - 2 \cos^2 \theta + \cos^4 \theta) \, d\theta$$

$$= \int (\sin \theta - 2 \sin \theta \cos^2 \theta + \sin \theta \cos^4 \theta) \, d\theta$$

$$= -\cos \theta + \frac{2 \cos^3 \theta}{3} - \frac{\cos^5 \theta}{5} + c$$

(Whenever a power of a cosine is multiplied by a sine of power 1, or vice-versa, the integral may be determined by inspection as shown.

In general, $\displaystyle\int \cos^n \theta \, \sin \theta \, d\theta = \frac{-\cos^{n+1}}{(n+1)} + c$

and $\displaystyle\int \sin^n \theta \, \cos \theta \, d\theta = \frac{\sin^{n+1} \theta}{(n+1)} + c)$

*Problem 17* Evaluate $\displaystyle\int_0^{\frac{\pi}{2}} \sin^2 x \cos^3 x \, dx$.

$$\int_0^{\frac{\pi}{2}} \sin^2 x \cos^3 x \, dx = \int_0^{\frac{\pi}{2}} \sin^2 x \cos^2 x \cos x \, dx$$

$$= \int_0^{\frac{\pi}{2}} (\sin^2 x)(1 - \sin^2 x)(\cos x) \, dx$$

$$= \int_0^{\frac{\pi}{2}} (\sin^2 x \cos x - \sin^4 x \cos x) \, dx$$

$$= \left[ \frac{\sin^3 x}{3} - \frac{\sin^5 x}{5} \right]_0^{\frac{\pi}{2}} = \left[ \frac{(\sin \frac{\pi}{2})^3}{3} - \frac{(\sin \frac{\pi}{2})^5}{5} \right] - [0 - 0]$$

$$= \frac{1}{3} - \frac{1}{5} = \mathbf{\frac{2}{15}} \ .$$

**Problem 18** Evaluate $\int_0^{\frac{\pi}{4}} 2 \cos^4 \theta \, d\theta$, correct to 4 significant figures.

$$\int_0^{\frac{\pi}{4}} 2 \cos^4 \theta \, d\theta = 2 \int_0^{\frac{\pi}{4}} (\cos^2 \theta)^2 \, d\theta = 2 \int_0^{\frac{\pi}{4}} \frac{1}{2}(1+\cos 2\theta)^2 \, d\theta$$

$$= \int_0^{\frac{\pi}{4}} (1+2\cos 2\theta+\cos^2 2\theta) \, d\theta$$

$$= \int_0^{\frac{\pi}{4}} [1+2\cos 2\theta + \frac{1}{2}(1+\cos 4\theta)] \, d\theta$$

$$= \int_0^{\frac{\pi}{4}} (\frac{3}{2}+2\cos 2\theta + \frac{1}{2}\cos 4\theta) \, d\theta$$

$$= \left[ \frac{3\theta}{2} + \sin 2\theta + \frac{\sin 4\theta}{8} \right]_0^{\frac{\pi}{4}}$$

$$= \left[ \frac{3}{2}(\frac{\pi}{4}) + \sin \frac{2\pi}{4} + \frac{\sin 4(\frac{\pi}{4})}{8} \right] - [0]$$

$$= \frac{3\pi}{8} + 1 = 2.1781, \text{ correct to 4 significant figures.}$$

**Problem 19** Find $\int \sin^2 t \cos^4 t \, dt$.

$$\int \sin^2 t \cos^4 t \, dt = \int \sin^2 t (\cos^2 t)^2 \, dt = \int \left( \frac{1-\cos 2t}{2} \right) \left( \frac{1+\cos 2t}{2} \right)^2 \, dt$$

$$= \frac{1}{8} \int (1-\cos 2t)(1+2\cos 2t+\cos^2 2t) \, dt$$

$$= \frac{1}{8} \int (1+2\cos 2t+\cos^2 2t-\cos 2t-2\cos^2 2t-\cos^3 2t) \, dt$$

$$= \frac{1}{8} \int (1+\cos 2t-\cos^2 2t-\cos^3 2t) \, dt$$

$$= \frac{1}{8} \int \left[ 1+\cos 2t- \left( \frac{1+\cos 4t}{2} \right) -\cos 2t (1-\sin^2 2t) \right] dt$$

$$= \frac{1}{8} \int \left( \frac{1}{2} - \frac{\cos 4t}{2} + \cos 2t \sin^2 2t \right) dt$$

$$= \frac{1}{8} \left( \frac{t}{2} - \frac{\sin 4t}{8} + \frac{\sin^3 2t}{6} \right) +c$$

**Problem 20** Determine $\int \sin 3t \cos 2t \, dt$

82

$$\int \sin 3t \cos 2t \, dt = \int \frac{1}{2} \left[ \sin (3t+2t) + \sin (3t-2t) \right] \, dt, \text{ from 6 of } \textit{Table 2,}$$
$$= \frac{1}{2} \int (\sin 5t + \sin t) \, dt$$
$$= \frac{1}{2} \left( \frac{-\cos 5t}{5} - \cos t \right) + c.$$

*Problem 21* Find $\int \frac{1}{3} \cos 5x \sin 2x \, dx.$

$$\int \frac{1}{3} \cos 5x \sin 2x \, dx = \frac{1}{3} \int \frac{1}{2} \left[ \sin (5x+2x) - \sin (5x-2x) \right] \, dx, \text{ from 7 of } \textit{Table 2,}$$
$$= \frac{1}{6} \int (\sin 7x - \sin 3x) \, dx$$
$$= \frac{1}{6} \left( \frac{-\cos 7x}{7} + \frac{\cos 3x}{3} \right) + c$$

*Problem 22* Evaluate $\int_0^1 2 \cos 6\theta \cos \theta \, d\theta$, correct to 4 decimal places.

$$\int_0^1 2 \cos 6\theta \cos \theta \, d\theta = 2 \int_0^1 \frac{1}{2} \left[ \cos (6\theta+\theta) + \cos (6\theta-\theta) \right] \, d\theta, \text{ from 8 of } \textit{Table 2,}$$
$$= \int_0^1 (\cos 7\theta + \cos 5\theta) \, d\theta$$
$$= \left[ \frac{\sin 7\theta}{7} + \frac{\sin 5\theta}{5} \right]_0^1 = \left( \frac{\sin 7}{7} + \frac{\sin 5}{5} \right) - \left( \frac{\sin 0}{7} + \frac{\sin 0}{5} \right)$$

'sin 7' means 'the sine of 7 radians' ($\equiv 401° \, 4'$) and sin $5 \equiv$ sin $286° \, 29'$.

Hence $\int_0^1 2 \cos 6\theta \cos \theta \, d\theta = (0.09386 + -0.19178) - (0)$
$$= -0.0979, \text{ correct to 4 decimal places.}$$

*Problem 23* Find $3 \int \sin 5x \sin 3x \, dx.$

$$3 \int \sin 5x \sin 3x \, dx = 3 \int - \frac{1}{2} \left[ \cos (5x+3x) - \cos (5x-3x) \right] \, dx, \text{ from 9 of } \textit{Table 2,}$$
$$= - \frac{3}{2} \int (\cos 8x - \cos 2x) \, dx$$
$$= - \frac{3}{2} \left( \frac{\sin 8x}{8} - \frac{\sin 2x}{2} \right) + c \text{ or } \frac{3}{16} (4 \sin 2x - \sin 8x) + c.$$

*Problem 24* Determine $\int \frac{1}{\sqrt{(a^2-x^2)}} \, dx.$

Let $x = a \sin \theta$, then $\frac{dx}{d\theta} = a \cos \theta$ and $dx = a \cos \theta \, d\theta$.

Hence $\int \frac{1}{\sqrt{(a^2-x^2)}} \, dx = \int \frac{1}{\sqrt{(a^2-a^2 \sin^2 \theta)}} a \cos \theta \, d\theta = \int \frac{a \cos \theta \, d\theta}{\sqrt{[a^2(1-\sin^2 \theta)]}}$

83

$$= \int \frac{a \cos \theta \, d\theta}{\sqrt{(a^2 \cos^2 \theta)}} \text{ , since } \sin^2 \theta + \cos^2 \theta = 1$$

$$= \int \frac{a \cos \theta \, d\theta}{a \cos \theta} = \int d\theta = \theta + c$$

Since $x = a \sin \theta$, then $\sin \theta = \frac{x}{a}$ and $\theta = \arcsin \frac{x}{a}$.

Hence $\int \frac{1}{\sqrt{(a^2 - x^2)}} \, dx = \arcsin \frac{x}{a} + c.$

*Problem 25* Evaluate $\int_0^3 \frac{1}{\sqrt{(9-x^2)}} \, dx.$

From *Problem 24*, $\int_0^3 \frac{1}{\sqrt{(9-x^2)}} \, dx = \left[ \arcsin \frac{x}{3} \right]_0^3$ , since $a = 3$,

$$= (\arcsin 1 - \arcsin 0) = \frac{\pi}{2} \text{ or } 1.5708$$

*Problem 26* Find $\int \sqrt{(a^2 - x^2)} \, dx.$

Let $x = a \sin \theta$ then $\frac{dx}{d\theta} = a \cos \theta$ and $dx = a \cos \theta \, d\theta$.

Hence $\int \sqrt{(a^2 - x^2)} \, dx = \int \sqrt{(a^2 - a^2 \sin^2 \theta)}(a \cos \theta \, d\theta)$

$$= \int \sqrt{[a^2(1 - \sin^2 \theta)]}(a \cos \theta \, d\theta)$$

$$= \int \sqrt{(a^2 \cos^2 \theta)}(a \cos \theta \, d\theta) = \int (a \cos \theta)(a \cos \theta \, d\theta)$$

$$= a^2 \int \cos^2 \theta \, d\theta = a^2 \int \left( \frac{1 + \cos 2\theta}{2} \right) d\theta$$

$$\text{(since } \cos 2\theta = 2 \cos^2 \theta - 1)$$

$$= \frac{a^2}{2} \left( \theta + \frac{\sin 2\theta}{2} \right) + c$$

In the compound angle addition formula: $\sin(A+B) = \sin A \cos B + \cos A \sin B$, let $B = A$, then $\sin 2A = 2 \sin A \cos A$.

Hence $\int \sqrt{(a^2 - x^2)} \, dx = \frac{a^2}{2} \left( \theta + \frac{2 \sin \theta \cos \theta}{2} \right) + c = \frac{a^2}{2} [\theta + \sin \theta \cos \theta] + c$

Since $x = a \sin \theta$, then $\sin \theta = \frac{x}{a}$ and $\theta = \arcsin \frac{x}{a}$.

Also, $\cos^2 \theta + \sin^2 \theta = 1$, from which, $\cos \theta = \sqrt{(1 - \sin^2 \theta)} = \sqrt{\left[ 1 - \left( \frac{x}{a} \right)^2 \right]}$

$$= \sqrt{\left( \frac{a^2 - x^2}{a^2} \right)} = \frac{\sqrt{(a^2 - x^2)}}{a}$$

Thus $\int \sqrt{(a^2 - x^2)} \, dx = \frac{a^2}{2} [\theta + \sin \theta \cos \theta] = \frac{a^2}{2} \left[ \arcsin \frac{x}{a} + \left( \frac{x}{a} \right) \frac{\sqrt{(a^2 - x^2)}}{a} \right] + c$

$$= \frac{a^2}{2} \arcsin \frac{x}{a} + \frac{x}{2} \sqrt{(a^2 - x^2)} + c.$$

**Problem 27** Evaluate $\int_0^4 \sqrt{(16-x^2)}\, dx$.

From *Problem 26*, $\int_0^4 \sqrt{(16-x^2)}\, dx = \left[\dfrac{16}{2} \arcsin \dfrac{x}{4} + \dfrac{x}{2} \sqrt{(16-x^2)}\right]_0^4$

$$= [8 \arcsin 1 + 2\sqrt{(0)}] - [8 \arcsin 0 + 0]$$

$$= 8 \arcsin 1 = 8 \left(\dfrac{\pi}{2}\right) = 4\pi \text{ or } 12.57$$

**Problem 28** Determine $\int \dfrac{1}{(a^2+x^2)}\, dx$

Let $x = a \tan \theta$ then $\dfrac{dx}{d\theta} = a \sec^2 \theta$ and $dx = a \sec^2 \theta\, d\theta$

Hence $\int \dfrac{1}{(a^2+x^2)}\, dx = \int \dfrac{1}{(a^2+a^2 \tan^2 \theta)}(a \sec^2 \theta\, d\theta) = \int \dfrac{a \sec^2 \theta\, d\theta}{a^2(1+\tan^2 \theta)}$

$$= \int \dfrac{a \sec^2 \theta\, d\theta}{a^2 \sec^2 \theta}, \text{ since } 1+\tan^2 \theta = \sec^2 \theta$$

$$= \int \dfrac{1}{a}\, d\theta = \dfrac{1}{a}(\theta)+c$$

Since $x = a \tan \theta$, $\theta = \arctan \dfrac{x}{a}$. Hence $\int \dfrac{1}{(a^2+x^2)}\, dx = \dfrac{1}{a} \arctan \dfrac{x}{a} + c$.

**Problem 29** Evaluate $\int_0^2 \dfrac{1}{(4+x^2)}\, dx$.

From *Problem 28*, $\int_0^2 \dfrac{1}{(4+x^2)}\, dx = \dfrac{1}{2}\left[\arctan \dfrac{x}{2}\right]_0^2$, since $a = 2$,

$$= \dfrac{1}{2}(\arctan 1 - \arctan 0) = \dfrac{1}{2}\left(\dfrac{\pi}{4} - 0\right) = \dfrac{\pi}{8} \text{ or } 0.3927$$

**Problem 30** Evaluate $\int_0^1 \dfrac{5}{(3+2x^2)}\, dx$, correct to 4 decimal places.

$$\int_0^1 \dfrac{5}{(3+2x^2)}\, dx = \int_0^1 \dfrac{5}{2(\frac{3}{2}+x^2)}\, dx = \dfrac{5}{2}\int_0^1 \dfrac{1}{\left\{[\sqrt{(\frac{3}{2})}]^2 + x^2\right\}}\, dx$$

$$= \dfrac{5}{2}\left[\dfrac{1}{\sqrt{(\frac{3}{2})}} \arctan \dfrac{x}{\sqrt{(\frac{3}{2})}}\right]_0^1 = \dfrac{5}{2}\sqrt{\left(\dfrac{2}{3}\right)}\left[\arctan \sqrt{\dfrac{2}{3}} - \arctan 0\right]$$

$$= (2.0412)[0.6847 - 0] = \mathbf{1.3976}, \text{ correct to 4 decimal places.}$$

**Problem 31** Determine $\int \dfrac{1}{\sqrt{(x^2+a^2)}}\, dx$

85

Let $x = a \sinh \theta$, then $\frac{dx}{d\theta} = a \cosh \theta$ and $dx = a \cosh \theta\, d\theta$.

Hence $\int \frac{1}{\sqrt{(x^2+a^2)}}\, dx = \int \frac{1}{\sqrt{(a^2 \sinh^2 \theta + a^2)}} (a \cosh \theta\, d\theta) = \int \frac{a \cosh \theta\, d\theta}{\sqrt{[a^2(\sinh^2 \theta + 1)]}}$

$$= \int \frac{a \cosh \theta\, d\theta}{\sqrt{(a^2 \cosh^2 \theta)}}\ ,\ \text{since } \cosh^2 \theta - \sinh^2 \theta = 1$$

$$= \int \frac{a \cosh \theta}{a \cosh \theta}\, d\theta = \int d\theta = \theta + c$$

$$= \operatorname{arsinh} \frac{x}{a} + c,\ \text{since } x = a \sinh \theta$$

It was shown in chapter 7 that $\operatorname{arsinh} \frac{x}{a} = \ln \left\{ \frac{x + \sqrt{(x^2 + a^2)}}{a} \right\}$, which provides an alternative solution to $\int \frac{1}{\sqrt{(x^2+a^2)}}\, dx$.

**Problem 32** Evaluate $\int_0^2 \frac{1}{\sqrt{(x^2+4)}}\, dx$, correct to 4 decimal places.

$\int_0^2 \frac{1}{\sqrt{(x^2+4)}}\, dx = \left[ \operatorname{arsinh} \frac{x}{2} \right]_0^2$ or $\left[ \ln \left\{ \frac{x+\sqrt{(x^2+4)}}{2} \right\} \right]_0^2$, from *Problem 31*. where $a = 2$.

Using the logarithmic form, $\int_0^2 \frac{1}{\sqrt{(x^2+4)}}\, dx = \left[ \ln \left( \frac{2+\sqrt{8}}{2} \right) - \ln \left( \frac{0+\sqrt{4}}{2} \right) \right]$

$$= \ln 2.4142 - \ln 1 = 0.8814,\ \text{correct to}$$
$$\text{4 decimal places.}$$

**Problem 33** Evaluate $\int_1^2 \frac{2}{x^2\sqrt{(1+x^2)}}\, dx$, correct to 3 significant figures.

Since the integral contains a term of the form $\sqrt{(a^2+x^2)}$, then let $x = \sinh \theta$, from which $dx/d\theta = \cosh \theta$ and $dx = \cosh \theta\, d\theta$.

Hence $\int \frac{2}{x^2\sqrt{(1+x^2)}}\, dx = \int \frac{2 (\cosh \theta\, d\theta)}{\sinh^2 \theta \sqrt{(1+\sinh^2 \theta)}} = 2 \int \frac{\cosh \theta\, d\theta}{\sinh^2 \theta \cosh \theta}\ ,$

$$\text{since } \cosh^2 \theta - \sinh^2 \theta = 1,$$

$$= 2 \int \frac{d\theta}{\sinh^2 \theta} = 2 \int \operatorname{cosech}^2 \theta\, d\theta = -2 \coth \theta + c.$$

$\coth \theta = \dfrac{\cosh \theta}{\sinh \theta} = \dfrac{\sqrt{(1+\sinh^2 \theta)}}{\sinh \theta} = \dfrac{\sqrt{(1+x^2)}}{x}$

Hence $\int_1^2 \frac{2}{x^2\sqrt{(1+x^2)}}\, dx = -\left[ 2 \coth \theta \right]_1^2 = -2\left[ \frac{\sqrt{(1+x^2)}}{x} \right]_1^2 = -2\left[ \frac{\sqrt{5}}{2} - \frac{\sqrt{2}}{1} \right]$

$$= 0.592,\ \text{correct to 3 significant figures}$$

**Problem 34** Find $\int \sqrt{(x^2+a^2)}\, dx$.

Let $x = a \sinh \theta$ then $\dfrac{dx}{d\theta} = a \cosh \theta$ and $dx = a \cosh \theta \, d\theta$.

Hence $\displaystyle\int \sqrt{(x^2 + a^2)} \, dx = \int \sqrt{(a^2 \sinh^2 \theta + a^2)}(a \cosh \theta \, d\theta)$

$$= \int \sqrt{[a^2(\sinh^2 \theta + 1)]}(a \cosh \theta \, d\theta)$$

$$= \int \sqrt{(a^2 \cosh^2 \theta)}(a \cosh \theta \, d\theta), \text{ since } \cosh^2 \theta - \sinh^2 \theta = 1,$$

$$= \int (a \cosh \theta)(a \cosh \theta) \, d\theta = a^2 \int \cosh^2 \theta \, d\theta$$

$$= a^2 \int \left( \frac{1 + \cosh 2\theta}{2} \right) d\theta = \frac{a^2}{2} \left( \theta + \frac{\sinh 2\theta}{2} \right) + c$$

$$= \frac{a^2}{2} [\theta + \sinh \theta \cosh \theta] + c, \text{ since } \sinh 2\theta = 2 \sinh \theta \cosh \theta.$$

Since $x = a \sinh \theta$, then $\sinh \theta = \dfrac{x}{a}$ and $\theta = \operatorname{arsinh} \dfrac{x}{a}$.

Also since $\cosh^2 \theta - \sinh^2 \theta = 1$ then $\cosh \theta = \sqrt{(1 + \sinh^2 \theta)} = \sqrt{\left[ 1 + \left( \dfrac{x}{a} \right)^2 \right]}$

$$= \sqrt{\left( \frac{a^2 + x^2}{a^2} \right)} = \frac{\sqrt{(a^2 + x^2)}}{a}$$

Hence $\displaystyle\int \sqrt{(x^2 + a^2)} \, dx = \frac{a^2}{2} \left[ \operatorname{arsinh} \frac{x}{a} + \left( \frac{x}{a} \right) \frac{\sqrt{(x^2 + a^2)}}{a} \right] + c$

$$= \frac{a^2}{2} \operatorname{arsinh} \frac{x}{a} + \frac{x}{2} \sqrt{(x^2 + a^2)} + c$$

**Problem 35** Determine $\displaystyle\int \frac{1}{\sqrt{(x^2 - a^2)}} \, dx$.

Let $x = a \cosh \theta$ then $\dfrac{dx}{d\theta} = a \sinh \theta$ and $dx = a \sinh \theta \, d\theta$.

Hence $\displaystyle\int \frac{1}{\sqrt{(x^2 - a^2)}} \, dx = \int \frac{1}{\sqrt{(a^2 \cosh^2 \theta - a^2)}} (a \sinh \theta \, d\theta) = \int \frac{a \sinh \theta \, d\theta}{\sqrt{[a^2(\cosh^2 \theta - 1)]}}$

$$= \int \frac{a \sinh \theta \, d\theta}{\sqrt{(a^2 \sinh^2 \theta)}}, \text{ since } \cosh^2 \theta - \sinh^2 \theta = 1$$

$$= \int \frac{a \sinh \theta \, d\theta}{a \sinh \theta} = \int d\theta = \theta + c = \operatorname{arcosh} \frac{x}{a} + c,$$

$$\text{since } x = a \cosh \theta$$

It was shown in chapter 7 that $\operatorname{arcosh} \dfrac{x}{a} = \ln \left\{ \dfrac{x + \sqrt{(x^2 - a^2)}}{a} \right\}$,

which provides an alternative solution to $\displaystyle\int \frac{1}{\sqrt{(x^2 - a^2)}} \, dx$.

**Problem 36** Determine $\displaystyle\int \frac{2x - 3}{\sqrt{(x^2 - 9)}} \, dx$.

$$\int \frac{2x - 3}{\sqrt{(x^2 - 9)}} \, dx = \int \frac{2x}{\sqrt{(x^2 - 9)}} \, dx - \int \frac{3}{\sqrt{(x^2 - 9)}} \, dx$$

The first integral is determined using the algebraic substitution $u = (x^2 - 9)$,

and the second integral is of the form $\displaystyle\int \frac{1}{\sqrt{(x^2 - a^2)}} \, dx$ (see *Problem 35*).

87

Hence $\int \frac{2x}{\sqrt{(x^2-9)}} dx - \int \frac{3}{\sqrt{(x^2-9)}} dx = 2\sqrt{(x^2-9)} - 3 \text{ arcosh} \frac{x}{3} + c.$

**Problem 37** Find $\int \sqrt{(x^2-a^2)} \, dx.$

Let $x = a \cosh \theta$ then $dx/d\theta = a \sinh \theta$ and $dx = a \sinh \theta \, d\theta.$

Hence $\int \sqrt{(x^2-a^2)} \, dx = \int \sqrt{(a^2 \cosh^2 \theta - a^2)}(a \sinh \theta \, d\theta)$

$$= \int \sqrt{[a^2(\cosh^2 \theta - 1)]} \, [a \sinh \theta \, d\theta]$$

$$= \int \sqrt{(a^2 \sinh^2 \theta)}(a \sinh \theta \, d\theta) = a^2 \int \sinh^2 \theta \, d\theta$$

$$= a^2 \int \left( \frac{\cosh 2\theta - 1}{2} \right) d\theta, \text{ since } \cosh 2\theta = 1 + 2 \sinh^2 \theta$$

$$= \frac{a^2}{2} \left[ \frac{\sinh 2\theta}{2} - \theta \right] + c$$

$$= \frac{a^2}{2} [\sinh \theta \cosh \theta - \theta] + c, \text{ since } \sinh 2\theta = 2 \sinh \theta \cosh \theta$$

Since $x = a \cosh \theta$ then $\cosh \theta = \frac{x}{a}$ and $\theta = \text{arcosh} \frac{x}{a}.$

Also, since $\cosh^2 \theta - \sinh^2 \theta = 1$, then $\sinh \theta = \sqrt{(\cosh^2 \theta - 1)}$

$$= \sqrt{\left[ \left( \frac{x}{a} \right)^2 - 1 \right]} = \frac{\sqrt{(x^2-a^2)}}{a}$$

Hence $\int \sqrt{(x^2-a^2)} \, dx = \frac{a^2}{2} \left[ \frac{\sqrt{(x^2-a^2)}}{a} \left( \frac{x}{a} \right) - \text{arcosh} \frac{x}{a} \right] + c$

$$= \frac{x}{2} \sqrt{(x^2-a^2)} - \frac{a^2}{2} \text{ arcosh} \frac{x}{a} + c.$$

**Problem 38** Evaluate $\int_2^3 \sqrt{(x^2-4)}$, correct to 4 significant figures.

$\int_2^3 \sqrt{(x^2-4)} \, dx = \left[ \frac{x}{2} \sqrt{(x^2-4)} - \frac{4}{2} \text{ arcosh} \frac{x}{2} \right]_2^3$, from *Problem 37*,

$$= \left( \frac{3}{2}\sqrt{5} - 2 \text{ arcosh} \frac{3}{2} \right) - (0 - 2 \text{ arcosh } 1)$$

Since $\text{arcosh} \frac{x}{a} = \ln \left\{ \frac{x + \sqrt{(x^2-a^2)}}{a} \right\}$ then $\text{arcosh} \frac{3}{2} = \ln \left\{ \frac{3 + \sqrt{(3^2-2^2)}}{2} \right\}$

$$= \ln 2.6180 = 0.9624$$

Similarly, $\text{arcosh } 1 = 0$

Hence $\int_2^3 \sqrt{(x^2-4)} \, dx = \left[ \frac{3}{2} \sqrt{5} - 2(0.9624) \right] - [0] = \mathbf{1.429}$, correct to 4 significant figures.

(c) INTEGRATION USING PARTIAL FRACTIONS

**Problem 39** Determine $\int \frac{11-3x}{x^2+2x-3} \, dx.$

As shown in *Mathematics 3 Checkbook*, page 47:

$$\frac{11-3x}{x^2+2x-3} \equiv \frac{11-3x}{(x-1)(x+3)} \equiv \frac{A}{(x-1)} + \frac{B}{(x+3)} \equiv \frac{A(x+3)+B(x-1)}{(x-1)(x+3)}$$

Hence $11-3x \equiv A(x+3)+B(x-1)$

Let $x = 1$, then $8 = 4A$, from which, $A = 2$

Let $x = -3$, then $20 = -4B$, from which, $B = -5$

Hence $\displaystyle\int \frac{11-3x}{x^2+2x-3}\,dx = \int \left\{\frac{2}{(x-1)} - \frac{5}{(x+3)}\right\} dx = 2\ln(x-1)-5\ln(x+3)+c$

or $\ln \left\{\dfrac{(x-1)^2}{(x+3)^5}\right\} +c$, by the laws of logarithms.

---

***Problem 40*** Find $\displaystyle\int \frac{2x^2-9x-35}{(x+1)(x-2)(x+3)}\,dx$.

It was shown in *Mathematics 3 Checkbook,* page 47 that:

$$\frac{2x^2-9x-35}{(x+1)(x-2)(x+3)} \equiv \frac{4}{(x+1)} - \frac{3}{(x-2)} + \frac{1}{(x+3)}$$

Hence $\displaystyle\int \frac{2x^2-9x-35}{(x+1)(x-2)(x+3)}\,dx \equiv \int \left\{\frac{4}{(x+1)} - \frac{3}{(x-2)} + \frac{1}{(x+3)}\right\} dx$

$$= 4\ln(x+1) -3\ln(x-2)+\ln(x+3)+c$$

$$= \ln \left\{\frac{(x+1)^4(x+3)}{(x-2)^3}\right\} +c$$

---

***Problem 41*** Determine $\displaystyle\int \frac{x^2+1}{x^2-3x+2}\,dx$.

By dividing out (since the numerator and denominator are of the same degree) and resolving into partial fractions it was shown in *Mathematics 3 Checkbook*, page 48, that

$$\frac{x^2+1}{x^2-3x+2} \equiv 1 - \frac{2}{(x-1)} + \frac{5}{(x-2)}$$

Hence $\displaystyle\int \frac{x^2+1}{(x^2-3x+2)}\,dx \equiv \int \left\{1 - \frac{2}{(x-1)} + \frac{5}{(x-2)}\right\} dx$

$$= x-2\ln(x-1)+5\ln(x-2)+c$$

or $x+\ln \left\{\dfrac{(x-2)^5}{(x-1)^2}\right\} +c.$

---

***Problem 42*** Evaluate $\displaystyle\int_2^3 \left(\frac{x^3-2x^2-4x-4}{x^2+x-2}\right) dx$, correct to 4 significant figures.

By dividing out and resolving into partial fractions it was shown in *Mathematics 3 Checkbook*, page 48 that

$$\frac{x^3-2x^2-4x-4}{x^2+x-2} \equiv x-3 + \frac{4}{(x+2)} - \frac{3}{(x-1)}$$

Hence $\int_2^3 \left(\frac{x^3-2x^2-4x-4}{x^2+x-2}\right) dx \equiv \int_2^3 \left\{x-3 + \frac{4}{(x+2)} - \frac{3}{(x-1)}\right\} dx$

$$= \left[\frac{x^2}{2} - 3x + 4 \ln (x+2) - 3 \ln (x-1)\right]_2^3$$

$$= (\tfrac{9}{2} - 9 + 4 \ln 5 - 3 \ln 2) - (2 - 6 + 4 \ln 4 - 3 \ln 1)$$

$$= -1.687, \text{ correct to 4 significant figures.}$$

**Problem 43** Determine $\int \frac{2x+3}{(x-2)^2} dx$.

It was shown in *Mathematics 3 Checkbook*, page 49 that:

$$\frac{2x+3}{(x-2)^2} \equiv \frac{A}{(x-2)} + \frac{B}{(x-2)^2} \equiv \frac{A(x-2)+B}{(x-2)^2}$$

Hence $2x+3 \equiv A(x-2)+B$, from which, $A = 2$ and $B = 7$.

Thus $\int \frac{2x+3}{(x-2)^2} dx \equiv \int \left\{\frac{2}{(x-2)} + \frac{7}{(x-2)^2}\right\} dx = 2 \ln (x-2) - \frac{7}{(x-2)} + c$

$\int \frac{7}{(x-2)^2} dx$ is determined using the algebraic substitution $u = (x-2)$

**Problem 44** Find $\int \frac{5x^2-2x-19}{(x+3)(x-1)^2} dx$.

It was shown in *Mathematics 3 Checkbook*, page 49, that:

$$\frac{5x^2-2x-19}{(x+3)(x-1)^2} \equiv \frac{2}{(x+3)} + \frac{3}{(x-1)} - \frac{4}{(x-1)^2}$$

Hence $\int \frac{5x^2-2x-19}{(x+3)(x-1)^2} dx \equiv \int \left\{\frac{2}{(x+3)} + \frac{3}{(x-1)} - \frac{4}{(x-1)^2}\right\} dx$

$$= 2 \ln (x+3) + 3 \ln (x-1) + \frac{4}{(x-1)} + c$$

or $\quad \ln\{(x+3)^2 (x-1)^3\} + \frac{4}{(x-1)} + c$

**Problem 45** Evaluate $\int_{-2}^1 \frac{3x^2+16x+15}{(x+3)^3} dx$, correct to 4 significant figures.

It was shown in *Mathematics 3 Checkbook*, page 50, that:

$$\frac{3x^2+16x+15}{(x+3)^3} \equiv \frac{3}{(x+3)} - \frac{2}{(x+3)^2} - \frac{6}{(x+3)^3}$$

Hence $\int_{-2}^1 \frac{3x^2+16x+15}{(x+3)^3} dx \equiv \int_{-2}^1 \left\{\frac{3}{(x+3)} - \frac{2}{(x+3)^2} - \frac{6}{(x+3)^3}\right\} dx$

$$= \left[3 \ln (x+3) + \frac{2}{(x+3)} + \frac{3}{(x+3)^2}\right]_{-2}^1$$

$$= (3 \ln 4 + \frac{2}{4} + \frac{3}{16}) - (3 \ln 1 + \frac{2}{1} + \frac{3}{1})$$

$$= -0.1536, \text{ correct to 4 significant figures.}$$

*Problem 46* Find $\int \dfrac{3+6x+4x^2-2x^3}{x^2(x^2+3)}\, dx.$

It was shown in *Mathematics 3 Checkbook,* page 50, that:

$$\frac{3+6x+4x^2-2x^3}{x^2(x^2+3)} \equiv \frac{A}{x} + \frac{B}{x^2} + \frac{Cx+D}{(x^2+3)} \equiv \frac{Ax(x^2+3)+B(x^2+3)+(Cx+D)x^2}{x^2(x^2+3)}$$

Hence $3+6x+4x^2-2x^3 \equiv Ax(x^2+3)+B(x^2+3)+(Cx+D)x^2$
from which $A = 2, B = 1, C = -4$ and $D = 3$.

Thus $\int \dfrac{3+6x+4x^2-2x^3}{x^2(x^2+3)}\, dx \equiv \int \left\{ \dfrac{2}{x} + \dfrac{1}{x^2} + \dfrac{3-4x}{(x^2+3)} \right\} dx$

$$= \int \left\{ \frac{2}{x} + \frac{1}{x^2} + \frac{3}{(x^2+3)} - \frac{4x}{(x^2+3)} \right\} dx$$

$\int \dfrac{3}{(x^2+3)}\, dx = 3 \int \dfrac{1}{x^2+(\sqrt{3})^2}\, dx = \dfrac{3}{\sqrt{3}} \arctan \dfrac{x}{\sqrt{3}}$ , from 12, *Table 2.*

$\int \dfrac{4x}{x^2+3}\, dx$ is determined using the algebraic substitution $u = (x^2+3)$.

Hence $\int \left\{ \dfrac{2}{x} + \dfrac{1}{x^2} + \dfrac{3}{(x^2+3)} - \dfrac{4x}{x^2+3} \right\} dx = 2 \ln x - \dfrac{1}{x} + \dfrac{3}{\sqrt{3}} \arctan \dfrac{x}{\sqrt{3}} - 2\ln(x^2+3) + c$

$$= \ln \left( \frac{x}{x^2+3} \right)^2 - \frac{1}{x} + \sqrt{3} \arctan \frac{x}{\sqrt{3}} + c$$

*Problem 47* Determine $\int \dfrac{1}{(x^2-a^2)}\, dx.$

Let $\dfrac{1}{(x^2-a^2)} \equiv \dfrac{A}{(x-a)} + \dfrac{B}{(x+a)} \equiv \dfrac{A(x+a)+B(x-a)}{(x+a)(x-a)}$

Equating the numerators gives: $1 \equiv A(x+a)+B(x-a)$

Let $x = a$, then $A = \dfrac{1}{2a}$ , and let $x = -a$, then $B = -\dfrac{1}{2a}$

Hence $\int \dfrac{1}{(x^2-a^2)}\, dx = \int \dfrac{1}{2a}\left[ \dfrac{1}{(x-a)} - \dfrac{1}{(x+a)} \right] dx = \dfrac{1}{2a}\left[ \ln(x-a) - \ln(x+a) \right] + c$

$$= \frac{1}{2a} \ln \left( \frac{x-a}{x+a} \right) + c.$$

*Problem 48* Evaluate $\displaystyle\int_3^4 \dfrac{3}{(x^2-4)}\, dx$, correct to 3 significant figures.

From *Problem 47,* $\displaystyle\int_3^4 \dfrac{3}{(x^2-4)}\, dx = 3\left[ \dfrac{1}{2(2)} \ln \left( \dfrac{x-2}{x+2} \right) \right]_3^4 = \dfrac{3}{4}\left[ \ln \dfrac{2}{6} - \ln \dfrac{1}{5} \right]$

$$= \frac{3}{4} \ln \frac{5}{3} = 0.383, \text{ correct to 3 significant figures.}$$

**Problem 49** Determine $\int \dfrac{1}{(a^2-x^2)}\, dx$.

There are two methods which may be used to determine this integral.

*Method 1*
Using partial fractions, let

$$\frac{1}{(a^2-x^2)} \equiv \frac{1}{(a-x)(a+x)} \equiv \frac{A}{(a-x)}+\frac{B}{(a+x)} \equiv \frac{A(a+x)+B(a-x)}{(a-x)(a+x)}$$

Then $1 \equiv A(a+x)+B(a-x)$

Let $x=a$ then $A = \dfrac{1}{2a}$. Let $x=-a$ then $B=\dfrac{1}{2a}$

Hence $\displaystyle\int \frac{1}{(a^2-x^2)}\, dx = \int \frac{1}{2a}\left[\frac{1}{(a-x)}+\frac{1}{(a+x)}\right]dx = \frac{1}{2a}\left[-\ln(a-x)+\ln(a+x)\right] + c$

$$= \frac{1}{2a}\ln\left(\frac{a+x}{a-x}\right)+c$$

*Method 2*
Let $x=a \tanh\theta$, then $\dfrac{dx}{d\theta} = a \,\text{sech}^2\,\theta$ and $dx = a\,\text{sech}^2\,\theta\, d\theta$.

Hence $\displaystyle\int \frac{1}{(a^2-x^2)}\, dx = \int \frac{1}{(a^2-a^2\tanh^2\theta)}(a\,\text{sech}^2\,\theta\, d\theta) = \int \frac{a\,\text{sech}^2\,\theta\, d\theta}{a^2(1-\tanh^2\theta)}$

$$= \int \frac{a\,\text{sech}^2\,\theta\, d\theta}{a^2\,\text{sech}^2\,\theta}$$

$$= \frac{1}{a}\int d\theta = \frac{1}{a}\,(\theta)+c = \frac{1}{a}\,\text{artanh}\,\frac{x}{a}+c;$$

**Problem 50** Evaluate $\displaystyle\int_0^2 \frac{5}{(9-x^2)}\, dx$, correct to 4 decimal places.

From *Problem 49*, $\displaystyle\int_0^2 \frac{5}{(9-x^2)}\, dx = 5\left[\frac{1}{2(3)}\,\ln\left(\frac{3+x}{3-x}\right)\right]_0^2 = \frac{5}{6}\left[\ln\frac{5}{1}-\ln 1\right]$

$$= 1.3412, \text{ correct to 4 decimal places.}$$

## C. FURTHER PROBLEMS ON INTEGRATION USING SUBSTITUTIONS AND PARTIAL FRACTIONS

(a) ALGEBRAIC SUBSTITUTIONS

In *Problems 1 to 15*, integrate with respect to the variable.

1  $2\sin(4x+9)$                              $[-\dfrac{1}{2}\cos(4x+9)+c]$

2  $3\cos(2\theta-5)$                           $[\dfrac{3}{2}\sin(2\theta-5)+c]$

3  $4\sec^2(3t+1)$                             $[\dfrac{4}{3}\tan(3t+1)+c]$

4  $\frac{1}{2}(5x-3)^6$ $\qquad\qquad$ $[\frac{1}{70}(5x-3)^7+c]$

5  $\dfrac{-3}{(2x-1)}$ $\qquad\qquad$ $[-\frac{3}{2}\ln(2x-1)+c]$

6  $3e^{3\theta+5}$ $\qquad\qquad$ $[e^{3\theta+5}+c]$

7  $2x\,(2x^2-3)^5$ $\qquad\qquad$ $[\frac{1}{12}(2x^2-3)^6+c]$

8  $\frac{1}{3}\sin^2\theta\,\cos\theta$ $\qquad\qquad$ $[\frac{1}{9}\sin^3\theta+c]$

9  $5\cos^5 t\,\sin t$ $\qquad\qquad$ $[-\frac{5}{6}\cos^6 t+c]$

10  $3\sec^2 3x\,\tan 3x$ $\qquad\qquad$ $[\frac{1}{2}\sec^2 3x+c$ or $\frac{1}{2}\tan^2 3x+c]$

11  $2t\surd(3t^2-1)$ $\qquad\qquad$ $[\frac{2}{9}\surd(3t^2-1)^3+c]$

12  $\dfrac{\ln\theta}{\theta}$ $\qquad\qquad$ $[\frac{1}{2}(\ln\theta)^2+c]$

13  $\dfrac{x}{\surd(x^2+4)}$ $\qquad\qquad$ $[\surd(x^2+4)+c]$

14  $3\tan 2t$ $\qquad\qquad$ $[\frac{3}{2}\ln(\sec 2t)+c]$

15  $\dfrac{2e^t}{\surd(e^t+4)}$ $\qquad\qquad$ $[4\surd(e^t+4)+c]$

In *Problems 16 to 22*, evaluate the definite integrals correct to 4 significant figures.

16  $\displaystyle\int_0^1 (3x+1)^5\,dx$ $\qquad\qquad$ [227.5]

17  $\displaystyle\int_0^2 x\surd(2x^2+1)\,dx$ $\qquad\qquad$ [4.333]

18  $\displaystyle\int_0^{\frac{\pi}{3}} 2\sin(3t+\frac{\pi}{4})\,dt$ $\qquad\qquad$ [0.9428]

19  $\displaystyle\int_0^1 3\cos(4x-3)\,dx$ $\qquad\qquad$ [0.7369]

20  $\displaystyle\int_0^1 3xe^{(2x^2-1)}dx$ $\qquad\qquad$ [1.763]

21  $\displaystyle\int_0^{\frac{\pi}{2}} 3\sin^4\theta\,\cos\theta\,d\theta$ $\qquad\qquad$ [0.6000]

22  $\displaystyle\int_0^1 \dfrac{3x}{(4x^2-1)^5}\,dx$ $\qquad\qquad$ [0.09259]

## (b) TRIGONOMETRIC AND HYPERBOLIC SUBSTITUTIONS

In *Problems 23 and 24*, integrate with respect to the variable.

23  (a) $\sin^2 2x$;  (b) $3\cos^2 t$ $\qquad\qquad$ $\left[(a)\ \frac{1}{2}\left(x-\frac{\sin 4x}{4}\right)+c;\ (b)\ \frac{3}{2}\left(t+\frac{\sin 2t}{2}\right)+c\right]$

93

24 (a) $5 \tan^2 3\theta$; (b) $2 \cot^2 2t$ $\qquad$ [(a) $5 (\frac{1}{3} \tan 3\theta - \theta) + c$; (b) $-(\cot 2t + 2t) + c$]

In *Problems 25 and 26*, evaluate the definite integrals, correct to 4 significant figures.

25 (a) $\int_0^{\frac{\pi}{3}} 3 \sin^2 3x \, dx$; (b) $\int_0^{\frac{\pi}{4}} \cos^2 4x \, dx$ $\qquad$ [(a) $\frac{\pi}{2}$ or 1.571; (b) $\frac{\pi}{8}$ or 0.3927]

26 (a) $\int_0^1 2 \tan^2 2t \, dt$; (b) $\int_{\frac{\pi}{6}}^{\frac{\pi}{3}} \frac{1}{2} \cot^2 \theta \, d\theta$ $\qquad$ [(a) −4.185; (b) 0.3156]

*Powers of sines and cosines*
In *Problems 27 to 29*, integrate with respect to the variable.

27 (a) $\sin^3 \theta$; (b) $2 \cos^3 2x$ $\qquad$ $\left[\text{(a)} -\cos \theta + \frac{\cos^3 \theta}{3} + c; \text{(b)} \sin 2x - \frac{\sin^3 2x}{3} + c\right]$

28 (a) $2 \sin^3 t \cos^2 t$; (b) $\sin^3 x \cos^4 x$

$\left[\text{(a)} \frac{-2}{3} \cos^3 t + \frac{2}{5} \cos^5 t + c; \text{(b)} \frac{-\cos^5 x}{5} + \frac{\cos^7 x}{7} + c\right]$

29 (a) $2 \sin^4 2\theta$; (b) $\sin^2 t \cos^2 t$

$\left[\text{(a)} \frac{3\theta}{4} - \frac{1}{4} \sin 4\theta + \frac{1}{32} \sin 8\theta + c; \text{(b)} \frac{t}{8} - \frac{1}{32} \sin 4t + c\right]$

30 Show that $\int_0^{\frac{\pi}{2}} \sin^4 2t \cos^2 2t \, dt = \frac{\pi}{32}$ .

*Products of sines and cosines*
In *Problems 31 and 32*, integrate with respect to the variable.
31 (a) $\sin 5t \cos 2t$; (b) $2 \sin 3x \sin x$.

$\left[\text{(a)} -\frac{1}{2}\left(\frac{\cos 7t}{7} + \frac{\cos 3t}{3}\right) + c; \text{(b)} \frac{\sin 2x}{2} - \frac{\sin 4x}{4} + c\right]$

32 (a) $3 \cos 6x \cos x$; (b) $\frac{1}{2} \cos 4\theta \sin 2\theta$

$\left[\text{(a)} \frac{3}{2}\left(\frac{\sin 7x}{7} + \frac{\sin 5x}{5}\right) + c; \text{(b)} \frac{1}{4}\left(\frac{\cos 2\theta}{2} - \frac{\cos 6\theta}{6}\right) + c\right]$

In *Problems 33 and 34*, evaluate the definite integrals.

33 (a) $\int_0^{\frac{\pi}{2}} \cos 4x \cos 3x \, dx$; (b) $\int_0^1 2 \sin 7t \cos 3t \, dt$
$\qquad$ [(a) $\frac{3}{7}$ or 0.4286; (b) 0.5973]

34 (a) $-4 \int_0^{\frac{\pi}{3}} \sin 5\theta \sin 2\theta \, d\theta$; (b) $\int_1^2 3 \cos 8t \sin 3t \, dt$ $\qquad$ [(a) 0.2474; (b) −0.1999]

*Sine θ substitution*

35 Determine (a) $\int \frac{5}{\sqrt{(4-t^2)}} \, dt$; (b) $\int \frac{3}{\sqrt{(9-x^2)}} \, dx$

$\qquad$ [(a) $5 \arcsin \frac{x}{2} + c$; (b) $3 \arcsin \frac{x}{3} + c$]

94

**36** Determine (a) $\int\sqrt{(4-x^2)}dx$; (b) $\int\sqrt{(16-9t^2)}dt$.

$$\left[(a)\; 2\arcsin\frac{x}{2}+\frac{x}{2}\sqrt{(4-x^2)}+c;\; (b)\; \frac{8}{3}\arcsin\frac{3t}{4}+\frac{t}{2}\sqrt{(16-9t^2)}+c\right]$$

**37** Evaluate (a) $\int_0^4 \frac{1}{\sqrt{(16-x^2)}}dx$; (b) $\int_0^1 \sqrt{(9-4x^2)}dx$

$$\left[(a)\; \frac{\pi}{2}\text{ or }1.571;\; (b)\; 2.760\right]$$

*tan θ substitution*

**38** Determine (a) $\int\frac{3}{4+t^2}dt$; (b) $\int\frac{5}{16+9\theta^2}d\theta$

$$\left[(a)\; \frac{3}{2}\arctan\frac{x}{2}+c;\; (b)\; \frac{5}{12}\arctan\frac{3\theta}{4}+c\right]$$

**39** Evaluate (a) $\int_0^1 \frac{3}{1+t^2}dt$; (b) $\int_0^3 \frac{5}{4+x^2}dx$     [(a) 2.356; (b) 2.457]

*sinh θ substitution*

**40** Find (a) $\int\frac{2}{\sqrt{(x^2+16)}}dx$; (b) $\int\frac{3}{\sqrt{(9+5x^2)}}dx$

$$\left[(a)\; 2\operatorname{arsinh}\frac{x}{4}+c;\; (b)\; \frac{3}{\sqrt{5}}\operatorname{arsinh}\frac{\sqrt{5}}{3}x+c\right]$$

**41** Find (a) $\int\sqrt{(x^2+9)}dx$; (b) $\int\sqrt{(4t^2+25)}dt$

$$\left[(a)\; \frac{9}{2}\operatorname{arsinh}\frac{x}{3}+\frac{x}{2}\sqrt{(x^2+9)}+c;\; (b)\; \frac{25}{4}\operatorname{arsinh}\frac{2t}{5}+\frac{t}{2}\sqrt{(4t^2+25)}+c\right]$$

**42** Evaluate (a) $\int_0^3 \frac{4}{\sqrt{(t^2+9)}}dt$; (b) $\int_0^1 \sqrt{(16+9\theta^2)}d\theta$     [(a) 3.525; (b) 4.348]

*cosh θ substitution*

**43** Find (a) $\int\frac{1}{\sqrt{(t^2-16)}}dt$; (b) $\int\frac{3}{\sqrt{(9-4x^2)}}dx$

$$\left[(a)\; \operatorname{arcosh}\frac{x}{4}+c;\; (b)\; \frac{3}{2}\operatorname{arcosh}\frac{2x}{3}+c\right]$$

**44** Find (a) $\int\sqrt{(\theta^2-9)}d\theta$; (b) $\int\sqrt{(4\theta^2-25)}d\theta$

$$\left[(a)\; \frac{\theta}{2}\sqrt{(\theta^2-9)}-\frac{9}{2}\operatorname{arcosh}\frac{\theta}{3}+c;\; (b)\; \theta\sqrt{(\theta^2-\frac{25}{4})}-\frac{25}{4}\operatorname{arcosh}\frac{2\theta}{5}+c\right]$$

**45** Evaluate (a) $\int_1^2 \frac{2}{\sqrt{(x^2-1)}}dx$; (b) $\int_2^3 \sqrt{(t^2-4)}dt$.     [(a) 2.634; (b) 1.429]

**46** Show that $\int\frac{3x-1}{\sqrt{(x^2-4)}}dx = 3\sqrt{(x^2-4)}-\operatorname{arcosh}\frac{x}{2}+c$.

### (c) INTEGRATION USING PARTIAL FRACTIONS

In *Problems 47 to 54*, integrate with respect to $x$

**47** $\int\frac{12}{(x^2-9)}dx$     $\left[2\ln(x-3)-2\ln(x+3)+c\text{ or }\ln\left(\frac{x-3}{x+3}\right)^2+c\right]$

48 $\int \dfrac{4(x-4)}{(x^2-2x-3)}dx$ $\qquad \left[ 5 \ln (x+1)-\ln (x-3)+c \text{ or } \ln \left\{ \dfrac{(x+1)^5}{(x-3)} \right\}+c \right]$

49 $\int \dfrac{3(2x^2-8x-1)}{(x+4)(x+1)(2x+1)}\,dx$ $\qquad \left[ \begin{array}{l} 7 \ln (x+4)-3 \ln (x+1)-\ln (2x-1)+c \\[2mm] \text{or } \ln \left\{ \dfrac{(x+4)^7}{(x+1)^3(2x-1)} \right\}+c \end{array} \right]$

50 $\int \dfrac{x^2+9x+8}{x^2+x-6}dx$ $\qquad [x+2 \ln (x+3)+6 \ln (x-2)+c \text{ or } x+\ln\{(x+3)^2(x-2)^6\}+c]$

51 $\int \dfrac{3x^3-2x^2-16x+20}{(x-2)(x+2)}\,dx$ $\qquad \left[ \dfrac{3x^2}{2} -2x+\ln (x-2)-5 \ln (x+2)+c \right]$

52 $\int \dfrac{4x-3}{(x+1)^2}dx$ $\qquad \left[ 4 \ln (x+1)+ \dfrac{7}{(x+1)} +c \right]$

53 $\int \dfrac{5x^2-30x+44}{(x-2)^3}\,dx$ $\qquad \left[ 5 \ln (x-2) + \dfrac{10}{(x-2)} - \dfrac{2}{(x-2)^2} +c \right]$

54 $\int \dfrac{x^2-x-13}{(x^2+7)(x-2)}\,dx$ $\qquad \left[ \ln (x^2+7)+ \dfrac{3}{\sqrt{7}} \arctan \dfrac{x}{\sqrt{7}} - \ln (x-2)+c \right]$

In *Problems 55 to 61*, evaluate the definite integrals correct to 4 significant figures.

55 $\displaystyle\int_3^4 \dfrac{x^2-3x+6}{x(x-2)(x-1)} dx$ $\qquad\qquad$ [0.6275]

56 $\displaystyle\int_4^6 \dfrac{x^2-x-14}{x^2-2x-3} dx$ $\qquad\qquad$ [0.8122]

57 $\displaystyle\int_1^2 \dfrac{x^2+7x+3}{x^2(x+3)} dx$ $\qquad\qquad$ [1.663]

58 $\displaystyle\int_6^7 \dfrac{18+21x-x^2}{(x-5)(x+2)^2} dx$ $\qquad\qquad$ [1.089]

59 $\displaystyle\int_5^6 \dfrac{6x-5}{(x-4)(x^2+3)} dx$ $\qquad\qquad$ [0.5799]

60 $\displaystyle\int_1^2 \dfrac{4}{(16-x^2)}dx$ $\qquad\qquad$ [0.2939]

61 $\displaystyle\int_4^5 \dfrac{2}{(x^2-9)} dx$ $\qquad\qquad$ [0.1865]

# 11 Integration by parts

## A. MAIN POINTS CONCERNED WITH INTEGRATION BY PARTS

1  From the product rule of differentiation: $\dfrac{d}{dx}(uv) = v\dfrac{du}{dx} + u\dfrac{dv}{dx}$,

where $u$ and $v$ are both functions of $x$.

Rearranging gives: $u\dfrac{dv}{dx} = \dfrac{d}{dx}(uv) - v\dfrac{du}{dx}$

Integrating both sides with respect to $x$ gives: $\displaystyle\int u\dfrac{dv}{dx}dx = \int \dfrac{d}{dx}(uv)dx - \int v\dfrac{du}{dx}dx$

i.e. $\displaystyle\int u\dfrac{dv}{dx}\,dx = uv - \int v\dfrac{du}{dx}dx$  or  $\displaystyle\int u\,dv = uv - \int v\,du$

This is known as the **integration by parts formula** and provides a method of integrating such products of simple functions as $\int xe^x\,dx$, $\int t\sin t\,dt$, $\int e^\theta \cos\theta\,d\theta$ and $\int x\ln x\,dx$. The method is used for determining certain integrals involved with Fourier series (see chapter 17).

2  Given a product of two terms to integrate the initial choice is: 'which part to make equal to $u$' and 'which part to make equal to $v$'. The choice must be such that the '$u$ part' becomes a constant after successive differentiation and the '$dv$ part' can be integrated from standard integrals. Invariably, the following rule holds: 'If a product to be integrated contains an algebraic term (such as $x$, $t^2$ or $3\theta$) then this term is chosen as the $u$ part'. The one exception to this rule is when a '$\ln x$' term is involved; in this case $\ln x$ is chosen as the '$u$ part'.

## B. WORKED PROBLEMS ON INTEGRATION BY PARTS

*Problem 1* Determine $\int x\cos x\,dx$.

From the integration by parts formula, $\displaystyle\int u\,dv = uv - \int v\,du$

Let $u = x$, from which $\dfrac{du}{dx} = 1$, i.e. $du = dx$

and let $dv = \cos x\,dx$, from which $v = \displaystyle\int \cos x\,dx = \sin x$.

Expressions for $u$, $du$, $v$ and $dv$ are now substituted into the parts formula as shown below.

$$\int \boxed{u} \quad \boxed{dv} \quad = \quad \boxed{u} \quad \boxed{v} \quad - \int \boxed{v} \quad \boxed{du}$$

$$\int \boxed{x} \quad \boxed{\cos x \, dx} \quad = \quad \boxed{(x)} \quad \boxed{(\sin x)} \quad - \int \boxed{(\sin x)} \quad \boxed{(dx)}$$

i.e. $\int x \cos x \, dx = x \sin x - (-\cos x) + c$
$$= x \sin x + \cos x + c.$$

This result may be checked by differentiating the right hand side,

i.e., $\dfrac{d}{dx} (x \sin x + \cos x + c) = [(x)(\cos x) + (\sin x)(1)] - \sin x + 0$
$$= x \cos x, \text{ which is the function being integrated.}$$

*Problem 2* Find $\int 3t \, e^{2t} \, dt$.

Let $u = 3t$, from which, $\dfrac{du}{dt} = 3$, i.e., $du = 3dt$,

and let $dv = e^{2t} \, dt$, from which, $v = \int e^{2t} \, dt = \dfrac{1}{2} e^{2t}$

Substituting into $\int u \, dv = uv - \int v \, du$ gives:

$$\int 3t e^{2t} dt = (3t) \left( \frac{1}{2} e^{2t} \right) - \int \left( \frac{1}{2} e^{2t} \right) (3dt)$$

$$= \frac{3}{2} te^{2t} - \frac{3}{2} \int e^{2t} \, dt$$

$$= \frac{3}{2} te^{2t} - \frac{3}{2} \left( \frac{e^{2t}}{2} \right) + c$$

Hence $\int 3te^{2t} dt = \dfrac{3}{2} e^{2t} \left( t - \dfrac{1}{2} \right) + c$, which may be checked by differentiation.

*Problem 3* Evaluate $\displaystyle\int_0^{\frac{\pi}{2}} 2\theta \sin \theta \, d\theta$.

Let $u = 2\theta$, from which, $\dfrac{du}{d\theta} = 2$, i.e., $du = 2d\theta$,

and let $dv = \sin \theta \, d\theta$, from which, $v = \int \sin \theta \, d\theta = -\cos \theta$.

Substituting into $\int u \, dv = uv - \int v \, du$ gives:

$$\int 2\theta \sin \theta \, d\theta = (2\theta)(-\cos \theta) - \int (-\cos \theta)(2 \, d\theta)$$

$$= -2\theta \cos \theta + 2 \int \cos \theta \, d\theta$$

$$= -2\theta \cos \theta + 2 \sin \theta + c$$

Hence $\displaystyle\int_0^{\frac{\pi}{2}} 2\theta \sin \theta \, d\theta = [-2\theta \cos \theta + 2 \sin \theta]_0^{\frac{\pi}{2}} = \left[ -2 \left( \frac{\pi}{2} \right) \cos \frac{\pi}{2} + 2 \sin \frac{\pi}{2} \right] - \left[ 0 + 2 \sin 0 \right]$

$$= (-0 + 2) - (0 + 0), \text{ (since } \cos \frac{\pi}{2} = 0 \text{ and } \sin \frac{\pi}{2} = 1)$$

$$= 2$$

*Problem 4* Evaluate $\displaystyle\int_0^1 5xe^{4x} dx$, correct to 3 significant figures.

Let $u = 5x$, from which $\dfrac{du}{dx} = 5$, i.e., $du = 5dx$,

and let $dv = e^{4x}\ dx$, from which, $v = \int e^{4x}\ dx = \dfrac{1}{4}e^{4x}$

Substituting into $\int u\ dv = uv - \int v\ du$ gives:

$$\int 5xe^{4x}\ dx = (5x)\left(\frac{e^{4x}}{4}\right) - \int\left(\frac{e^{4x}}{4}\right)(5dx)$$

$$= \frac{5}{4}xe^{4x} - \frac{5}{4}\int e^{4x}\ dx$$

$$= \frac{5}{4}xe^{4x} - \frac{5}{4}\left(\frac{e^{4x}}{4}\right) + c = \frac{5}{4}e^{4x}\left(x - \frac{1}{4}\right) + c$$

Hence $\displaystyle\int_0^1 5xe^{4x}\ dx = \left[\frac{5}{4}e^{4x}\left(x - \frac{1}{4}\right)\right]_0^1 = \left[\frac{5}{4}e^4\left(1 - \frac{1}{4}\right)\right] - \left[\frac{5}{4}e^0\left(0 - \frac{1}{4}\right)\right]$

$$= \left(\frac{15}{16}e^4\right) - \left(-\frac{5}{16}\right) = 51.186 + 0.313 = 51.499 = \mathbf{51.5},$$

correct to 3 significant figures.

---

*Problem 5* Determine $\int x^2\ \sin x\ dx$.

Let $u = x^2$, from which, $\dfrac{du}{dx} = 2x$, i.e., $du = 2x\ dx$,

and let $dv = \sin x\ dx$, from which, $v = \int \sin x\ dx = -\cos x$.

Substituting into $\int u\ dv = uv - \int v\ du$ gives:

$$\int x^2\ \sin x\ dx = (x^2)(-\cos x) - \int(-\cos x)(2x\ dx)$$

$$= -x^2\ \cos x + 2\left[\int x\ \cos x\ dx\right]$$

The integral, $\int x\ \cos x\ dx$, is not a 'standard integral' and it can only be determined by using the integration by parts formula again.

From *Problem 1*, $\int x\ \cos x\ dx = x\ \sin x + \cos x$

Hence $\int x^2\ \sin x\ dx = -x^2\ \cos x + 2[x\ \sin x + \cos x] + c$

$$= -x^2\ \cos x + 2x\ \sin x + 2\ \cos x + c$$

$$= (2 - x^2)\ \cos x + 2x\ \sin x + c.$$

In general, if the algebraic term of a product is of power $n$, then the integration
by parts formula is applied $n$ times.

---

*Problem 6* Find $\int x\ \ln x\ dx$.

The logarithmic function is chosen as the '$u$ part'.

Thus when $u = \ln x$, then $\dfrac{du}{dx} = \dfrac{1}{x}$, i.e., $du = \dfrac{dx}{x}$.

Letting $dv = x\ dx$ gives $v = \int x\ dx = \dfrac{x^2}{2}$

Substituting into $\int u\ dv = uv - \int v\ du$ gives:

$$\int x\ \ln x\ dx = (\ln x)\left(\frac{x^2}{2}\right) - \int\left(\frac{x^2}{2}\right)\frac{dx}{x}$$

$$= \frac{x^2}{2}\ \ln x - \frac{1}{2}\int x\ dx = \frac{x^2}{2}\ \ln x - \frac{1}{2}\left(\frac{x^2}{2}\right) + c$$

Hence $\int x\ \ln x\ dx = \dfrac{x^2}{2}\left(\ln x - \dfrac{1}{2}\right) + c$ or $\dfrac{x^2}{4}(2\ \ln x - 1) + c$

*Problem 7* Determine $\int \ln x \, dx$.

$\int \ln x \, dx$ is the same as $\int (1) \ln x \, dx$

Let $u = \ln x$, from which, $\dfrac{du}{dx} = \dfrac{1}{x}$, i.e., $du = \dfrac{dx}{x}$

and let $dv = 1 dx$, from which, $v = \int 1 dx = x$.

Substituting into $\int u \, dv = uv - \int v \, du$ gives:

$$\int \ln x \, dx = (\ln x)(x) - \int x \, \frac{dx}{x} = x \ln x - \int dx = x \ln x - x + c$$

Hence $\int \ln x \, dx = x (\ln x - 1) + c$

*Problem 8* Evaluate $\int_1^9 \sqrt{x} \ln x \, dx$, correct to 3 significant figures.

Let $u = \ln x$, from which, $du = \dfrac{dx}{x}$

and let $dv = \sqrt{x} \, dx = x^{\frac{1}{2}} dx$, from which, $v = \int x^{\frac{1}{2}} dx = \dfrac{2}{3} x^{\frac{3}{2}}$

Substituting into $\int u \, dv = uv - \int v \, du$ gives:

$$\int \sqrt{x} \ln x \, dx = (\ln x) \left[ \frac{2}{3} x^{\frac{3}{2}} \right] - \int \left( \frac{2}{3} x^{\frac{3}{2}} \right) \frac{dx}{x}$$

$$= \frac{2}{3} \sqrt{x^3} \ln x - \frac{2}{3} \int x^{\frac{1}{2}} \, dx$$

$$= \frac{2}{3} \sqrt{x^3} \ln x - \frac{2}{3} \left( \frac{2}{3} x^{\frac{3}{2}} \right) + c = \frac{2}{3} \sqrt{x^3} \left[ \ln x - \frac{2}{3} \right] + c$$

Hence $\int_1^9 \sqrt{x} \ln x \, dx = \left[ \frac{2}{3} \sqrt{x^3} \left( \ln x - \frac{2}{3} \right) \right]_1^9 = \left[ \frac{2}{3} \sqrt{9^3} \left( \ln 9 - \frac{2}{3} \right) \right] - \left[ \frac{2}{3} \sqrt{1^3} \left( \ln 1 - \frac{2}{3} \right) \right]$

$$= \left[ 18 \left( \ln 9 - \frac{2}{3} \right) \right] - \left[ \frac{2}{3} \left( 0 - \frac{2}{3} \right) \right]$$

$= 27.550 + 0.444 = 27.994 = \mathbf{28.0}$, correct to
3 significant figures.

*Problem 9* Find $\int e^{ax} \cos bx \, dx$

When integrating a product of an exponential and a sine or cosine function it is immaterial which part is made equal to '$u$'.

Let $u = e^{ax}$, from which, $\dfrac{du}{dx} = ae^{ax}$, i.e., $du = ae^{ax} \, dx$,

and let $dv = \cos bx \, dx$, from which, $v = \int \cos bx \, dx = \dfrac{1}{b} \sin bx$.

Substituting into $\int u \, dv = uv - \int v \, du$ gives:

$$\int e^{ax} \cos bx \, dx = (e^{ax}) \left( \frac{1}{b} \sin bx \right) - \int \left( \frac{1}{b} \sin bx \right) (ae^{ax} dx)$$

$$= \frac{1}{b} e^{ax} \sin bx - \frac{a}{b} \left[ \int e^{ax} \sin bx \, dx \right] \qquad (1)$$

$\int e^{ax} \sin bx \, dx$ is now determined separately using integration by parts again:
Let $u = e^{ax}$ then $du = ae^{ax} \, dx$,
and let $dv = \sin bx \, dx$, from which, $v = \int \sin bx \, dx = -\dfrac{1}{b} \cos bx$
Substituting into the parts formula gives:

$$\int e^{ax} \sin bx \, dx = (e^{ax}) \left( -\frac{1}{b} \cos bx \right) - \int \left( -\frac{1}{b} \cos bx \right) (ae^{ax} \, dx)$$

$$= -\frac{1}{b} e^{ax} \cos bx + \frac{a}{b} \int e^{ax} \cos bx \, dx$$

Substituting this result into equation (1) gives:

$$\int e^{ax} \cos bx \, dx = \frac{1}{b} e^{ax} \sin bx - \frac{a}{b} \left[ -\frac{1}{b} e^{ax} \cos bx + \frac{a}{b} \int e^{ax} \cos bx \, dx \right]$$

$$= \frac{1}{b} e^{ax} \sin bx + \frac{a}{b^2} e^{ax} \cos bx - \frac{a^2}{b^2} \int e^{ax} \cos bx \, dx$$

The integral on the far right of this equation is the same as the integral on the left hand side and thus they may be combined.

$$\int e^{ax} \cos bx \, dx + \frac{a^2}{b^2} \int e^{ax} \cos bx \, dx = \frac{1}{b} e^{ax} \sin bx + \frac{a}{b^2} e^{ax} \cos bx$$

i.e. $\left( 1 + \dfrac{a^2}{b^2} \right) \int e^{ax} \cos bx \, dx = \dfrac{1}{b} e^{ax} \sin bx + \dfrac{a}{b^2} e^{ax} \cos bx$

i.e. $\left( \dfrac{b^2 + a^2}{b^2} \right) \int e^{ax} \cos bx \, dx = \dfrac{e^{ax}}{b^2} (b \sin bx + a \cos bx)$

Hence $\int e^{ax} \cos bx \, dx = \left( \dfrac{b^2}{a^2 + b^2} \right) \left( \dfrac{e^{ax}}{b^2} \right) (b \sin bx + a \cos bx)$

$$= \frac{e^{ax}}{a^2 + b^2} (b \sin bx + a \cos bx) + c$$

Using a similar method to above, that is, integrating by parts twice, the following result may be proved:

$$e^{ax} \sin bx \, dx = \frac{e^{ax}}{a^2 + b^2} (a \sin bx - b \cos bx) + c \qquad (2)$$

*Problem 10* Evaluate $\displaystyle\int_0^{\frac{\pi}{4}} e^t \sin 2t \, dt$, correct to 4 decimal places.

Comparing $\int e^t \sin 2t \, dt$ with $\int e^{ax} \sin bx \, dx$ shows that $x = t$, $a = 1$ and $b = 2$. Hence, substituting into equation (2) gives:

$$\int_0^{\frac{\pi}{4}} e^t \sin 2t \, dt = \left[ \frac{e^t}{1^2 + 2^2} (1 \sin 2t - 2 \cos 2t) \right]_0^{\frac{\pi}{4}}$$

$$= \left[ \frac{e^{\frac{\pi}{4}}}{5} \left( \sin 2(\tfrac{\pi}{4}) - 2 \cos 2(\tfrac{\pi}{4}) \right) \right] - \left[ \frac{e^0}{5} (\sin 0 - 2 \cos 0) \right]$$

$$= \left[ \frac{e^{\frac{\pi}{4}}}{5} (1 - 0) \right] - \left[ \frac{1}{5} (0 - 2) \right] = \frac{e^{\frac{\pi}{4}}}{5} + \frac{2}{5}$$

$$= 0.8387, \text{ correct to 4 decimal places.}$$

## C. FURTHER PROBLEMS ON INTEGRATION BY PARTS

Determine the integrals in *Problems 1 to 10* using integration parts.

1   $\int x e^{2x}\ dx$                             $\left[\dfrac{e^{2x}}{2}\left(x-\dfrac{1}{2}\right)+c\right]$

2   $\int \dfrac{4x}{e^{3x}}\ dx$                      $\left[-\dfrac{4}{3}e^{-3x}\left(x+\dfrac{1}{3}\right)+c\right]$

3   $\int x \sin x\ dx$                   $[-x \cos x+\sin x+c]$

4   $\int 2x^2 \ln x\ dx$                 $\left[\dfrac{2}{3}x^3\left(\ln x-\dfrac{1}{3}\right)+c\right]$

5   $\int 2 \ln 3x\ dx$                  $[2x\ (\ln 3x-1)+c]$

6   $\int 5\theta \cos 2\theta\ d\theta$          $\left[\dfrac{5}{2}\left(\theta \sin 2\theta + \dfrac{1}{2}\cos 2\theta\right)+c\right]$

7   $\int 3t^2\ e^{2t}\ dt$             $\left[\dfrac{3}{2}e^{2t}\left(t^2-t+\dfrac{1}{2}\right)+c\right]$

8   $\int x^2 \sin 3x\ dx$      $\left[\dfrac{\cos 3x}{27}(2-9x^2)+\dfrac{2}{9}x \sin 3x+c\right]$

9   $\int 2e^{5x} \cos 2x\ dx$     $\left[\dfrac{2}{29}e^{5x}\ (2 \sin 2x+5 \cos 2x)+c\right]$

10   $\int 2\theta \sec^2 \theta\ d\theta$          $[2[\theta \tan \theta-\ln (\sec \theta)]+c]$

Evaluate the integrals in *Problems 11 to 18*, correct to 4 significant figures.

11   $\displaystyle\int_0^2 2xe^x\ dx$                             $[16.78]$

12   $\displaystyle\int_0^{\frac{\pi}{4}} x \sin 2x\ dx$                       $[0.2500]$

13   $\displaystyle\int_1^2 x \ln x\ dx$                          $[0.6363]$

14   $\displaystyle\int_0^{\frac{\pi}{2}} t^2 \cos t\ dt$                       $[0.4674]$

15   $\displaystyle\int_1^2 3x^2 e^{\frac{x}{2}}\ dx$                      $[15.78]$

16   $\displaystyle\int_0^1 2e^{3x} \sin 2x\ dx$                   $[11.31]$

17   $\displaystyle\int_0^{\frac{\pi}{2}} e^t \cos 3t\ dt$                    $[-1.543]$

18   $\displaystyle\int_1^4 \sqrt{x^3}\ \ln x\ dx$                    $[12.78]$

19 In determining a Fourier series to represent $f(x) = x$ in the range $-\pi$ to $\pi$, Fourier coefficients are given by:

$$a_n = \frac{1}{\pi} \int_{-\pi}^{\pi} x \cos nx \, dx \quad \text{and} \quad b_n = \frac{1}{\pi} \int_{-\pi}^{\pi} x \sin nx \, dx$$

where $n$ is a positive integer. Show by using integration by parts that $a_n = 0$ and $b_n = -\frac{2}{n} \cos n\pi$.

# 12 First order differential equations by separation of variables

## A. MAIN POINTS CONCERNED WITH FIRST ORDER DIFFERENTIAL EQUATIONS BY SEPARATION OF THE VARIABLES

1  A **differential equation** is one that contains differential coefficients.

Examples include: (i) $\dfrac{dy}{dx} = 4x$; (ii) $\dfrac{d^2 y}{dt^2} + 2\dfrac{dy}{dt} - 5y = 0$.

2  Differential equations are **classified** according to the highest derivative which occurs in them. Thus example (i) above is a **first order differential equation** and example (ii) is a **second order differential equation**.

3  (i) Given a differential equation it is possible, by integration and by being given enough data to determine the constants, to obtain the original function. This process is called 'solving the differential equation'.

  (ii) A solution to a differential equation which contains one or more arbitrary constants of integration is called the **general solution** of the differential equation.

  (iii) When additional information is given (called 'boundary conditions'), so that constants may be determined, the **particular solution** of the differential equation is obtained.

4  A differential equation of the form $\dfrac{dy}{dx} = f(x)$ is solved by direct integration, i.e., $y = \int f(x)\, dx$.

(See *Problems 1 to 5*)

5  A differential equation of the form $\dfrac{dy}{dx} = f(y)$ is initially rearranged to give $dx = \dfrac{dy}{f(y)}$, and then the solution is obtained by direct integration, i.e., $\int dx = \int \dfrac{dy}{f(y)}$ .

(See *Problems 6 to 9*)

6  A differential equation of the form $\dfrac{dy}{dx} = f(x) \cdot g(y)$, where $f(x)$ is a function of $x$ only and $g(y)$ is a function of $y$ only, may be rearranged as $\dfrac{dy}{g(y)} = f(x)\, dx$, and then the solution is obtained by direct integration, i.e., $\int \dfrac{dy}{g(y)} = \int f(x)\, dx$.

When two variables are rearranged into two separate groups as shown above, each containing only one variable, the variables are said to be separable. (See *Problems 10 to 15*)

Differential equations of the forms $\frac{dy}{dx} = f(x)$ and $\frac{dy}{dx} = f(y)$ are merely special cases of 'separating the variables'.

## B WORKED PROBLEMS ON FIRST ORDER DIFFERENTIAL EQUATIONS BY SEPARATING THE VARIABLES

*Problem 1* Find the general solution of the differential equation $\frac{dy}{dx} = 3x^2 - \sin 2x$.

Integrating both sides of $\frac{dy}{dx} = 3x^2 - \sin 2x$ gives: $\int \frac{dy}{dx} = \int (3x^2 - \sin 2x)\, dx$

i.e. $y = x^3 + \frac{1}{2}\cos 2x + c$, which is the general solution.

*Problem 2* Determine the general solution of $x\frac{dy}{dx} = 2 - 4x^3$.

Rearranging $x\frac{dy}{dx} = 2 - 4x^3$ gives: $\frac{dy}{dx} = \frac{2-4x^3}{x} = \frac{2}{x} - \frac{4x^3}{x} = \frac{2}{x} - 4x^2$.

Integrating both sides gives: $y = \int (\frac{2}{x} - 4x^2)\, dx$

i.e. $y = 2\ln x - \frac{4}{3}x^3 + c$, which is the general solution.

*Problem 3* Find the particular solution of the differential equation $5\frac{dy}{dx} + 2x = 3$, given the boundary conditions $y = 1\frac{2}{5}$ when $x = 2$.

Since $5\frac{dy}{dx} + 2x = 3$ then $\frac{dy}{dx} = \frac{3-2x}{5} = \frac{3}{5} - \frac{2x}{5}$

Hence $y = \int (\frac{3}{5} - \frac{2x}{5})\, dx$

i.e. $y = \frac{3x}{5} - \frac{x^2}{5} + c$, which is the general solution.

Substituting the boundary conditions $y = 1\frac{2}{5}$ and $x = 2$ to evaluate $c$ gives:

$1\frac{2}{5} = \frac{6}{5} - \frac{4}{5} + c$, from which, $c = 1$.

**Hence the particular solution is $y = \frac{3x}{5} - \frac{x^2}{5} + 1$.**

*Problem 4* Solve the equation $2t\left(t - \frac{d\theta}{dt}\right) = 5$, given $\theta = 2$ when $t = 1$.

Rearranging gives: $t - \frac{d\theta}{dt} = \frac{5}{2t}$ and $\frac{d\theta}{dt} = t - \frac{5}{2t}$

Integrating gives: $\theta = \int \left(t - \frac{5}{2t}\right)\, dt$

i.e. $\qquad \theta = \dfrac{t^2}{2} - \dfrac{5}{2}\ln t + c$, which is the general solution.

When $\theta = 2$, $t = 1$, thus $2 = \dfrac{1}{2} - \dfrac{5}{2}\ln 1 + c$ from which, $c = \dfrac{3}{2}$

Hence the particular solution is: $\qquad \theta = \dfrac{t^2}{2} - \dfrac{5}{2}\ln t + \dfrac{3}{2}$

i.e. $\qquad\qquad\qquad\qquad \theta = \dfrac{1}{2}(t^2 - \ln t + 3)$

---

*Problem 5* The angular velocity $\omega$ of a flywheel of moment of inertia $I$ is given by $I(d\omega/dt) + N = 0$, where $N$ is a constant. Determine $\omega$ in terms of $t$ given that $\omega = \omega_0$ when $t = 0$.

---

Rearranging $I\dfrac{d\omega}{dt} + N = 0$ gives $\dfrac{d\omega}{dt} = \dfrac{-N}{I}$

Integrating gives: $\omega = \displaystyle\int \dfrac{-N}{I}\, dt$, i.e. $\omega = \dfrac{-Nt}{I} + c$.

When $t = 0$, $\omega = \omega_0$ thus $\omega_0 = \dfrac{-N(0)}{I} + c$, from which, $c = \omega_0$.

Hence $\omega = \dfrac{-Nt}{I} + \omega_0$ or $\omega = \omega_0 - \dfrac{Nt}{I}$

---

*Problem 6* Find the general solution of $dy/dx = 3 + 2y$.

---

Rearranging $\dfrac{dy}{dx} = 3 + 2y$ gives: $dx = \dfrac{dy}{3 + 2y}$

Integrating both sides gives: $\displaystyle\int dx = \int \dfrac{dy}{(3 + 2y)}$

Thus, by using the substitution $u = (3 + 2y)$, $x = \dfrac{1}{2}\ln(3 + 2y) + c$ $\qquad$ (1)

It is possible to give the general solution of a differential equation in a different form. For example, if $c = \ln k$, where $k$ is a constant, then:

$x = \dfrac{1}{2}\ln(3 + 2y) + \ln k$, i.e. $x = \ln(3 + 2y)^{\frac{1}{2}} + \ln k$

$x = \ln[k\sqrt{(3 + 2y)}]$ $\qquad\qquad\qquad\qquad\qquad\qquad\qquad$ (2)

by the laws of logarithms, from which, $e^x = k\sqrt{(3 + 2y)}$ $\qquad\qquad$ (3)

Equations (1), (2) and (3) are all acceptable general solutions of the differential equation $\dfrac{dy}{dx} = 3 + 2y$.

---

*Problem 7* Solve the differential equation $5\,(dy/dx) = \sin^2 3y$.

---

Rearranging $5\dfrac{dy}{dx} = \sin^2 3y$ gives: $dx = \dfrac{5}{\sin^2 3y}\, dy = 5\,\text{cosec}^2\, 3y\, dy$.

Integrating both sides gives: $\displaystyle\int dx = \int 5\,\text{cosec}^2\, 3y\, dy$

i.e., $\qquad x = 5\left(-\dfrac{1}{3}\cot 3y\right) + c$, from standard integrals.

Hence the general solution is $x = c - \dfrac{5}{3}\cot 3y$.

*Problem 8* Determine the particular solution of $(y^2-1)\dfrac{dy}{dx} = 3y$ given that $y = 1$ when $x = 2\dfrac{1}{6}$.

Rearranging gives: $dx = \left(\dfrac{y^2-1}{3y}\right)dy = \left(\dfrac{y}{3} - \dfrac{1}{3y}\right)dy$

Integrating gives: $\displaystyle\int dx = \int\left(\dfrac{y}{3} - \dfrac{1}{3y}\right)dy$

i.e. $x = \dfrac{y^2}{6} - \dfrac{1}{3}\ln y + c$, which is the general solution.

When $y = 1$, $x = 2\dfrac{1}{6}$, thus $2\dfrac{1}{6} = \dfrac{1}{6} - \dfrac{1}{3}\ln 1 + c$, from which, $c = 2$.

Hence the particular solution is: $x = \dfrac{y^2}{6} - \dfrac{1}{3}\ln y + 2$.

*Problem 9* (a) The variation of resistance, $R$ ohms, of an aluminium conductor with temperature $\theta°C$ is given by $dR/d\theta = \alpha R$, where $\alpha$ is the temperature coefficient of resistance of aluminium. If $R = R_0$ when $\theta = 0°C$, solve the equation for $R$. (b) If $\alpha = 38 \times 10^{-4}/°C$, determine the resistance of an aluminium conductor at $50°C$, correct to 3 significant figures, when its resistance at $0°C$ is $24.0\ \Omega$.

(a) $\dfrac{dR}{d\theta} = \alpha R$ is of the form $\dfrac{dy}{dx} = f(y)$.

Rearranging gives: $d\theta = \dfrac{dR}{\alpha R}$

Integrating both sides gives: $\displaystyle\int d\theta = \int \dfrac{dR}{\alpha R}$

i.e. $\theta = \dfrac{1}{\alpha}\ln R + c$, which is the general solution.

Substituting the boundary conditions $R = R_0$ when $\theta = 0$ gives: $0 = \dfrac{1}{\alpha}\ln R_0 + c$

from which $c = -\dfrac{1}{\alpha}\ln R_0$

Hence the particular solution is $\theta = \dfrac{1}{\alpha}\ln R - \dfrac{1}{\alpha}\ln R_0 = \dfrac{1}{\alpha}(\ln R - \ln R_0)$

i.e. $\theta = \dfrac{1}{\alpha}\ln\left(\dfrac{R}{R_0}\right)$ or $\alpha\theta = \ln\left(\dfrac{R}{R_0}\right)$

Hence $e^{\alpha\theta} = \dfrac{R}{R_0}$ from which, $R = R_0 e^{\alpha\theta}$

(b) Substituting $\alpha = 38 \times 10^{-4}$, $R_0 = 24.0$ and $\theta = 50$ into $R = R_0 e^{\alpha\theta}$ gives the resistance at $50°C$, i.e., $R_{50} = 24.0e^{(38\times 10^{-4}\times 50)} = $ **29.0 ohms**.

*Problem 10* Solve: $\dfrac{dy}{dx} = \dfrac{(2x^3-1)}{(3-2y)}$.

Separating the variables gives: $(3-2y)dy = (2x^3-1)dx$
Integrating both sides gives: $\displaystyle\int(3-2y)dy = \int(2x^3-1)dx$

Hence the general solution is: $3y-y^2 = \dfrac{x^4}{2} - x + c$.

*Problem 11* Solve the equation $4xy\dfrac{dy}{dx} = y^2 - 1$.

Separating the variables gives: $\left(\dfrac{4y}{y^2 - 1}\right) dy = \dfrac{1}{x} dx$

Integrating both sides gives: $\displaystyle\int\left(\dfrac{4y}{y^2 - 1}\right) dy = \int\left(\dfrac{1}{x}\right)dx$

Hence the general solution is: $2\ln(y^2 - 1) = \ln x + c$      (1)

or    $\ln(y^2 - 1)^2 - \ln x = c$

from which,    $\ln\left\{\dfrac{(y^2 - 1)^2}{x}\right\} = c$

and    $\dfrac{(y^2 - 1)^2}{x} = e^c$      (2)

If in equation (1), $c = \ln A$, where $A$ is a different constant,
then   $\ln(y^2 - 1)^2 = \ln x + \ln A$
i.e.    $\ln(y^2 - 1)^2 = \ln Ax$
i.e.    $(y^2 - 1)^2 = Ax$      (3)
Equations (1) to (3) are thus three valid solutions of the differential equation

$4xy\dfrac{dy}{dx} = y^2 - 1$.

*Problem 12* Determine the particular solution of $d\theta/dt = 2e^{3t - 2\theta}$, given that $t = 0$ when $\theta = 0$.

$d\theta/dt = 2e^{3t - 2\theta} = 2(e^{3t})(e^{-2\theta})$, by the laws of indices.

Separating the variables gives: $\dfrac{d\theta}{e^{-2\theta}} = 2e^{3t} dt$   i.e.   $e^{2\theta} d\theta = 2e^{3t} dt$

Integrating both sides gives: $\displaystyle\int e^{2\theta} d\theta = \int 2e^{3t} dt$

Thus the general solution is:    $\dfrac{1}{2}e^{2\theta} = \dfrac{2}{3}e^{3t} + c$

When $t = 0$, $\theta = 0$, thus:    $\dfrac{1}{2}e^0 = \dfrac{2}{3}e^0 + c$

from which,    $c = \dfrac{1}{2} - \dfrac{2}{3} = -\dfrac{1}{6}$

Hence the particular solution is:    $\dfrac{1}{2}e^{2\theta} = \dfrac{2}{3}e^{3t} - \dfrac{1}{6}$   or   $3e^{2\theta} = 4e^{3t} - 1$

*Problem 13* Find the curve which satisfies the equation $xy = (1 + x^2)\dfrac{dy}{dx}$ and passes through the point $(0, 1)$.

Separating the variables gives: $\dfrac{x}{(1 + x^2)} dx = \dfrac{dy}{y}$

Integrating both sides gives: $\dfrac{1}{2}\ln(1 + x^2) = \ln y + c$

When $x = 0$, $y = 1$, thus $\dfrac{1}{2}\ln 1 = \ln 1 + c$, from which, $c = 0$.

Hence the particular solution is $\dfrac{1}{2}\ln(1 + x^2) = \ln y$

i.e. $\ln (1+x^2)^{\frac{1}{2}} = \ln y$, from which, $(1+x^2)^{\frac{1}{2}} = y$.

**Hence the equation of the curve is $y = \sqrt{(1+x^2)}$.**

*Problem 14* The current $i$ in an electric circuit containing resistance $R$ and inductance $L$ in series with a constant voltage source $E$ is given by the differential equation $E - L(di/dt) = Ri$. Solve the equation and find $i$ in terms of time $t$ given that when $t = 0, i = 0$.

Rearranging $E - L\dfrac{di}{dt} = Ri$ gives $\dfrac{di}{dt} = \dfrac{E-Ri}{L}$

and separating the variables gives: $\dfrac{di}{E-Ri} = \dfrac{dt}{L}$

Integrating both sides gives: $\displaystyle\int \dfrac{di}{E-Ri} = \int \dfrac{dt}{L}$

Hence the general solution is: $-\dfrac{1}{R} \ln (E-Ri) = \dfrac{t}{L} + c$

When $t = 0, i = 0$, thus $-\dfrac{1}{R} \ln E = c$.

Thus the particular solution is: $-\dfrac{1}{R} \ln (E-Ri) = \dfrac{t}{L} - \dfrac{1}{R} \ln E$

Transposing gives: $-\dfrac{1}{R} \ln(E-Ri) + \dfrac{1}{R} \ln E = \dfrac{t}{L}, \dfrac{1}{R} [\ln E - \ln(E-Ri)] = \dfrac{t}{L},$

$\ln\left(\dfrac{E}{E-Ri}\right) = \dfrac{Rt}{L}$ ; $\dfrac{E}{E-Ri} = e^{Rt/L}, \dfrac{E-Ri}{E} = e^{-Rt/L}, Ri = E - Ee^{-Rt/L}$

Hence current, $i = \dfrac{E}{R} (1 - e^{-Rt/L})$, which represents the law of growth of current in an inductive circuit.

*Problem 15* For an adiabatic expansion of a gas $C_v \dfrac{dp}{p} + C_p \dfrac{dV}{V} = 0$, where $C_p$ and $C_v$ are constants. Given $n = \dfrac{C_p}{C_v}$, show that $pV^n = $ constant.

Separating the variables gives: $C_v \dfrac{dp}{p} = -C_p \dfrac{dV}{V}$

Integrating both sides gives: $C_v \displaystyle\int \dfrac{dp}{p} = -C_p \int \dfrac{dV}{V}$

i.e. $C_v \ln p = -C_p \ln V + k$

Dividing throughout by constant $C_v$ gives: $\ln p = -\dfrac{C_p}{C_v} \ln V + \dfrac{k}{C_v}$.

Since $\dfrac{C_p}{C_v} = n$, then $\ln p + n \ln V = K$, where $K = \dfrac{k}{C_v}$

i.e. $\ln p + \ln V^n = K$ or $\ln pV^n = K$, by the laws of logarithms.

Hence $pV^n = e^K$, i.e. $\mathbf{pV^n = }$ **constant**.

## C. FURTHER PROBLEMS ON FIRST ORDER DIFFERENTIAL EQUATIONS BY SEPARATION OF THE VARIABLES

DIFFERENTIAL EQUATIONS OF THE FORM $\frac{dy}{dx} = f(x)$

In *Problems 1 to 7*, solve the differential equations.

1   $\frac{dy}{dx} = 3x^5$.                                            $[y = \frac{1}{2}x^6 + c]$

2   $\frac{dy}{dx} = \cos 4x - 2x$.                           $\left[y = \frac{\sin 4x}{4} - x^2 + c\right]$

3   $2x\frac{dy}{dx} = 3 - x^3$.                           $\left[y = \frac{3}{2}\ln x - \frac{x^3}{6} + c\right]$

4   $3\frac{dy}{dx} - 2x^2 = e^{4x}$                        $\left[y = \frac{2}{9}x^3 + \frac{e^{4x}}{12} + c\right]$

5   $\frac{dy}{dx} + x = 3$, given $y = 2$ when $x = 1$.        $\left[y = 3x - \frac{x^2}{2} - \frac{1}{2}\right]$

6   $3\frac{dy}{d\theta} + \sin\theta = 0$, given $y = \frac{2}{3}$ when $\theta = \frac{\pi}{2}$.    $\left[y = \frac{1}{3}\cos\theta + \frac{1}{2}\right]$

7   $\frac{1}{e^x} + 2 = x - 3\frac{dy}{dx}$, given $y = 1$ when $x = 0$.    $\left[y = \frac{1}{6}(x^2 - 4x + \frac{2}{e^x} + 4)\right]$

8   The gradient of a curve is given by $\frac{dy}{dx} + \frac{x^2}{2} = 3x$. Find the equation of the curve if it passes through the point $(1, \frac{1}{3})$.    $\left[y = \frac{3}{2}x^2 - \frac{x^3}{6} - 1\right]$

9   The acceleration, $a$, of a body is equal to its rate of change of velocity, $\frac{dv}{dt}$. Find an equation for $v$ in terms of $t$, given that when $t = 0$, velocity $v = u$.    $[v = u + at]$

10   The bending moment, $M$, of a beam is given by the equation $\frac{dM}{dx} = w(x - l)$, where $x$ is the distance from one end of a beam of length $l$ and $w$ is a constant. Solve the equation and show that $M = \frac{1}{2}w(l - x)^2$ given $M = \frac{wl^2}{2}$ when $x = 0$.

11   The velocity, $v$, of a body is equal to its rate of change of distance $\frac{dx}{dt}$. Find an equation for $x$ in terms of $t$ given $v = u + at$, where $u$ and $a$ are constants and $x = 0$ when $t = 0$.

$$[x = ut + \frac{1}{2}at^2]$$

DIFFERENTIAL EQUATIONS OF THE FORM $\frac{dy}{dx} = f(y)$

In *Problems 12 to 17*, solve the differential equations.

12   $\frac{dy}{dx} = 2 + 3y$                              $[x = \frac{1}{3}\ln(2 + 3y) + c]$

13   $2\frac{dy}{dx} = \cot 4y$                          $\left[\frac{1}{4}\ln(\sec 4y) = \frac{x}{2} + c\right]$

14   $3y\frac{dy}{dt} = 2 - y^2$                       $\left[-\frac{1}{2}\ln(2 - y^2) = \frac{t}{3} + c\right]$

15   $\frac{dy}{dx} = 2\cos^2 y$                             $[\tan y = 2x + c]$

16   $\left(\frac{y}{2}\right)\frac{dy}{dx} = 3 - y$, given $y = 2$ when $x = -1$.    $[2x + y + 3\ln(3 - y) = 0]$

110

17  $(y^2 + 2)\dfrac{dy}{dx} = 5y$, given $y = 1$ when $x = \dfrac{1}{2}$.   $\left[\dfrac{y^2}{2} + 2\ln y = 5x - 2\right]$

18  The current in an electric circuit is given by the equation $Ri + L\dfrac{di}{dt} = 0$, where $L$

and $R$ are constants. Show that $i = Ie^{-\frac{Rt}{L}}$, given that $i = I$ when $t = 0$.

19  The velocity of a chemical reaction is given by $\dfrac{dx}{dt} = k(a - x)$, where $x$ is the amount
transferred in time $t$, $k$ is a constant and $a$ is the concentration at time $t = 0$ when
$x = 0$. Solve the equation and determine $x$ in terms of $t$.   $[x = a(1 - e^{-kt})]$

20  (a) Charge $Q$ coulombs at time $t$ seconds is given by the differential equation

$R\dfrac{dQ}{dt} + \dfrac{Q}{C} = 0$, where $C$ is the capacitance in farads and $R$ the resistance in

ohms. Solve the equation for $Q$ given that $Q = Q_0$ when $t = 0$.

(b) A circuit possesses a resistance of $250 \times 10^3$ ohms and a capacitance of
$8.5 \times 10^{-6}$ farads, and after $0.32$ seconds the charge falls to $8.0$ C. Determine
the initial charge and the charge after 1 second, each correct to 3 significant
figures.

$\left[\text{(a) } Q = Q_0 e^{-\frac{t}{CR}} ; \text{(b) } 9.30 \text{ C}; 5.81 \text{ C}\right]$

21  The rate of decay of a radioactive substance is given by $dN/dt = -\lambda N$, where $\lambda$ is
the decay constant and $\lambda N$ the number of radioactive atoms disintegrating per
second. Determine the half-life of a mercury isotope, in hours, (i.e. the time for $N$
to become one half of its original value), assuming the decay constant for mercury
to be $2.917 \times 10^{-6}$ atoms per second.   [66 hours]

22  A differential equation relating the difference in tension $T$, pulley contact angle $\theta$
and coefficient of friction $\mu$ is $dT/d\theta = \mu T$. When $\theta = 0$, $T = 150$ N, and $\mu = 0.30$
as slipping starts. Determine the tension at the point of slipping when $\theta = 2$ radians.
Determine also the value of $\theta$ when $T$ is 300 N.   [273 N; 2.31 rads]

23  The rate of cooling of a body is given by $d\theta/dt = k\theta$, where $k$ is a constant.
If $\theta = 60°$C when $t = 2$ minutes and $\theta = 50°$C when $t = 5$ minutes, determine the
time taken for $\theta$ to fall to $40°$C, correct to the nearest second.
   [8 minutes 40 seconds]


VARIABLE-SEPARABLE TYPES OF DIFFERENTIAL EQUATIONS

In *Problem 24 to 30*, solve the differential equations.

24  $\dfrac{dy}{dx} = yx^3$.   $\left[\ln y = \dfrac{x^4}{4} + c\right]$

25  $\dfrac{dy}{dx} = \dfrac{3x^2 - 2}{2y + 1}$   $[x^3 - 2x = y^2 + y + c]$

26  $\dfrac{dy}{dx} = 2y\cos x$   $[\ln y = 2\sin x + c]$

27  $2xy\dfrac{dy}{dx} = y^2 + 3$   $[\ln (y^2 + 3) = \ln x + c \text{ or } y^2 + 3 = Ax]$

28  $(2y - 1)\dfrac{dy}{dx} = (3x^2 + 1)$, given $x = 1$ when $y = 2$.   $[y^2 - y = x^3 + x]$

29  $\dfrac{dy}{dx} = e^{2x - y}$, given $x = 0$ when $y = 0$.   $\left[e^y = \dfrac{1}{2}e^{2x} + \dfrac{1}{2}\right]$

111

30  $2y(1-x) + x(1+y)\dfrac{dy}{dx} = 0$, given $x = 1$ when $y = 1$.  $\left[\ln(x\sqrt{y}) = x - \dfrac{y}{2} - \dfrac{1}{2}\right]$

31  Show that the solution of the equation $\dfrac{y^2+1}{x^2+1} = \dfrac{y}{x}\dfrac{dy}{dx}$ is of the form

$\sqrt{\left(\dfrac{y^2+1}{x^2+1}\right)}$ = constant.

32  Solve $xy = (1-x^2)\dfrac{dy}{dx}$ for $y$, given $x = 0$ when $y = 1$.  $\left[y = \dfrac{1}{\sqrt{(1-x^2)}}\right]$

33  Determine the equation of the curve which satisfies the equation $xy\dfrac{dy}{dx} = x^2 - 1$, and which passes through the point $(1, 2)$.

$[x^2 - 2\ln x = y^2 - 3]$

34  Solve the equation $y\cos^2\theta\dfrac{dy}{d\theta} = 3 + \sin\theta$, given that $y = \sqrt{6}$ when $x = \pi/4$.

$[y^2 = 2(3\tan\theta + \sec\theta - \sqrt{2})]$

35  The p.d., $V$, between the plates of a capacitor $C$ charged by a steady voltage $E$ through a resistor $R$ is given by the equation $CR\dfrac{dV}{dt} + V = E$.

(a) Solve the equation for $V$ given that at $t = 0$, $V = 0$.

(b) Calculate $V$, correct to 3 significant figures, when $E = 25$ volts, $C = 20 \times 10^{-6}$ farads, $R = 200 \times 10^3$ ohms and $t = 3.0$ seconds.

$[$(a) $V = E(1 - e^{-\frac{t}{CR}})$; (b) 13.2 volts$]$

# 13 Homogeneous first order differential equations

## A. MAIN POINTS CONCERNED WITH HOMOGENEOUS FIRST ORDER DIFFERENTIAL EQUATIONS

1 Certain first order differential equations are not of the 'variable–separable' type but can be made separable by changing the variable.

2 An equation of the form $P\dfrac{dy}{dx} = Q$, where $P$ and $Q$ are functions of both $x$ and $y$ of the same degree throughout, is said to be **homogeneous** in $y$ and $x$. For example, $f(x, y) = x^2 + 3xy + y^2$ is a homogeneous function since each of the three terms are of degree 2. Similarly, $f(x, y) = \dfrac{x - 3y}{2x + y}$ is homogeneous in $x$ and $y$ since each of the four terms are of degree 1. However, $f(x, y) = \dfrac{x^2 - y}{2x^2 + y^2}$ is not homogeneous since the term in $y$ in the numerator is of degree 1 and the other three terms are of degree 2.

3 **Procedure to solve differential equations of the form $P\dfrac{dy}{dx} = Q$.**

(i) Rearrange $P\dfrac{dy}{dx} = Q$ into the form $\dfrac{dy}{dx} = \dfrac{Q}{P}$.

(ii) Make the substitution $y = vx$ (where $v$ is a function of $x$), from which, $\dfrac{dy}{dx} = v(1) + x\dfrac{dv}{dx}$, by the product rule.

(iii) Substitute for both $y$ and $\dfrac{dy}{dx}$ in the equation $\dfrac{dy}{dx} = \dfrac{Q}{P}$. Simplify, by cancelling, and an equation results in which the variables are separable.

(iv) Separate the variables and solve using the method shown in chapter 12.

(v) Substitute $v = \dfrac{y}{x}$ to solve in terms of the original variables.

## B. WORKED PROBLEMS ON HOMOGENEOUS FIRST ORDER DIFFERENTIAL EQUATIONS

*Problem 1* Solve the differential equation $y - x = x\dfrac{dy}{dx}$, given $x = 1$ when $y = 2$.

Using the procedure of para. 3:

(i) Rearranging $y - x = x\dfrac{dy}{dx}$ gives: $\dfrac{dy}{dx} = \dfrac{y - x}{x}$, which is homogeneous in $x$ and $y$.

113

(ii) Let $y = vx$, then $\dfrac{dy}{dx} = v + x\dfrac{dv}{dx}$.

(iii) Substituting for $y$ and $\dfrac{dy}{dx}$ gives: $v + x\dfrac{dv}{dx} = \dfrac{vx-x}{x} = \dfrac{v-1}{1} = v-1$.

(iv) Separating the variables gives: $x\dfrac{dv}{dx} = v-1-v = -1$, i.e. $dv = -\dfrac{1}{x}\,dx$.

Integrating both sides gives: $\displaystyle\int dv = \int -\dfrac{1}{x}\,dx$

Hence, $v = -\ln x + c$.

(v) Replacing $v$ by $\dfrac{y}{x}$ gives: $\dfrac{y}{x} = -\ln x + c$, which is the general solution.

When $x = 1$, $y = 2$, thus: $\dfrac{2}{1} = -\ln 1 + c$,

from which, $c = 2$.

Thus the particular solution is: $\dfrac{y}{x} = -\ln x + 2$

or   $y = -x\,(\ln x - 2)$.

---

*Problem 2* Find the particular solution of the equation:

$$x\dfrac{dy}{dx} = \dfrac{x^2+y^2}{y},$$

given the boundary conditions that $x = 1$ when $y = 4$.

---

Using the procedure of para. 3:

(i) Rearranging $x\dfrac{dy}{dx} = \dfrac{x^2+y^2}{y}$ gives $\dfrac{dy}{dx} = \dfrac{x^2+y^2}{xy}$ which is homogeneous in

$x$ and $y$ since each of the three terms on the right hand side are of the same degree (i.e. degree 2).

(ii) Let $y = vx$ then $\dfrac{dy}{dx} = v + x\dfrac{dv}{dx}$.

(iii) Substituting for $y$ and $\dfrac{dy}{dx}$ in the equation $\dfrac{dy}{dx} = \dfrac{x^2+y^2}{xy}$ gives:

$$v + x\dfrac{dv}{dx} = \dfrac{x^2+(vx)^2}{x(vx)} = \dfrac{x^2+v^2x^2}{vx^2} = \dfrac{1+v^2}{v}$$

(iv) Separating the variables gives: $x\dfrac{dv}{dx} = \dfrac{1+v^2}{v} - v = \dfrac{1+v^2-v^2}{v} = \dfrac{1}{v}$

Hence, $v\,dv = \dfrac{1}{x}\,dx$.

Integrating both sides gives: $\displaystyle\int v\,dv = \int \dfrac{1}{x}\,dx$, i.e. $\dfrac{v^2}{2} = \ln x + c$

(v) Replacing $v$ by $\dfrac{y}{x}$ gives: $\dfrac{y^2}{2x^2} = \ln x + c$, which is the general solution

When $x = 1$, $y = 4$, thus: $\dfrac{16}{2} = \ln 1 + c$ from which, $c = 8$

Hence the particular solution is $\dfrac{y^2}{2x^2} = \ln x + 8$

or   $y^2 = 2x^2\,(\ln x + 8)$.

*Problem 3* Solve the equation $7x\,(x-y)\,dy = 2(x^2+6xy-5y^2)\,dx$ given that $x = 1$ when $y = 0$.

Using the procedure of para. 3:

(i) Rearranging gives: $\dfrac{dy}{dx} = \dfrac{2x^2+12xy-10y^2}{7x^2-7xy}$ , which is homogeneous in $x$ and $y$ since each of the terms on the right hand side is of degree 2.

(ii) Let $y = vx$, then $\dfrac{dy}{dx} = v+x\,\dfrac{dv}{dx}$ .

(iii) Substituting for $y$ and $\dfrac{dy}{dx}$ gives: $v+x\,\dfrac{dv}{dx} = \dfrac{2x^2-12x(vx)-10(vx)^2}{7x^2-7x(vx)}$

$$= \frac{2+12v-10v^2}{7-7v}$$

(iv) Separating the variables gives: $x\,\dfrac{dv}{dx} = \dfrac{2+12v-10v^2}{7-7v} -v$

$$= \frac{(2+12v-10v^2)-v(7-7v)}{7-7v}$$

$$= \frac{2+5v-3v^2}{7-7v}$$

Hence, $\dfrac{7-7v}{2+5v-3v^2}\,dv = \dfrac{dx}{x}$

Integrating both sides gives: $\displaystyle\int\!\left(\frac{7-7v}{2+5v-3v^2}\right)\,dv = \int\frac{dx}{x}$

Resolving $\dfrac{7-7v}{2+5v-3v^2}$ into partial fractions gives $\dfrac{4}{(1+3v)} - \dfrac{1}{(2-v)}$

Hence $\displaystyle\int\!\left(\frac{4}{(1+3v)} - \frac{1}{(2-v)}\right)\,dv = \int\frac{dx}{x}$ .

i.e., $\dfrac{4}{3}\ln(1+3v) + \ln(2-v) = \ln x+c.$

(v) Replacing $v$ by $\dfrac{y}{x}$ gives: $\dfrac{4}{3}\ln\left(1 + \dfrac{3y}{x}\right) + \ln\left(2-\dfrac{y}{x}\right) = \ln x+c$

or $\dfrac{4}{3}\ln\left(\dfrac{x+3y}{x}\right) + \ln\left(\dfrac{2x-y}{x}\right) = \ln x+c$

When $x = 1$, $y = 0$, thus $\dfrac{4}{3}\ln 1 + \ln 2 = \ln 1 + c$ from which $c = \ln 2$

Hence the particular solution is: $\dfrac{4}{3}\ln\left(\dfrac{x+3y}{x}\right) + \ln\left(\dfrac{2x-y}{x}\right) = \ln x + \ln 2$

i.e. $\ln\left(\dfrac{x+3y}{x}\right)^{\frac{4}{3}}\left(\dfrac{2x-y}{x}\right) = \ln(2x)$, from the laws of logarithms,

i.e. $\left(\dfrac{x+3y}{x}\right)^{\frac{4}{3}}\left(\dfrac{2x-y}{x}\right) = 2x$

*Problem 4* Show that the solution of the differential equation:

$x^2-3y^2+2xy\,\dfrac{dy}{dx} = 0$ is $y = x\sqrt{(8x+1)}$ given that $y = 3$ when $x = 1$.

Using the procedure or para. 3:

(i) Rearranging gives: $2xy \dfrac{dy}{dx} = 3y^2 - x^2$

and $\dfrac{dy}{dx} = \dfrac{3y^2 - x^2}{2xy}$

(ii) Let $y = vx$ then $\dfrac{dy}{dx} = v + x\dfrac{dv}{dx}$

(iii) Substituting for $y$ and $\dfrac{dy}{dx}$ gives: $v + x\dfrac{dv}{dx} = \dfrac{3(vx)^2 - x^2}{2x(vx)} = \dfrac{3v^2 - 1}{2v}$

(iv) Separating the variables gives: $x\dfrac{dv}{dx} = \dfrac{3v^2 - 1}{2v} - v = \dfrac{3v^2 - 1 - 2v^2}{2v} = \dfrac{v^2 - 1}{2v}$

Hence $\dfrac{2v}{v^2 - 1}\,dv = \dfrac{1}{x}\,dx$

Integrating both sides gives: $\displaystyle\int \dfrac{2v}{v^2 - 1}\,dv = \int \dfrac{1}{x}\,dx$, i.e., $\ln(v^2 - 1) = \ln x + c$

(v) Replacing $v$ by $\dfrac{y}{x}$ gives: $\ln\left(\dfrac{y^2}{x^2} - 1\right) = \ln x + c$

When $y = 3, x = 1$, thus $\ln\left(\dfrac{9}{1} - 1\right) = \ln 1 + c$, from which, $c = \ln 8$

Hence the particular integral is: $\ln\left(\dfrac{y^2}{x^2} - 1\right) = \ln x + \ln 8$
$= \ln 8x$, by the laws of logarithms.

Hence $\dfrac{y^2}{x^2} - 1 = 8x$; $\dfrac{y^2}{x^2} = 8x + 1$; $y^2 = x^2\,(8x+1)$,

i.e. $y = x\sqrt{(8x+1)}$.

## C. FURTHER PROBLEMS ON HOMOGENEOUS FIRST ORDER DIFFERENTIAL EQUATIONS

In *Problems 1 to 5*, find the general solution of the differential equations.

1 $x^2 = y^2\dfrac{dy}{dx}$  $\qquad\left[-\dfrac{1}{3}\ln\left(\dfrac{x^3 - y^3}{x^3}\right) = \ln x + c\right]$

2 $x - y + x\dfrac{dy}{dx} = 0$  $\qquad[y = x\,(c - \ln x)]$

3 $\dfrac{x+y}{y-x} = \dfrac{dy}{dx}$  $\qquad\left[-\dfrac{1}{2}\ln\left(\dfrac{-y^2}{x^2} + \dfrac{2y}{x} + 1\right) = \ln x + c \text{ or } x^2 + 2xy - y^2 = k\right]$

4 $xy^3\,dy = (x^4 + y^4)\,dx$  $\qquad[y^4 = 4x^4\,(\ln x + c)]$

5 $(9xy - 11xy)\dfrac{dy}{dx} = 11y^2 - 16xy + 3x^2$  $\quad\left[\dfrac{1}{5}\left[\dfrac{3}{13}\ln\left(\dfrac{13y - 3x}{x}\right) - \ln\left(\dfrac{y-x}{x}\right)\right] = \ln x + c\right]$

6 Show that the solution of the equation $2xy\dfrac{dy}{dx} = x^2 + y^2$ can be expressed as $x = K\,(x^2 - y^2)$, where $K$ is a constant.

7 Find the particular solution of the differential equation: $(x^2 + y^2)\,dy = xy\,dx$, given that $x = 1$ when $y = 1$.  $\qquad[x^2 = 2y^2\,(\ln y + \tfrac{1}{2})]$

8 Determine the solution of $\dfrac{dy}{dx} = \dfrac{x^3 + y^3}{xy^2}$ given $x = 1$ when $y = 4$.  $\qquad[y^3 = x^3\,(3\ln x + 64)]$

9  Show that the solution of the differential equation:

$\dfrac{dy}{dx} = \dfrac{y^3-xy^2-x^2y-5x^3}{xy^2-x^2y-2x^3}$ is of the form:

$\dfrac{y^2}{2x^2} + \dfrac{4y}{x} + 18 \ln\left(\dfrac{y-5x}{x}\right) = \ln x + 42$, when $x = 1$ and $y = 6$.

# 14 Linear first order differential equations

## A. MAIN POINTS CONCERNED WITH LINEAR FIRST ORDER DIFFERENTIAL EQUATIONS

1 An equation of the form $dy/dx + Py = Q$, where $P$ and $Q$ are functions of $x$ only is called a **linear differential equation** since $y$ and its derivatives are of the first degree.

2 (i) The solution of $dy/dx + Py = Q$ is obtained by multiplying throughout by what is termed an **integrating factor**.

(ii) Multiplying $dy/dx + Py = Q$ by say $R$, a function of $x$ only, gives

$$R\frac{dy}{dx} + RPy = RQ \tag{1}$$

(iii) The differential coefficient of a product $Ry$ is obtained using the product rule, i.e., $\frac{d}{dx}(Ry) = R\frac{dy}{dx} + y\frac{dR}{dx}$, which is the same as the left-hand side of equation (1), when $R$ is chosen such that $RP = \frac{dR}{dx}$.

(iv) If $dR/dx = RP$, then separating the variables gives $dR/R = P\,dx$. Integrating both sides gives $\int \frac{dR}{R} = \int P\,dx$, i.e. $\ln R = \int P\,dx + c$
from which, $R = e^{\int P\,dx + c} = e^{\int P\,dx}e^{c}$
i.e. $R = Ae^{\int P\,dx}$, where $A = e^{c} = $ a constant.

(v) Substituting $R = Ae^{\int P\,dx}$ in equation (1) gives:

$$Ae^{\int P\,dx}(\frac{dy}{dx}) + Ae^{\int P\,dx}Py = Ae^{\int P\,dx}Q$$

i.e. $e^{\int P\,dx}(\frac{dy}{dx}) + e^{\int P\,dx}Py = e^{\int P\,dx}Q$ $\tag{2}$

(vi) The left-hand side of equation (2) is $\frac{d}{dx}(ye^{\int P\,dx})$, which may be checked by differentiation $ye^{\int P\,dx}$ with respect to $x$, using the product rule.

(vii) From equation (2), $\frac{d}{dx}(ye^{\int P\,dx}) = e^{\int P\,dx}Q$
Integrating both sides gives: $ye^{\int P\,dx} = \int e^{\int P\,dx}Q\,dx$ $\tag{3}$

(viii) $e^{\int P\,dx}$ is the integrating factor.

118

3  **Procedure to solve differential equations of the form $\dfrac{dy}{dx} + Py = Q$.**

(i)  Rearrange the differential equation into the form $\dfrac{dy}{dx} + Py = Q$, where $P$ and $Q$ are functions of $x$.

(ii)  Determine $\int P\, dx$.

(iii)  Determine the integrating factor $e^{\int P dx}$.

(iv)  Substitute $e^{\int P dx}$ into equation (3).

(v)  Integrate the right-hand side of equation (3) to give the general solution of the differential equation. Given boundary conditions, the particular solution may be determined.

## B.  WORKED PROBLEMS ON LINEAR FIRST ORDER DIFFERENTIAL EQUATIONS

*Problem 1* Solve $\dfrac{1}{x}\dfrac{dy}{dx} + 4y = 2$ given the boundary conditions $x = 0$ when $y = 4$.

Using the procedure of para. 3:

(i)  Rearranging gives $\dfrac{dy}{dx} + 4xy = 2x$, which is of the form $\dfrac{dy}{dx} + Py = Q$.

where $P = 4x$ and $Q = 2x$.

(ii)  $\int P\, dx = \int 4x\, dx = 2x^2$.

(iii)  Integrating factor $e^{\int P dx} = e^{2x^2}$.

(iv)  Substituting into equation (3) gives: $ye^{2x^2} = \int e^{2x^2}(2x)\, dx$

(v)  Hence the general solution is: $ye^{2x^2} = \dfrac{1}{2}e^{2x^2} + c$, by using the substitution $u = 2x^2$.

When $x = 0$, $y = 4$, thus $4e^0 = \dfrac{1}{2}e^0 + c$, from which, $c = \dfrac{7}{2}$

Hence the particular solution is $ye^{2x^2} = \dfrac{1}{2}e^{2x^2} + \dfrac{7}{2}$

or $y = \dfrac{1}{2} + \dfrac{7}{2}e^{-2x^2}$  or  $y = \dfrac{1}{2}(1 + 7e^{-2x^2})$.

*Problem 2*  Show that the solution of the equation $\dfrac{dy}{dx} + 1 = -\dfrac{y}{x}$ is given by $y = \dfrac{3 - x^2}{2x}$, given $x = 1$ when $y = 1$.

Using the procedure of para. 3:

(i)  Rearranging gives $\dfrac{dy}{dx} + \left(\dfrac{1}{x}\right)y = -1$, which is of the form $\dfrac{dy}{dx} + Py = Q$,

where $P = \dfrac{1}{x}$ and $Q = -1$. (Note that $Q$ can be considered to be $-1x^0$, i.e. a function of $x$.)

(ii)  $\displaystyle\int P\, dx = \int \dfrac{1}{x}\, dx = \ln x$.

(iii)  Integrating factor $e^{\int P dx} = e^{\ln x} = x$ (from the definition of a logarithm).

(iv)  Substituting into equation (3) gives: $yx = \displaystyle\int x\,(-1)\, dx$.

119

(v) Hence the general solution is $\qquad yx = \dfrac{-x^2}{2} + c$

When $x = 1$, $y = 1$, thus $1 = \dfrac{-1}{2} + c$, from which, $c = \dfrac{3}{2}$.

Hence the particular solution is: $yx = \dfrac{-x^2}{2} + \dfrac{3}{2}$

i.e. $2yx = 3 - x^2$ and $y = \dfrac{3 - x^2}{2x}$

---

*Problem 3* Determine the particular solution of $(dy/dx) - x + y = 0$, given that $x = 0$ when $y = 2$.

---

Using the procedure of para. 3:

(i) Rearranging gives $\dfrac{dy}{dx} + y = x$, which is of the form $\dfrac{dy}{dx} + Py = Q$, where $P = 1$

and $Q = \dot{x}$. (In this case $P$ can be considered to be $1x^0$, i.e., a function of $x$.)

(ii) $\int P\,dx = \int 1\,dx = x$

(iii) Integrating factor $e^{\int P\,dx} = e^x$

(iv) Substituting in equation (3) gives: $ye^x = \int e^x(x)\,dx$ $\qquad\qquad$ (4)

(v) $\int e^x(x)\,dx$ is determined using integration by parts (see chapter 11).

$\int xe^x\,dx = xe^x - e^x + c$

Hence from equation (4): $ye^x = xe^x - e^x + c$, which is the general solution.

When $x = 0$, $y = 2$ thus $2e^0 = 0 - e^0 + c$, from which, $c = 3$.

Hence the particular solution is: $ye^x = xe^x - e^x + 3$ or $y = x - 1 + 3e^{-x}$.

---

*Problem 4* Solve the differential equation $dy/d\theta = \sec\theta + y\tan\theta$ given the boundary conditions $y = 1$ when $\theta = 0$.

---

Using the procedure of para. 3:

(i) Rearranging gives $\dfrac{dy}{d\theta} - (\tan\theta)\,y = \sec\theta$, which is of the form $\dfrac{dy}{d\theta} + Py = Q$,

where $P = -\tan\theta$ and $Q = \sec\theta$.

(ii) $\int P\,dx = \int -\tan\theta\,d\theta = -\ln(\sec\theta) = \ln(\sec\theta)^{-1} = \ln(\cos\theta)$.

(iii) Integrating factor $e^{\int P\,dx} = e^{\ln(\cos\theta)} = \cos\theta$ (from the definition of a logarithm).

(iv) Substituting in equation (3) gives: $y\cos\theta = \int \cos\theta\,(\sec\theta)\,d\theta$

i.e., $y\cos\theta = \int d\theta$

(v) Integrating gives: $y\cos\theta = \theta + c$, which is the general solution.

When $\theta = 0$, $y = 1$, thus $1\cos 0 = 0 + c$, from which, $c = 1$.

Hence the particular solution is: $y\cos\theta = \theta + 1$ or $y = \sec\theta\,(\theta + 1)$.

---

*Problem 5* (a) Find the general solution of the equation $(x-2)\dfrac{dy}{dx} + \dfrac{3(x-1)}{(x+1)}y = 1$.

(b) Given the boundary conditions that $y = 5$ when $x = -1$, find the particular solution of the equation given in (a).

---

(a) Using the procedure of para. 3:

(i) Rearranging gives $\dfrac{dy}{dx} + \dfrac{3(x-1)}{(x+1)(x-2)}\,y = \dfrac{1}{(x-2)}$, which is of the form

$$\frac{dy}{dx} + Py = Q, \text{ where } P = \frac{3(x-1)}{(x+1)(x-2)} \text{ and } Q = \frac{1}{(x-2)}.$$

(ii) $\int P\,dx = \int \frac{3(x-1)}{(x+1)(x-2)}\,dx$, which is integrated using partial fractions.

Let $\dfrac{3x-3}{(x+1)(x-2)} \equiv \dfrac{A}{(x+1)} + \dfrac{B}{(x-2)} \equiv \dfrac{A(x-2)+B(x+1)}{(x+1)(x-2)}$,

from which, $3x-3 \equiv A(x-2)+B(x+1)$

When $x = -1$, $-6 = -3A$, from which, $A = 2$

When $x = 2$, $3 = 3B$, from which, $B = 1$

Hence $\int \dfrac{3x-3}{(x+1)(x-2)}\,dx \equiv \int \left[ \dfrac{2}{(x+1)} + \dfrac{1}{(x-2)} \right] dx = 2\ln(x+1)+\ln(x-2)$

$$= \ln(x+1)^2 + \ln(x-2)$$
$$= \ln\{(x+1)^2(x-2)\}$$

(iii) Integrating factor $e^{\int P\,dx} = e^{\ln\{(x+1)^2(x-2)\}} = (x+1)^2(x-2)$

(iv) Substituting in equation (3) gives: $y(x+1)^2(x-2) = \int (x+1)^2(x-2)\dfrac{1}{(x-2)}dx$

$$= \int (x+1)^2\,dx$$

(v) **Hence the general solution is:** $y(x+1)^2(x-2) = \dfrac{1}{3}(x+1)^3 + c.$

(b) When $x = -1$, $y = 5$ thus $5(0)(-3) = 0+c$, from which, $c = 0$.

Hence $y(x+1)^2(x-2) = \dfrac{1}{3}(x+1)^3$

i.e., $y = \dfrac{(x+1)^3}{3(x+1)^2(x-2)}$, and hence **the particular solution is** $y = \dfrac{x+1}{3(x-2)}$.

## C. FURTHER PROBLEMS ON LINEAR FIRST ORDER DIFFERENTIAL EQUATIONS

In *Problems 1 to 8*, solve the differential equations.

1   $x\dfrac{dy}{dx} = 3-y.$      $\left[ y = 3 + \dfrac{c}{x} \right]$

2   $\dfrac{dy}{dx} = x(1-2y)$      $\left[ y = \dfrac{1}{2} + ce^{-x^2} \right]$

3   $t\dfrac{dy}{dt} - 5t = -y$      $\left[ y = \dfrac{5t}{2} + \dfrac{c}{t} \right]$

4   $x\left( \dfrac{dy}{dx} + 1 \right) = x^3 - 2y$, given $x = 1$ when $y = 3$.      $\left[ y = \dfrac{x^3}{5} - \dfrac{x}{3} + \dfrac{47}{15x^2} \right]$

5   $\dfrac{1}{x}\dfrac{dy}{dx} + y = 1$      $\left[ y = 1 + ce^{-\frac{x^2}{2}} \right]$

6   $\dfrac{dy}{dx} + x = 2y$      $\left[ y = \dfrac{x}{2} + \dfrac{1}{4} + ce^{2x} \right]$

7   $\cot x\dfrac{dy}{dx} = 1-2y$, given $y = 1$ when $x = \dfrac{\pi}{4}$.      $\left[ y = \dfrac{1}{2} + \cos^2 x \right]$

8   $t\dfrac{d\theta}{dt} + \sec t\,(t\sin t + \cos t)\,\theta = \sec t$, given $t = \pi$ when $\theta = 1$.      $\left[ \theta = \dfrac{1}{t}(\sin t - \pi) \right]$

9   Given the equation $x\dfrac{dy}{dx} = \dfrac{2}{x+2} - y$ show that the particular solution is $y = \dfrac{2}{x}\ln(x+2)$, given the boundary conditions that $x = -1$ when $y = 0$.

10 Show that the solution of the differential equation $\dfrac{dy}{dx} - 2(x+1)^3 = \dfrac{4}{(x+1)}y$

is $y = (x+1)^4 \ln (x+1)^2$, given that $x = 0$ when $y = 0$.

11 Show that the solution of the differential equation $\dfrac{dy}{dx} + ky = a \sin bx$ is given by:

$y = \left(\dfrac{a}{k^2+b^2}\right)$ $(k \sin bx - b \cos bx) + \left(\dfrac{k^2+b^2+ab}{k^2+b^2}\right) e^{-kx}$, given $y = 1$ when $x = 0$.

12 The equation $\dfrac{dv}{dt} = -(av+bt)$, where $a$ and $b$ are constants, represents an equation

of motion when a particle moves in a resisting medium. Solve the equation for $v$
given that $v = u$ when $t = 0$.

$$\left[ v = \dfrac{b}{a^2} - \dfrac{bt}{a} + \left(u - \dfrac{b}{a^2}\right) e^{-at} \right]$$

13 In an alternating current circuit containing resistance $R$ and inductance $L$ the

current $i$ is given by: $Ri + L\dfrac{di}{dt} = E_0 \sin \omega t$. Given $i = 0$ when $t = 0$, show that the

solution of the equation is given by:

$$i = \left(\dfrac{E_0}{R^2+\omega^2 L^2}\right) (R \sin \omega t - \omega L \cos \omega t) + \left(\dfrac{E_0 \omega L}{R^2+\omega^2 L^2}\right) e^{-\frac{Rt}{L}}$$

14 The concentration, $C$, of impurities of an oil purifier varies with time $t$ and is

described by the equation $a\dfrac{dC}{dt} = b + dm - Cm$, where $a$, $b$, $d$ and $m$ are constants.

Given $C = c_0$ when $t = 0$, solve the equation and show that:

$$C = \left(\dfrac{b}{m} + d\right) \left(1 - e^{-\frac{mt}{a}}\right) + c_0 e^{-\frac{mt}{a}}.$$

15 The equation of motion of a train is given by: $m\dfrac{dv}{dt} = mk(1-e^{-t}) - mcv$, where $v$

is the speed, $t$ is the time and $m$, $k$ and $c$ are constants. Determine the speed, $v$,
given $v = 0$ at $t = 0$.

$$\left[ v = k\left\{\dfrac{1}{c} - \dfrac{e^{-t}}{(c-1)} + \dfrac{e^{-ct}}{c(c-1)}\right\} \right]$$

# 15 Second order differential equations (1)

## A. MAIN POINTS CONCERNED WITH DIFFERENTIAL EQUATIONS OF THE FORM $a\dfrac{d^2y}{dx^2} + b\dfrac{dy}{dx} + cy = 0$

1  An equation of the form $a\dfrac{d^2y}{dx^2} + b\dfrac{dy}{dx} + cy = 0$, where $a$, $b$ and $c$ are constants, is called a **linear second order differential equation with constant coefficients.**

2  If $D$ represents $\dfrac{d}{dx}$ and $D^2$ represents $\dfrac{d^2}{dx^2}$, then the equation in para. 1 may be stated as $(aD^2 + bD + c)\,y = 0$. This equation is said to be in 'D-operator' form.

3  If $y = Ae^{mx}$, then $\dfrac{dy}{dx} = Ame^{mx}$ and $\dfrac{d^2y}{dx^2} = Am^2e^{mx}$.

Substituting these values into $a\dfrac{d^2y}{dx^2} + b\dfrac{dy}{dx} + cy = 0$ gives:

$a\,(Am^2e^{mx}) + b\,(Ame^{mx}) + cy = 0$
i.e., $Ae^{mx}(am^2 + bm + c) = 0$.
Thus $y = Ae^{mx}$ is a solution of the given equation provided that $(am^2 + bm + c) = 0$.

4  $am^2 + bm + c = 0$ is called the **auxiliary equation,** and since the equation is a quadratic, $m$ may be obtained either by factorising or by using the quadratic formula.

5  Since, in the auxiliary equation, $a$, $b$ and $c$ are real values, then the equation may have either

        (i)  two different real roots  (when $b^2 > 4ac$)

or     (ii)  two real equal roots    (when $b^2 = 4ac$)

or    (iii)  two complex roots     (when $b^2 < 4ac$)

6  **Procedure to solve differential equations of the form** $a\dfrac{d^2y}{dx^2} + b\dfrac{dy}{dx} + cy = 0$

(a) Rewrite the differential equation $a\dfrac{d^2y}{dx^2} + b\dfrac{dy}{dx} + cy = 0$ as $(aD^2 + bD + c)y = 0$.

(b) Substitute $m$ for $D$ and solve the auxiliary equation $am^2 + bm + c = 0$ for $m$.

(c) If the roots of the auxiliary equation are:
   (i)   **real and different,** say $m = \alpha$ and $m = \beta$, then the general solution is
      $y = Ae^{\alpha x} + Be^{\beta x}$.
   (ii)  **real and equal,** say $m = \alpha$ twice, then the general solution is
      $y = (Ax + B)\,e^{\alpha x}$

123

**complex,** say $m = \alpha \pm j\beta$, then the general solution is
$$y = e^{\alpha x}\{C \cos \beta x + D \sin \beta x\}.$$

(d) Given boundary conditions, constants $A$ and $B$, or $C$ and $D$, may be determined and the **particular solution** of the differential equation obtained.

7   The particular solutions obtained in the worked problems of section B may each be **verified** by substituting expressions for $y$, $\dfrac{dy}{dx}$ and $\dfrac{d^2y}{dx^2}$ into the original equation.

## B.  WORKED PROBLEMS ON DIFFERENTIAL EQUATIONS OF THE FORM $\dfrac{d^2y}{dx^2} + b\dfrac{dy}{dx} + cy = 0$

*Problem 1* Determine the general solution of $2\dfrac{d^2y}{dx^2} + 5\dfrac{dy}{dx} - 3y = 0$. Find also the particular solution given that when $x = 0$, $y = 4$ and $\dfrac{dy}{dx} = 9$.

Using the procedure of para. 6:

(a) $2\dfrac{d^2y}{dx^2} + 5\dfrac{dy}{dx} - 3y = 0$ in $D$-operator form is $(2D^2 + 5D - 3)\,y = 0$, where $D \equiv \dfrac{d}{dx}$.

(b) Substituting $m$ for $D$ gives the auxiliary equation $2m^2 + 5m - 3 = 0$.

Factorising gives: $(2m-1)(m+3) = 0$, from which, $m = \dfrac{1}{2}$ or $m = -3$.

(c) Since the roots are real and different the **general solution is**
$$y = Ae^{\frac{1}{2}x} + Be^{-3x}.$$

(d) When $x = 0$, $y = 4$, hence $4 = A + B$         (1)

Since $y = Ae^{\frac{1}{2}x} + Be^{-3x}$ then $\dfrac{dy}{dx} = \dfrac{1}{2}Ae^{\frac{1}{2}x} - 3Be^{-3x}$

When $x = 0$, $\dfrac{dy}{dx} = 9$, thus $9 = \dfrac{1}{2}A - 3B$         (2)

Solving the simultaneous equations (1) and (2) gives $A = 6$ and $B = -2$.
**Hence the particular solution is** $y = 6e^{\frac{1}{2}x} - 2e^{-3x}$.

*Problem 2* Find the general solution of $9\dfrac{d^2y}{dt^2} - 24\dfrac{dy}{dt} + 16y = 0$ and also the particular solution given the boundary conditions that when $t = 0$, $y = \dfrac{dy}{dt} = 3$.

Using the procedure of para. 6:

(a) $9\dfrac{d^2y}{dt^2} - 24\dfrac{dy}{dt} + 16y = 0$ in $D$-operator form is $(9D^2 - 24D + 16)\,y = 0$ where $D \equiv \dfrac{d}{dt}$.

(b) Substituting $m$ for $D$ gives the auxiliary equation $9m^2 - 24m + 16 = 0$.

Factorising gives: $(3m-4)(3m-4) = 0$, i.e., $m = \dfrac{4}{3}$ twice.

(c) Since the roots are real and equal, **the general solution is** $y = (At + B)e^{\frac{4}{3}t}$.

(d) When $t = 0$, $y = 3$ hence $3 = (0+B)e^0$, i.e., $B = 3$.

Since $y = (At+B)e^{\frac{4}{3}t}$ then $\dfrac{dy}{dt} = (At+B)\left(\dfrac{4}{3}e^{\frac{4}{3}t}\right)+Ae^{\frac{4}{3}t}$, by the product rule.

When $t = 0$, $\dfrac{dy}{dt} = 3$ thus $3 = (0+B)\dfrac{4}{3}e^0+Ae^0$

i.e., $3 = \dfrac{4}{3}B+A$

from which, $A = -1$, since $B = 3$.

**Hence the particular solution is $y = (-t+3)e^{\frac{4}{3}t}$ or $y = (3-t)e^{\frac{4}{3}t}$.**

*Problem 3* Solve the differential equation $\dfrac{d^2y}{dx^2} +6\dfrac{dy}{dx} + 13y = 0$, given that when $x = 0$, $y = 3$ and $\dfrac{dy}{dx} = 7$.

Using the procedure of para. 6:

(a) $\dfrac{d^2y}{dx^2} + 6\dfrac{dy}{dx} + 13y = 0$ in $D$-operator form is $(D^2 +6D+13)y = 0$, where $D \equiv \dfrac{d}{dx}$.

(b) Substituting $m$ for $D$ gives the auxiliary equation $m^2 +6m+13 = 0$.

Using the quadratic formula: $m = \dfrac{-6\pm\sqrt{[(6)^2 -4(1)(13)]}}{2(1)} = \dfrac{-6\pm\sqrt{(-16)}}{2}$

i.e. $m = \dfrac{-6\pm j4}{2} = -3 \pm j2$.

(c) Since the roots are complex, **the general solution is $y = e^{-3x}(C \cos 2x+D \sin 2x)$**

(d) When $x = 0$, $y = 3$ hence $3 = e^0$ $(C \cos 0+D \sin 0)$, i.e., $C = 3$.

Since $y = e^{-3x}(C \cos 2x+D \sin 2x)$,

then $\dfrac{dy}{dx} = e^{-3x} (-2C \sin 2x+2D \cos 2x)-3e^{-3x} (C \cos 2x+D \sin 2x)$,

by the product rule,

$= e^{-3x}\{ (2D-3C) \cos 2x-(2C+3D) \sin 2x\}$

When $x = 0$, $\dfrac{dy}{dx} = 7$, hence $7 = e^0\{ (2D-3C) \cos 0-(2C+3D) \sin 0\}$

i.e. $7 = 2D-3C$, from which, $D = 8$, since $C = 3$.

**Hence the particular solution is $y = e^{-3x} (3 \cos 2x+8 \sin 2x)$.**

Since $a \cos \omega t+b \sin \omega t = R \sin (\omega t+\alpha)$, where $R = \sqrt{(a^2 +b^2)}$ and

$\alpha = \arctan \dfrac{a}{b}$, then

$3 \cos 2x+8 \sin 2x = \sqrt{(3^2 +8^2)} \sin (2x+\arctan \dfrac{3}{8})$

$= \sqrt{73} \sin (2x+20°\ 33') = \sqrt{73} \sin (2x+0.359)$

**Thus the particular solution may also be expressed as**

$y = \sqrt{73}e^{-3x} \sin (2x+0.359)$

*Problem 4* The equation of motion of a body oscillating on the end of a spring is $(d^2x/dt^2)+100x = 0$, where $x$ is the displacement in metres of the body from its equilibrium position after time $t$ seconds. Determine $x$ in terms of $t$ given that at time $t = 0$, $x = 2m$ and $dx/dt = 0$

An equation of the form $\dfrac{d^2x}{dt^2} + m^2x = 0$ is a differential equation representing simple harmonic motion (S.H.M.). Using the procedure of para. 6:

(a) $\dfrac{d^2x}{dt^2} + 100x = 0$ in $D$-operator form is $(D^2+100)x = 0$.

(b) The auxiliary equation is $m^2+100 = 0$, i.e., $m^2 = -100$ and $m = \sqrt{(-100)}$, i.e., $m = \pm j10$.

(c) Since the roots are complex, the general solution is
$x = e^0 \,(C\cos 10t + D\sin 10t)$, i.e., $x = (C\cos 10t + D\sin 10t)$ metres.

(d) When $t = 0, x = 2$ thus $2 = C$

$\dfrac{dx}{dt} = -10C\sin 10t + 10D\cos 10t$

When $t = 0, \dfrac{dx}{dt} = 0$ thus $0 = -10C\sin 0 + 10D\cos 0$, i.e., $D = 0$.

**Hence the particular solution is $x = 2\cos 10t$ metres.**

---

**Problem 5** Given the differential equation $\dfrac{d^2V}{dt^2} = \omega^2 V$, where $\omega$ is a constant, show that its solution may be expressed as $V = 7\cosh \omega t + 3\sinh \omega t$ given the boundary conditions that when $t = 0$, $V = 7$ and $dV/dt = 3\omega$.

---

Using the procedure of para. 6:

(a) $\dfrac{d^2V}{dt^2} = \omega^2 V$, i.e., $\dfrac{d^2V}{dt^2} - \omega^2 V = 0$ in $D$-operator form is $(D^2-\omega^2)y = 0$, where $D \equiv \dfrac{d}{dt}$.

(b) The auxiliary equation is $m^2-\omega^2 = 0$, from which, $m^2 = \omega^2$ and $m = \pm\omega$.

(c) Since the roots are real and different, the general solution is $V = Ae^{\omega t} + Be^{-\omega t}$.

(d) When $t = 0$, $V = 7$ hence $7 = A+B$ $\qquad\qquad$ (1)

$\dfrac{dV}{dt} = A\omega e^{\omega t} - B\omega e^{-\omega t}$.

When $t = 0$, $\dfrac{dV}{dt} = 3\omega$, thus $3\omega = A\omega - B\omega$ $\quad$ i.e., $3 = A-B$ $\qquad$ (2)

From equations (1) and (2), $A = 5$ and $B = 2$.

Hence the particular solution is $V = 5e^{\omega t} + 2e^{-\omega t}$.

Since $\sinh \omega t = \dfrac{1}{2}(e^{\omega t} - e^{-\omega t})$ and $\cosh \omega t = \dfrac{1}{2}(e^{\omega t} + e^{-\omega t})$

then $\sinh \omega t + \cosh \omega t = e^{\omega t}$

and $\cosh \omega t - \sinh \omega t = e^{-\omega t}$.

Hence the particular solution may also be written as
$\qquad V = 5(\sinh \omega t + \cosh \omega t) + 2(\cosh \omega t - \sinh \omega t)$

i.e. $\quad V = (5+2)\cosh \omega t + (5-2)\sinh \omega t$

i.e. $\quad V = 7\cosh \omega t + 3\sinh \omega t$.

---

**Problem 6** The equation $\dfrac{d^2i}{dt^2} + \dfrac{R}{L}\dfrac{di}{dt} + \dfrac{1}{LC}i = 0$ represents a current $i$ flowing in an electrical circuit containing resistance $R$, inductance $L$ and capacitance $C$ connected in series. If $R = 200$ ohms, $L = 0.20$ henry and $C = 20 \times 10^{-6}$ farads, solve the equation for $i$ given the boundary conditions that when $t = 0$, $i = 0$ and $\dfrac{di}{dt} = 100$.

Using the procedure of para. 6:

(a) $\dfrac{d^2 i}{dt^2} + \dfrac{R}{L}\dfrac{di}{dt} + \dfrac{1}{LC}i = 0$ in $D$-operator form is $\left(D^2 + \dfrac{R}{L}D + \dfrac{1}{LC}\right)i = 0$ where $D \equiv \dfrac{d}{dt}$.

(b) The auxiliary equation is $m^2 + \dfrac{R}{L}m + \dfrac{1}{LC} = 0$.

Hence $m = \dfrac{-\dfrac{R}{L} \pm \sqrt{\left[\left(\dfrac{R}{L}\right)^2 - 4(1)\left(\dfrac{1}{LC}\right)\right]}}{2}$

When $R = 200$, $L = 0.2$ and $C = 20 \times 10^{-6}$,

then $m = \dfrac{-\dfrac{200}{0.2} \pm \sqrt{\left[\left(\dfrac{200}{0.2}\right)^2 - \dfrac{4}{(0.2)(20 \times 10^{-6})}\right]}}{2} = \dfrac{-1000 \pm \sqrt 0}{2} = -500$.

(c) Since the two roots are real and equal (i.e., $-500$ twice, since for a second order differential equation there must be two solutions), the general solution is $i = (At+B)e^{-500t}$.

(d) When $t = 0$, $i = 0$ hence $B = 0$

$\dfrac{di}{dt} = (At+B)(-500e^{-500t}) + (e^{-500t})(A)$, by the product rule.

When $t = 0$, $\dfrac{di}{dt} = 100$, thus $100 = -500B + A$
i.e., $A = 100$, since $B = 0$.
**Hence the particular solution is $i = 100te^{-500t}$.**

*Problem 7* The oscillations of a heavily damped pendulum satisfy the differential equation $\dfrac{d^2 x}{dt^2} + 6\dfrac{dx}{dt} + 8x = 0$, where $x$ cm is the displacement of the bob at time $t$ seconds. The initial displacement is equal to $+4$ cm and the initial velocity (i.e. $\dfrac{dx}{dt}$) is 8 cm/s. Solve the equation for $x$.

Using the procedure of para. 6:

(a) $\dfrac{d^2 x}{dt^2} + 6\dfrac{dx}{dt} + 8x = 0$ in $D$-operator form is $(D^2 + 6D + 8)x = 0$, where $D \equiv \dfrac{d}{dt}$.

(b) The auxiliary equation is $m^2 + 6m + 8 = 0$.
Factorising gives: $(m+2)(m+4) = 0$, from which, $m = -2$ or $m = -4$.

(c) Since the roots are real and different, the general solution is $x = Ae^{-2t} + Be^{-4t}$.

(d) Initial displacement means that time $t = 0$. At this instant, $x = 4$.
Thus $4 = A + B$          (1)

Velocity, $\dfrac{dx}{dt} = -2Ae^{-2t} - 4Be^{-4t}$.

$\dfrac{dx}{dt} = 8$ cm/s when $t = 0$, thus $8 = -2A - 4B$     (2)

From equations (1) and (2), $A = 12$ and $B = -8$.
Hence the particular solution is $x = 12e^{-2t} - 8e^{-4t}$
i.e., **displacement, $x = 4(3e^{-2t} - 2e^{-4t})$ cm.**

## C. FURTHER PROBLEMS ON DIFFERENTIAL EQUATIONS OF THE FORM

$$a\frac{d^2y}{dx^2} + b\frac{dy}{dx} + cy = 0$$

In *Problems 1 to 6*, determine the general solution of the given differential equations.

1  $\dfrac{d^2y}{dx^2} + 3\dfrac{dy}{dx} - 4y = 0.$ $\qquad\qquad\qquad$ $[y = Ae^x + Be^{-4x}]$

2  $6\dfrac{d^2y}{dt^2} - \dfrac{dy}{dt} - 2y = 0$ $\qquad\qquad\qquad$ $\left[y = Ae^{\frac{2}{3}t} + Be^{-\frac{1}{2}t}\right]$

3  $\dfrac{d^2y}{dx^2} - 4\dfrac{dy}{dx} + 4y = 0$ $\qquad\qquad\qquad$ $[y = (Ax+B)e^{2x}]$

4  $4\dfrac{d^2\theta}{dt^2} + 4\dfrac{d\theta}{dt} + \theta = 0$ $\qquad\qquad\qquad$ $[\theta = (At+B)e^{-\frac{1}{2}t}]$

5  $\dfrac{d^2y}{dx^2} + 2\dfrac{dy}{dx} + 5y = 0$ $\qquad\qquad\qquad$ $[y = e^{-x}(C\cos 2x + D\sin 2x)]$

6  $\dfrac{d^2y}{dx^2} + y = 0$ $\qquad\qquad\qquad$ $[y = C\cos x + D\sin x]$

In *Problems 7 to 12*, find the particular solution of the given differential equations for the stated boundary conditions.

7  $6\dfrac{d^2y}{dx^2} + 5\dfrac{dy}{dx} - 6y = 0$; when $x = 0$, $y = 5$ and $\dfrac{dy}{dx} = -1$. $\quad$ $[y = 3e^{\frac{2}{3}x} + 2e^{-\frac{3}{2}x}]$

8  $4\dfrac{d^2y}{dt^2} - 5\dfrac{dy}{dt} + y = 0$, when $t = 0$, $y = 1$ and $\dfrac{dy}{dt} = -2$. $\qquad$ $[y = 4e^{\frac{1}{4}t} - 3e^t]$

9  $(9D^2 + 30D + 25)y = 0$, where $D \equiv \dfrac{d}{dx}$; when $x = 0$, $y = 0$ and $\dfrac{dy}{dx} = 2$.

$\qquad\qquad\qquad\qquad\qquad\qquad\qquad\qquad\qquad\qquad$ $[y = 2xe^{-\frac{5}{3}x}]$

10  $\dfrac{d^2x}{dt^2} - 6\dfrac{dx}{dt} + 9x = 0$; when $t = 0$, $x = 2$ and $\dfrac{dx}{dt} = 0$. $\qquad$ $[x = 2(1-3t)e^{3t}]$

11  $\dfrac{d^2y}{dx^2} + 6\dfrac{dy}{dx} + 13y = 0$; when $x = 0$, $y = 4$ and $\dfrac{dy}{dx} = 0$.

$\qquad\qquad\qquad\qquad\qquad\qquad\qquad\qquad\qquad$ $[y = 2e^{-3x}(2\cos 2x + 3\sin 2x)]$

12  $(4D^2 + 20D + 125)\theta = 0$, where $D \equiv \dfrac{d}{dt}$; when $t = 0$, $\theta = 3$ and $\dfrac{d\theta}{dt} = 2.5$.

$\qquad\qquad\qquad\qquad\qquad\qquad\qquad\qquad$ $[\theta = e^{-2.5t}(3\cos 5t + 2\sin 5t)]$

13  A body moves in a straight line so that its distance $s$ metres from the origin after time $t$ seconds is given by $\dfrac{d^2s}{dt^2} + a^2s = 0$, where $a$ is a constant. Solve the equation for $s$ given that $s = c$ and $\dfrac{ds}{dt} = 0$ when $t = \dfrac{2\pi}{a}$. $\qquad$ $[s = c\cos at]$

14  Determine an expression for $x$ for a differential equation $\dfrac{d^2x}{dt^2} + 2n\dfrac{dx}{dt} + n^2x = 0$ which represents a critically damped oscillator, given that at time $t = 0$, $x = s$ and $\dfrac{dx}{dt} = u.$ $\qquad\qquad\qquad\qquad\qquad\qquad$ $[x = \{s + (u+ns)t\}e^{-nt}]$

15  The charge, $q$, on a capacitor in a certain electrical circuit satisfies the differential equation $\dfrac{d^2q}{dt^2} + 4\dfrac{dq}{dt} + 5q = 0$. Initially, (i.e., when $t = 0$), $q = Q$ and $\dfrac{dq}{dt} = 0$. Show that the charge in the circuit can be expressed as: $q = \sqrt{5}Qe^{-2t}\sin(t+0.464)$.

16  The motion of the pointer of a galvanometer about its position of equilibrium is represented by the equation $I\dfrac{d^2\theta}{dt^2} + K\dfrac{d\theta}{dt} + F\theta = 0$. If $I$, the moment of inertia of the pointer about its pivot, is $5 \times 10^{-3}$, $K$, the resistance due to friction at unit angular velocity, is $2 \times 10^{-2}$ and $F$, the force on the spring necessary to produce unit displacement, is 0.20, solve the equation for $\theta$ in terms of $t$ given that when $t = 0$, $\theta = 0.3$ and $\dfrac{d\theta}{dt} = 0$.

$\qquad\qquad\qquad\qquad\qquad\qquad [\theta = e^{-2t}(0.3\cos 6t + 0.1\sin 6t)]$

17  Solve the differential equation $\dfrac{d^2y}{dx^2} - n^2 y = 0$, where $n$ is a constant, and show that $y = y_0\cosh nx + \dfrac{3}{n}\sinh nx$, given that when $x = 0$, $y = y_0$ and $\dfrac{dy}{dx} = 3$.

18  $L\dfrac{d^2i}{dt^2} + R\dfrac{di}{dt} + \dfrac{1}{C}i = 0$ is an equation representing current $i$ in an electric circuit. If inductance $L$ is 0.25 henry, capacitance $C$ is $29.76 \times 10^{-6}$ farads and $R$ is 250 ohms, solve the equation for $i$ given the boundary conditions that when $t = 0$, $i = 0$ and $\dfrac{di}{dt} = 34$.

$\qquad\qquad\qquad\qquad\qquad\qquad [i = \dfrac{1}{20}(e^{-160t} - e^{-840t})]$

## A. MAIN POINTS CONCERNED WITH DIFFERENTIAL EQUATIONS OF THE

FORM $a\dfrac{d^2y}{dx^2} + b\dfrac{dy}{dx} + cy = f(x)$

1    If in the differential equation $a\dfrac{d^2y}{dx^2} + b\dfrac{dy}{dx} + cy = f(x)$      (1)

the substitution $y = u+v$ is made then:

$$a\frac{d^2(u+v)}{dx^2} + b\frac{d(u+v)}{dx} + c(u+v) = f(x)$$

Rearranging gives: $\left(a\dfrac{d^2u}{dx^2} + b\dfrac{du}{dx} + cu\right)+\left(a\dfrac{d^2v}{dx^2} + b\dfrac{dv}{dx} + cv\right) = f(x)$

If we let $a\dfrac{d^2v}{dx^2} + b\dfrac{dv}{dx} + cv = f(x)$      (2)

then $a\dfrac{d^2u}{dx^2} + b\dfrac{du}{dx} + cu = 0$      (3)

2    The general solution, $u$, of equation (3) will contain two unknown constants, as required for the general solution of equation (1). The method of solution of equation (3) is shown in chapter 15. The function $u$ is called the **complementary function**, **(C.F.)**.

3    If the particular solution, $v$, of equation (2) can be determined without containing any unknown constants then $y = u+v$ will give the general solution of equation (1). The function $v$ is called the **particular integral**, **(P.I.)**. Hence the general solution of equation (1) is given by: $y = $ **C.F.+P.I.**

4    **Procedure to solve differential equations of the form** $a\dfrac{d^2y}{dx^2} + b\dfrac{dy}{dx} + cy = f(x)$.

 (i)   Rewrite the given differential equation as $(aD^2+bD+c)y = f(x)$.

 (ii)  Substitute $m$ for $D$, and solve the auxiliary equation $am^2+bm+c = 0$ for $m$.

 (iii) Obtain the complementary function, $u$, which is achieved using the same procedure as in para. 6(c) of chapter 15.

 (iv)  To determine the particular integral, $v$, firstly assume a particular integral which is suggested by $f(x)$, but which contains undetermined coefficients. *Table 1* gives some suggested substitutions for different functions $f(x)$.

 (v)   Substitute the suggested P.I. into the differential equation $(aD^2+bD+c)v = f(x)$ and equate relevant coefficients to find the constants introduced.

**TABLE 1.** Form of particular integral for different functions

| Type | Straightforward cases<br>Try as particular integral: | 'Snag' cases<br>Try as particular integral: | See problem |
|---|---|---|---|
| (a) $f(x) = $ a constant | $v = k$ | $v = kx$ (used when C.F. contains a constant) | 1, 2 |
| (b) $f(x) = $ polynomial (i.e., $f(x) = L+Mx+Nx^2+\dots$ where any of the coefficients may be zero) | $v = a+bx+cx^2+\dots\dots$ | | 3 |
| (c) $f(x) = $ an exponential function (i.e., $f(x) = Ae^{\alpha x}$ | $v = ke^{\alpha x}$ | (i) $v = kxe^{\alpha x}$ (used when $e^{\alpha x}$ appears in the C.F.).<br>(ii) $v = kx^2 e^{\alpha x}$ (used when $e^{\alpha x}$ and $xe^{\alpha x}$ both appear in the C.F.), etc. | 4, 5<br><br>6 |
| (d) $f(x) = $ a sine or cosine function (i.e., $f(x) = a \sin px + b \cos px$, where $a$ or $b$ may be zero). | $v = A \sin px + B \cos px$ | $v = x (A \sin px + B \cos px)$ (used when $\sin px$ and/or $\cos px$ appears in the C.F.). | 7, 8 |
| (e) $f(x) = $ a sum e.g.<br>(i) $f(x) = 4x^2 - 3 \sin 2x$<br>(ii) $f(x) = 2 - x + e^{3x}$ | (i) $v = ax^2 + bx + c + d \sin 2x + e \cos 2x$<br>(ii) $v = ax + b + ce^{3x}$ | | 9 |
| (f) $f(x) = $ a product e.g. $f(x) = 2e^x \cos 2x$ | $v = e^x (A \sin 2x + B \cos 2x)$ | | 10 |

(vi) The general solution is given by $y = $ C.F.+P.I., i.e. $y = u+v$.

(vii) Given boundary conditions, arbitrary constants in the C.F. may be determined and the particular solution of the differential equation obtained.

## B. WORKED PROBLEMS ON DIFFERENTIAL EQUATIONS OF THE FORM

$$a \frac{d^2 y}{dx^2} + b \frac{dy}{dx} + cy = f(x)$$

*Problem 1* Solve the differential equation $\dfrac{d^2 y}{dx^2} + \dfrac{dy}{dx} - 2y = 4$.

Using the procedure of para. 4:

(i) $\dfrac{d^2y}{dx^2} + \dfrac{dy}{dx} - 2y = 4$ in $D$-operator form is $(D^2+D-2)y = 4$.

(ii) Substituting $m$ for $D$ gives the auxiliary equation $m^2+m-2 = 0$.
Factorising gives $(m-1)(m+2) = 0$, from which $m = 1$ or $m = -2$.

(iii) Since the roots are real and different, the C.F., $u = Ae^x + Be^{-2x}$.

(iv) Since the term on the right-hand side of the given equation is a constant, i.e., $f(x) = 4$, let the P.I. also be a constant, say $v = k$ (see *Table 1(a)*).

(v) Substituting $v = k$ into $(D^2+D-2)v = 4$ gives $(D^2+D-2)k = 4$.
Since $D(k) = 0$ and $D^2(k) = 0$ then $-2k = 4$, from which, $k = -2$.
Hence the P.I., $v = -2$.

(vi) The general solution is given by $y = u+v$, i.e. $y = Ae^x + Be^{-2x} - 2$.

---

**Problem 2** Determine the particular solution of the equation $\dfrac{d^2y}{dx^2} - 3\dfrac{dy}{dx} = 9$, given the boundary conditions that when $x = 0$, $y = 0$ and $\dfrac{dy}{dx} = 0$.

Using the procedure of para. 4:

(i) $\dfrac{d^2y}{dx^2} - 3\dfrac{dy}{dx} = 9$ in $D$-operator form is $(D^2-3D)y = 9$.

(ii) Substituting $m$ for $D$ gives the auxiliary equation $m^2-3m = 0$.
Factorising gives: $m(m-3) = 0$, from which, $m = 0$ or $m = 3$.

(iii) Since the roots are real and different, the C.F.
$u = Ae^0 + Be^{3x}$, i.e. $u = A+Be^{3x}$.

(iv) Since the C.F. contains a constant (i.e. $A$) then let the P.I., $v=kx$ (see *Table 1(a)*).

(v) Substituting $v = kx$ into $(D^2-3D)v = 9$ gives $(D^2-3D)kx = 9$.
$D(kx) = k$ and $D^2(kx) = 0$.
Hence $(D^2-3D)kx = 0-3k = 9$, from which, $k = -3$.
Hence the P.I., $v = -3x$.

(vi) The general solution is given by $y = u+v$, i.e. $y = A+Be^{3x} - 3x$.

(vii) When $x = 0$, $y = 0$ then $0 = A+Be^0 - 0$, i.e. $0 = A+B$ \hfill (1)
$\dfrac{dy}{dx} = 3Be^{3x} - 3$; $\dfrac{dy}{dx} = 0$ when $x = 0$, thus $0 = 3Be^0 - 3$,

from which, $B = 1$. From equation (1), $A = -1$
Hence the particular solution is $y = -1+1e^{3x} - 3x$, i.e., $y = e^{3x} - 3x - 1$.

---

**Problem 3** Solve the differential equation $2\dfrac{d^2y}{dx^2} - 11\dfrac{dy}{dx} + 12y = 3x-2$.

Using the procedure of para. 4:

(i) $2\dfrac{d^2y}{dx^2} - 11\dfrac{dy}{dx} + 12y = 3x-2$ in $D$-operator form is $(2D^2-11D+12)y = 3x-2$.

(ii) Substituting $m$ for $D$ gives the auxiliary equation $2m^2-11m+12 = 0$.
Factorising gives: $(2m-3)(m-4) = 0$, from which, $m = \dfrac{3}{2}$ or $m = 4$.

(iii) Since the roots are real and different, the C.F., $u = Ae^{\frac{3}{2}x} + Be^{4x}$

(iv) Since $f(x) = 5x-2$ is a polynomial let the P.I., $v = ax+b$, (see *Table 1(b)*).

132

(v) Substituting $v = ax+b$ into $(2D^2 -11D+12)\,v = 3x-2$ gives:

$(2D^2 -11D+12)(ax+b) = 3x-2$,

i.e. $2D^2(ax+b)-11D(ax+b)+12(ax+b) = 3x-2$

i.e. $0-11a+12ax+12b = 3x-2$

Equating the coefficients of $x$ gives: $12a = 3$, from which, $a = 4$.

Equating the constant terms gives: $-11a+12b = -2$, i.e. $-44+12b = -2$

from which, $12b = 42$, i.e., $b = \dfrac{7}{2}$.

Hence the P.I., $v = ax+b = 4x+\dfrac{7}{2}$.

(vi) The general solution is given by $y = u+v$, i.e., $y = Ae^{\frac{3}{2}x} +Be^{4x} +4x + \dfrac{7}{2}$.

---

*Problem 4* Solve the equation $\dfrac{d^2 y}{dx^2} - 2\dfrac{dy}{dx} + y = 3e^{4x}$ given the boundary conditions that when $x = 0$, $y = -\dfrac{2}{3}$ and $\dfrac{dy}{dx} = 4\dfrac{1}{3}$.

---

Using the procedure of para. 4:

(i) $\dfrac{d^2 y}{dx^2} -2\dfrac{dy}{dx} +y = 3e^{4x}$ in D-operator form is $(D^2 -2D+1)y = 3e^{4x}$.

(ii) Substituting $m$ for $D$ gives the auxiliary equation $m^2 -2m+1 = 0$.
Factorising gives: $(m-1)(m-1) = 0$, from which, $m = 1$ twice.

(iii) Since the roots are real and equal the C.F., $u = (Ax+B)e^x$.

(iv) Let the particular integral, $v = ke^{4x}$ (see *Table 1(c)*).

(v) Substituting $v = ke^{4x}$ into $(D^2 -2D+1)v = 3e^{4x}$ gives

$(D^2 -2D+1)ke^{4x} = 3e^{4x}$,

i.e. $D^2 (ke^{4x})-2D(ke^{4x})+1(ke^{4x}) = 3e^{4x}$

i.e. $16ke^{4x} -8ke^{4x} +ke^{4x} = 3e^{4x}$

Hence $9ke^{4x} = 3e^{4x}$, from which, $k = \dfrac{1}{3}$

Hence the P.I., $v = ke^{4x} = \dfrac{1}{3}e^{4x}$

(vi) The general solution is given by $y = u+v$, i.e., $y = (Ax+B)e^x + \dfrac{1}{3}e^{4x}$

(vii) When $x = 0$, $y = -\dfrac{2}{3}$, thus $-\dfrac{2}{3}= (0+B)e^0 + \dfrac{1}{3}e^0$, from which, $B = -1$.

$\dfrac{dy}{dx} = (Ax+B)\,e^x +e^x(A)+ \dfrac{4}{3}e^{4x}$

When $x = 0$, $\dfrac{dy}{dx} = 4\dfrac{1}{3}$, thus $\dfrac{13}{3} = B+A + \dfrac{4}{3}$

from which, $A = 4$, since $B = -1$.

**Hence the particular solution is $y = (4x-1)e^x + \dfrac{1}{3}e^{4x}$.**

---

*Problem 5* Solve the differential equation $2\dfrac{d^2 y}{dx^2} - \dfrac{dy}{dx} -3y = 5e^{\frac{3}{2}x}$.

Using the procedure of para. 4:

(i) $2\dfrac{d^2y}{dx^2} - \dfrac{dy}{dx} - 3y = 5e^{\frac{3}{2}x}$ in D-operator form is $(2D^2 - D - 3)y = 5e^{\frac{3}{2}x}$.

(ii) Substituting $m$ for $D$ gives the auxiliary equation $2m^2 - m - 3 = 0$.

Factorising gives: $(2m-3)(m+1) = 0$, from which, $m = \dfrac{3}{2}$ or $m = -1$.

(iii) Since the roots are real and different then the C.F., $u = Ae^{\frac{3}{2}x} + Be^{-x}$.

(iv) Since $e^{\frac{3}{2}x}$ appears in the C.F. **and** in the right-hand side of the differential equation, let the P.I., $v = kxe^{\frac{3}{2}x}$ (see *Table 1(c)*, snag case (i)).

(v) Substituting $v = kxe^{\frac{3}{2}x}$ into $(2D^2 - D - 3)v = 5e^{\frac{3}{2}x}$ gives:

$(2D^2 - D - 3)\, kxe^{\frac{3}{2}x} = 5e^{\frac{3}{2}x}$

$D(kxe^{\frac{3}{2}x}) = (kx)\left(\dfrac{3}{2}e^{\frac{3}{2}x}\right) + (e^{\frac{3}{2}x})(k)$, by the product rule,

$\qquad = ke^{\frac{3}{2}x}\left(\dfrac{3}{2}x + 1\right)$

$D^2(kxe^{\frac{3}{2}x}) = D\left[ke^{\frac{3}{2}x}\left(\dfrac{3}{2}x+1\right)\right] = (ke^{\frac{3}{2}x})\left(\dfrac{3}{2}\right) + \left(\dfrac{3}{2}x+1\right)\left(\dfrac{3}{2}ke^{\frac{3}{2}x}\right)$

$\qquad\qquad\qquad\qquad = ke^{\frac{3}{2}x}\left(\dfrac{9}{4}x+3\right)$

Hence $(2D^2 - D - 3)(kxe^{\frac{3}{2}x}) = 2\left[ke^{\frac{3}{2}x}\left(\dfrac{9}{4}x+3\right)\right] - \left[ke^{\frac{3}{2}x}\left(\dfrac{3}{2}x+1\right)\right] - 3\left[kxe^{\frac{3}{2}x}\right] = 5e^{\frac{3}{2}x}$

i.e., $\dfrac{9}{2}kxe^{\frac{3}{2}x} + 6ke^{\frac{3}{2}x} - \dfrac{3}{2}xke^{\frac{3}{2}x} - ke^{\frac{3}{2}x} - 3kxe^{\frac{3}{2}x} = 5e^{\frac{3}{2}x}$

Equating coefficients of $e^{\frac{3}{2}x}$ gives: $5k = 5$, from which, $k = 1$.

Hence the P.I., $v = kxe^{\frac{3}{2}x} = xe^{\frac{3}{2}x}$

(vi) The general solution is $y = u + v$, i.e. $y = Ae^{\frac{3}{2}x} + Be^{-x} + xe^{\frac{3}{2}x}$

---

*Problem 6* Solve $\dfrac{d^2y}{dx^2} - 4\dfrac{dy}{dx} + 4y = 3e^{2x}$.

Using the procedure of para. 4:

(i) $\dfrac{d^2y}{dx^2} - 4\dfrac{dy}{dx} + 4y = 3e^{2x}$ in D-operator form is $(D^2 - 4D + 4)y = 3e^{2x}$

(ii) Substituting $m$ for $D$ gives the auxiliary equation $m^2 - 4m + 4 = 0$.

Factorising gives: $(m-2)(m-2) = 0$, from which, $m = 2$ twice.

(iii) Since the roots are real and equal, the C.F., $u = (Ax+B)e^{2x}$.

(iv) Since $e^{2x}$ and $xe^{2x}$ both appear in the C.F. let the P.I., $v = kx^2e^{2x}$ (see *Table 1(c)*, snag case (ii)).

(v) Substituting $v = kx^2e^{2x}$ into $(D^2 - 4D + 4)v = 3e^{2x}$ gives:

$(D^2 - 4D + 4)(kx^2e^{2x}) = 3e^{2x}$.

$D(kx^2e^{2x}) = (kx^2)(2e^{2x}) + (e^{2x})(2kx) = 2ke^{2x}(x^2 + x)$

$D^2(kx^2e^{2x}) = D[2ke^{2x}(x^2 + x)] = (2ke^{2x})(2x+1) + (x^2+x)(4ke^{2x})$

$\qquad\qquad\qquad\qquad = 2ke^{2x}(4x+1+2x^2)$

Hence $(D^2-4D+4)(kx^2e^{2x}) = [2ke^{2x}(4x+1+2x^2)]-4[2ke^{2x}(x^2+x)]$
$$+4[kx^2e^{2x}] = 3e^{2x}$$

from which, $2ke^{2x} = 3e^{2x}$ and $k = \dfrac{3}{2}$

Hence the P.I., $v = kx^2e^{2x} = \dfrac{3}{2}x^2e^{2x}$.

(vi) The general solution, $y = u+v$, i.e. $y = (Ax+B)e^{2x} + \dfrac{3}{2}x^2e^{2x}$.

---

*Problem 7* Solve the differential equation $2\dfrac{d^2y}{dx^2} +3\dfrac{dy}{dx} -5y = 6\sin 2x$.

Using the procedure of para. 4:

(i) $2\dfrac{d^2y}{dx^2} +3\dfrac{dy}{dx} -5y = 6\sin 2x$ in $D$-operator form is $(2D^2+3D-5)y = 6\sin 2x$.

(ii) The auxiliary equation is $2m^2+3m-5 = 0$, from which,

$(m-1)(2m+5) = 0$, i.e., $m = 1$ or $m = -\dfrac{5}{2}$.

(iii) Since the roots are real and different the C.F., $u = Ae^x +Be^{-\frac{5}{2}x}$.

(iv) Let the P.I., $v = C\sin 2x+D\cos 2x$ (see *Table 1(d)*).

(v) Substituting $v = C\sin 2x+D\cos 2x$ into $(2D^2+3D-5)v = 6\sin 2x$ gives:

$(2D^2+3D-5)(C\sin 2x+D\cos 2x) = 6\sin 2x$.

$D(C\sin 2x+D\cos 2x) = 2C\cos 2x-2D\sin 2x$

$D^2(C\sin 2x+D\cos 2x) = D(2C\cos 2x-2D\sin 2x)$

$\qquad = -4C\sin 2x-4D\cos 2x$

Hence $(2D^2+3D-5)(C\sin 2x+D\cos 2x) = -8C\sin 2x-8D\cos 2x$

$\qquad\qquad +6C\cos 2x-6D\sin 2x-5C\sin 2x-5D\cos 2x$

$\qquad\qquad = 6\sin 2x$

Equating coefficient of $\sin 2x$ gives:　　　$-13C-6D = 6$　　　　(1)

Equating coefficients of $\cos 2x$ gives:　　　$6C-13D = 0$　　　　(2)

$6 \times (1)$ gives:　　$-78C-36D = 36$　　　　(3)

$13 \times (2)$ gives:　　$78C-169D = 0$　　　　(4)

$(3)+(4)$ gives:　　$-205D = 36$

from which,　　$D = \dfrac{-36}{205}$

Substituting $D = \dfrac{-36}{205}$ into equation (1) or (2) gives $C = \dfrac{-78}{205}$

Hence the P.I., $v = \dfrac{-78}{205}\sin 2x-\dfrac{36}{205}\cos 2x$.

(vi) The general solution, $y = u+v$, i.e. $y = Ae^x +Be^{-\frac{5}{2}x} -\dfrac{2}{205}(39\sin 2x+18\cos 2x)$

---

*Problem 8* Solve $\dfrac{d^2y}{dx^2} +16y = 10\cos 4x$ given $y = 3$ and $\dfrac{dy}{dx} = 4$ when $x = 0$.

Using the procedure of para. 4:

(i) $\dfrac{d^2y}{dx^2} +16y = 10\cos 4x$ in $D$-operator form is $(D^2+16)y = 10\cos 4x$.

(ii) The auxiliary equation is $m^2+16 = 0$, from which $m = \sqrt{-16} = \pm j4$.

135

(iii) Since the roots are complex the C.F., $u = e^0 (C \cos 4x + D \sin 4x)$

i.e., $u = C \cos 4x + D \sin 4x$.

(iv) Since $\sin 4x$ occurs in the C.F. and in the right-hand side of the given differential equation, let the P.I., $v = x(A \sin 4x + B \cos 4x)$, (see *Table 1(d)*, snag case).

(v) Substituting $v = x(A \sin 4x + B \cos 4x)$ into $(D^2 + 16)v = 10 \cos 4x$ gives:

$(D^2 + 16)[x(A \sin 4x + B \cos 4x)] = 10 \cos 4x$.

$D[x(A \sin 4x + B \cos 4x)] = x(4A \cos 4x - 4B \sin 4x)$

$\qquad + (A \sin 4x + B \cos 4x)(1)$, by the product rule.

$D^2[x(A \sin 4x + B \cos 4x)] = x(-16A \sin 4x - 16B \cos 4x)$

$\qquad + (4A \cos 4x - 4B \sin 4x) + (4A \cos 4x - 4B \sin 4x)$

Hence $(D^2 + 16)[x(A \sin 4x + B \cos 4x)]$

$= -16 Ax \sin 4x - 16Bx \cos 4x + 4A \cos 4x - 4B \sin 4x + 4A \cos 4x$

$\quad -4B \sin 4x + 16Ax \sin 4x + 16Bx \cos 4x = 10 \cos 4x$,

i.e., $-8B \sin 4x + 8A \cos 4x = 10 \cos 4x$.

Equating coefficients of $\cos 4x$ gives: $8A = 10$, from which $A = \dfrac{10}{8} = \dfrac{5}{4}$

Equating coefficients of $\sin 4x$ gives: $-8B = 0$, from which $B = 0$.

Hence the P.I., $v = x \left(\dfrac{5}{4} \sin 4x\right)$.

(vi) The general solution, $y = u + v$, i.e., $y = C \cos 4x + D \sin 4x + \dfrac{5}{4}x \sin 4x$.

(vii) When $x = 0$, $y = 3$, thus $3 = C \cos 0 + D \sin 0 + 0$, i.e., $C = 3$.

$\dfrac{dy}{dx} = -4C \sin 4x + 4D \cos 4x + \dfrac{5}{4}x(4 \cos 4x) + \dfrac{5}{4} \sin 4x$.

When $x = 0$, $\dfrac{dy}{dx} = 4$, thus $4 = -4C \sin 0 + 4D \cos 0 + 0 + \dfrac{5}{4} \sin 0$

i.e. $4 = 4D$, from which $D = 1$.

**Hence the particular solution is $y = 3 \cos 4x + \sin 4x + \dfrac{5}{4}x \sin 4x$.**

*Problem 9* Solve $\dfrac{d^2 y}{dx^2} + \dfrac{dy}{dx} - 6y = 12x - 50 \sin x$.

Using the procedure of para. 4:

(i) $\dfrac{d^2 y}{dx^2} + \dfrac{dy}{dx} - 6y = 12x - 50 \sin x$ in $D$-operator form is $(D^2 + D - 6)y = 12x - 50 \sin x$.

(ii) The auxiliary equation is $(m^2 + m - 6) = 0$, from which, $(m - 2)(m + 3) = 0$ i.e. $m = 2$ or $m = -3$.

(iii) Since the roots are real and different, the C.F., $u = Ae^{2x} + Be^{-3x}$.

(iv) Since the right-hand side of the given differential equation is the sum of a polynomial and a sine function let the P.I.

$v = ax + b + C \sin x + D \cos x$ (see *Table 1(e)*).

(v) Substituting $v$ into $(D^2 + D - 6)v = 12x - 50 \sin x$ gives:

$(D^2 + D - 6)(ax + b + C \sin x + D \cos x) = 12x - 50 \sin x$

$D(ax + b + C \sin x + D \cos x) = a + C \cos x - D \sin x$

$D^2(ax + b + C \sin x + D \cos x) = -C \sin x - D \cos x$

Hence $(D^2 + D - 6)(v) = (-C \sin x - D \cos x) + (a + C \cos x - D \sin x)$

$\qquad\qquad -6(ax + b + C \sin x + D \cos x) = 12x - 50 \sin x$

Equating constant terms gives: $a - 6b = 0$ $\qquad\qquad\qquad$ (1)

Equating coefficients of $x$ gives: $-6a = 12$, from which $a = -2$.

136

Hence, from (1), $b = -\dfrac{1}{3}$

Equating the coefficients of $\cos x$ gives: $-D+C-6D = 0$

$$\text{i.e.} \quad C-7D = 0 \tag{2}$$

Equating the coefficients of $\sin x$ gives: $-C-D-6C = -50$

$$\text{i.e.} \quad -7C-D = -50 \tag{3}$$

Solving equations (2) and (3) gives: $\quad C = 7$ and $D = 1$

Hence the P.I., $v = -2x - \dfrac{1}{3} + 7 \sin x + \cos x$

(vi) The general solution, $y = u+v$, i.e., $y = Ae^{2x} + Be^{-3x} - 2x - \dfrac{1}{3} + 7 \sin x + \cos x$.

---

**Problem 10** Solve the differential equation $\dfrac{d^2 y}{dx^2} - 2\dfrac{dy}{dx} + 2y = 3e^x \cos 2x$, given that when $x = 0$, $y = 2$ and $\dfrac{dy}{dx} = 5$.

---

Using the procedure of para. 4:

(i) $\dfrac{d^2 y}{dx^2} - 2\dfrac{dy}{dx} + 2y = 3e^x \cos 2x$ in $D$-operator form is $(D^2 - 2D + 2)y = 3e^x \cos 2x$.

(ii) The auxiliary equation is $m^2 - 2m + 2 = 0$.

Using the quadratic formula, $m = \dfrac{2 \pm \sqrt{[4-4(1)(2)]}}{2} = \dfrac{2 \pm \sqrt{-4}}{2} = \dfrac{2 \pm j2}{2}$

$$\text{i.e., } m = 1 \pm j1.$$

(iii) Since the roots are complex, the C.F., $u = e^x(C \cos x + D \sin x)$.

(iv) Since the right-hand side of the given differential equation is a product of an exponential and a cosine function, let the P.I., $v = e^x(A \sin 2X + B \cos 2x)$ (see *Table 1(f)*).

(v) Substituting $v$ into $(D^2 - 2D + 2)v = 3e^x \cos 2x$ gives:

$(D^2 - 2D + 2)[e^x(A \sin 2x + B \cos 2x)] = 3e^x \cos 2x$

$D(v) = e^x(2A \cos 2x - 2B \sin 2x) + e^x(A \sin 2x + B \cos 2x)$

$\qquad (\equiv e^x\{(2A+B) \cos 2x + (A-2B) \sin 2x\})$

$D^2(v) = e^x(-4A \sin 2x - 4B \cos 2x) + e^x(2A \cos 2x - 2B \sin 2x)$

$\qquad\qquad + e^x(2A \cos 2x - 2B \sin 2x) + e^x(A \sin 2x + B \cos 2x)$

$\qquad \equiv e^x\{(-3A-4B) \sin 2x + (4A-3B) \cos 2x\}$

Hence $(D^2 - 2D + 2)v = e^x\{(-3A-4B) \sin 2x + (4A-3B) \cos 2x\}$
$\qquad -2e^x\{(2A+B) \cos 2x + (A-2B) \sin 2x\} + 2e^x(A \sin 2x + B \cos 2x)$
$\qquad = 3e^x \cos 2x.$

Equating coefficients of $e^x \sin 2x$ gives: $-3A-4B-2A+4B+2A = 0$

$\qquad\qquad \text{i.e. } -3A = 0$, from which, $A = 0$

Equating coefficients of $e^x \cos 2x$ gives: $4A-3B-4A-2B+2B = 3$

$\qquad\qquad \text{i.e. } -3B = 3$, from which, $B = -1$

Hence the P.I., $v = e^x(-\cos 2x)$.

(vi) The general solution, $y = u+v$, i.e. $y = e^x(C \cos x + D \sin x) - e^x \cos 2x$.

(vii) When $x = 0$, $y = 2$ thus $2 = e^0(C \cos 0 + D \sin 0) - e^0 \cos 0$

i.e. $2 = C-1$, from which, $C = 3$

$\dfrac{dy}{dx} = e^x(-C \sin x + D \cos x) + e^x(C \cos x + D \sin x) - e^x(-2 \sin 2x) - e^x \cos 2x$

When $x = 0$, $\dfrac{dy}{dx} = 3$ thus $3 = e^0 \, (-C \sin 0 + D \cos 0)$

$\qquad\qquad\qquad\qquad + e^0 (C \cos 0 + D \sin 0) - e^0 (-2 \sin 0) - e^0 \cos 0$

i.e., $3 = D + C - 1$, from which, $D = 1$, since $C = 3$

**Hence the particular solution is $y = e^x (3 \cos x + \sin x) - e^x \cos 2x$**

## C. FURTHER PROBLEMS ON DIFFERENTIAL EQUATIONS OF THE FORM

$a \dfrac{d^2 y}{dx^2} + b \dfrac{dy}{dx} + cy = f(x)$

In *Problems 1 to 18*, find the general solutions of the given differential equations.

1   $2 \dfrac{d^2 y}{dx^2} + 5 \dfrac{dy}{dx} - 3y = 6$      $[y = A e^{\frac{1}{2}x} + B e^{-3x} - 2]$

2   $\dfrac{d^2 y}{dx^2} + 2 \dfrac{dy}{dx} + 2y = 8$      $[y = e^{-x}(C \cos x + D \sin x) + 4]$

3   $4 \dfrac{d^2 y}{dx^2} - 4 \dfrac{dy}{dx} + y = x$      $[y = (Ax + B) e^{\frac{1}{2}x} + 4 + x]$

4   $6 \dfrac{d^2 y}{dx^2} + 4 \dfrac{dy}{dx} - 2y = 3x - 2$      $[y = A e^{\frac{1}{3}x} + B e^{-x} - 2 - \frac{3}{2}x]$

5   $\dfrac{d^2 y}{dx^2} + 4y = 2x^2 + x - 5$      $[y = C \cos 2x + D \sin 2x - \frac{3}{2} + \frac{1}{4}x + \frac{1}{2}x^2]$

6   $\dfrac{d^2 y}{dx^2} - \dfrac{dy}{dx} - 6y = 2e^x$      $[y = A e^{3x} + B e^{-2x} - \frac{1}{3} e^x]$

7   $\dfrac{d^2 y}{dx^2} - 3 \dfrac{dy}{dx} - 4y = 3e^{-x}$      $[y = A e^{4x} + B e^{-x} - \frac{3}{5} x e^{-x}]$

8   $\dfrac{d^2 y}{dx^2} + 9y = 26e^{2x}$      $[y = C \cos 3x + D \sin 3x + 2 e^{2x}]$

9   $9 \dfrac{d^2 y}{dt^2} - 6 \dfrac{dy}{dt} + y = 12 e^{\frac{t}{3}}$      $[y = (At + B) e^{\frac{1}{3}t} + \frac{2}{3} t^2 e^{\frac{1}{3}t}]$

10   $2 \dfrac{d^2 y}{dx^2} - \dfrac{dy}{dx} - 3y = 25 \sin 2x$      $[y = A e^{\frac{3}{2}x} + B e^{-x} - \frac{1}{5}(11 \sin 2x - 2 \cos 2x)]$

11   $\dfrac{d^2 y}{dx^2} - 4 \dfrac{dy}{dx} + 4y = 5 \cos x$      $[y = (Ax + B) e^{2x} - \frac{4}{5} \sin x + \frac{3}{5} \cos x]$

12   $\dfrac{d^2 y}{dx^2} + 25y = 10 \sin 5x$      $[y = C \cos 5x + D \sin 5x - x \cos 5x]$

13   $\dfrac{d^2 y}{dx^2} + y = 4 \cos x$      $[y = C \cos x + D \sin x + 2x \sin x]$

14   $8 \dfrac{d^2 y}{dx^2} - 6 \dfrac{dy}{dx} + y = 2x + 40 \sin x$      $[y = A e^{\frac{x}{4}} + B e^{\frac{x}{2}} + 2x + 12 + \frac{8}{17}(6 \cos x - 7 \sin x)]$

15   $\dfrac{d^2 y}{d\theta^2} - 3 \dfrac{dy}{d\theta} + 2y = 2 \sin 2\theta - 4 \cos 2\theta$.      $[y = A e^{2\theta} + B e^{\theta} + \frac{1}{2}(\sin 2\theta + \cos 2\theta)]$

16   $\dfrac{d^2 y}{dx^2} + \dfrac{dy}{dx} - 2y = x^2 + e^{2x}$.      $\left[ y = A e^x + B e^{-2x} - \frac{3}{4} - \frac{1}{2}x - \frac{1}{2}x^2 + \frac{1}{4} e^{2x} \right]$

17   $\dfrac{d^2 y}{dx^2} - \dfrac{dy}{dx} - 2y = e^x - 2 \cos x$.      $\left[ y = A e^{2x} + B e^{-x} - \frac{1}{2} e^x + \frac{1}{5}(\sin x + 3 \cos x) \right]$

18   $\dfrac{d^2 y}{dt^2} - 2 \dfrac{dy}{d.} + 2y = e^t \sin t$      $\left[ y = e^t (C \cos t + D \sin t) - \frac{t}{2} e^t \cos t \right]$

In *Problems 19 to 25* find the particular solutions of the given differential equations.

19 $3\dfrac{d^2y}{dx^2} + \dfrac{dy}{dx} - 4y = 8$; when $x = 0$, $y = 0$ and $\dfrac{dy}{dx} = 0$. $\left[ y = \dfrac{2}{7}(3e^{-\frac{4}{3}x} + 4e^x) - 2 \right]$

20 $9\dfrac{d^2y}{dx^2} - 12\dfrac{dy}{dx} + 4y = 3x - 1$; when $x = 0$, $y = 0$ and $\dfrac{dy}{dx} = -\dfrac{4}{3}$

$$\left[ y = -(2 + \tfrac{3}{4}x)e^{\frac{2}{3}x} + 2 + \tfrac{3}{4}x \right]$$

21 $5\dfrac{d^2y}{dx^2} + 9\dfrac{dy}{dx} - 2y = 3e^x$; when $x = 0$, $y = \dfrac{1}{4}$ and $\dfrac{dy}{dx} = 0$.

$$\left[ y = \dfrac{5}{44}(e^{-2x} - e^{\frac{1}{5}x}) + \tfrac{1}{4}e^x \right]$$

22 $\dfrac{d^2y}{dt^2} - 6\dfrac{dy}{dt} + 9y = 4e^{3t}$; when $t = 0$, $y = 2$ and $\dfrac{dy}{dt} = 0$.  $[y = 2e^{3t}(1 - 3t + t^2)]$

23 $\dfrac{d^2y}{dx^2} - 3\dfrac{dy}{dx} - 4y = 3\sin x$; when $x = 0$, $y = 0$ and $\dfrac{dy}{dx} = 0$.

$$\left[ y = \dfrac{1}{170}(6e^{4x} - 51e^{-x}) - \dfrac{1}{34}(15\sin x - 9\cos x) \right]$$

24 $\dfrac{d^2y}{dx^2} - 7\dfrac{dy}{dx} + 10y = e^{2x} + 20$; when $x = 0$, $y = 0$ and $\dfrac{dy}{dx} = -\dfrac{1}{3}$

$$\left[ y = \dfrac{4}{3}e^{5x} - \dfrac{10}{3}e^{2x} - \dfrac{1}{3}xe^{2x} + 2 \right]$$

25 $2\dfrac{d^2y}{dx^2} - \dfrac{dy}{dx} - 6y = 6e^x\cos x$; when $x = 0$, $y = -\dfrac{21}{29}$ and $\dfrac{dy}{dx} = -6\dfrac{20}{29}$

$$\left[ y = 2e^{-\frac{3}{2}x} - 2e^{2x} + \dfrac{3e^x}{29}(3\sin x - 7\cos x) \right]$$

26 The equation of motion of a forced oscillation is $\dfrac{d^2y}{dt^2} + 2\dfrac{dy}{dt} + 5y = 4\sin 5t$.

Solve the equation given that when $t = 0$, $y = 0$ and $\dfrac{dy}{dt} = 0$.

$$\left[ y = \dfrac{e^{-t}}{25}(11\sin 2t + 2\cos 2t) - \dfrac{2}{25}(2\sin 5t + \cos 5t) \right]$$

27 The charge $q$ is an electric circuit at time $t$ satisfies the equation

$L\dfrac{d^2q}{dt^2} + R\dfrac{dq}{dt} + \dfrac{1}{C}q = E$, where $L$, $R$, $C$ and $E$ are constants. Solve the equation given $L = 2$, $C = 200 \times 10^{-6}$ and $E = 250$, when (a) $R = 200$ and (b) $R$ is

negligible. Assume that when $t = 0$, $q = 0$ and $\dfrac{dq}{dt} = 0$.

$$\left[ \text{(a) } q = \left( \dfrac{5}{2}t - \dfrac{1}{20} \right)e^{-50t} + \dfrac{1}{20} \text{; (b) } q = \dfrac{1}{20}(1 - \cos 50t) \right]$$

28 A differential equation representing the motion of a body is $\dfrac{d^2y}{dt^2} + n^2y = k\sin pt$,

where $k$, $n$ and $p$ are constants. Solve the equation (given $n \neq 0$ and $p^2 \neq n^2$) given

that when $t = 0$, $y = \dfrac{dy}{dt} = 0$.  $\left[ y = \dfrac{k}{n^2 - p^2}\left( \sin pt - \dfrac{p}{n}\sin nt \right) \right]$

29 In a galvanometer the deflection $\theta$ satisfies the differential equation

$\dfrac{d^2\theta}{dt^2} + 4\dfrac{d\theta}{dt} + 4\theta = 8$. Solve the equation for $\theta$ given that when $t = 0$, $\theta = \dfrac{d\theta}{dt} = 2$.

$$[\theta = 2(te^{-2t} + 1)]$$

30 The motion of a vibrating mass is given by $\frac{d^2y}{dt^2} + 8\frac{dy}{dt} + 20y = 300 \sin 4t$. Show

that the general solution of the differential equation is given by:

$$y = e^{-4t} (C \cos 2t + D \sin 2t) + \frac{15}{13} (\sin 4t - 8 \cos 4t).$$

31 $L\frac{d^2q}{dt^2} + R\frac{dq}{dt} + \frac{1}{C}q = V_0 \sin \omega t$ represents the variation of capacitor charge in

an electric circuit. Determine and expression for $q$ at time $t$ seconds given that
$R = 40, L = 0.02, C = 50 \times 10^{-6}, V_0 = 540.8$ and $\omega = 200$ and given the
boundary conditions that when $t = 0, q = 0$ and $\frac{dq}{dt} = 4.8$

$$[q = (10t + 0.01)e^{-1000t} + 0.024 \sin 200t - 0.010 \cos 200t]$$

# 17 Fourier series for periodic functions of period $2\pi$

## A. MAIN POINTS CONCERNED WITH FOURIER SERIES FOR PERIODIC FUNCTIONS OF PERIOD $2\pi$

1   **Fourier series** provides a method of analysing periodic functions into their constituent components. Alternating currents and voltages, displacement, velocity and acceleration of slider-crank mechanisms and acoustic waves are typical practical examples in engineering and science where periodic functions are involved and often requiring analysis.

2   A function $f(x)$ is said to be **periodic** if $f(x+T) = f(x)$ for all values of $x$, where $T$ is some positive number. $T$ is the interval between two successive repetitions and is called the **period** of the functions $f(x)$. For example, $y = \sin x$ is periodic in $x$ with period $2\pi$ since $\sin x = \sin(x+2\pi) = \sin(x+4\pi)$, and so on. Similarly, $y = \cos x$ is a periodic function with period $2\pi$ since $\cos x = \cos(x+2\pi) = \cos(x+4\pi)$, and so on. In general, if $y = \sin \omega t$ or $y = \cos \omega t$ then the period of the wave form is $2\pi/\omega$. The function shown in *Fig 1* is also periodic of period $2\pi$ and is defined by:

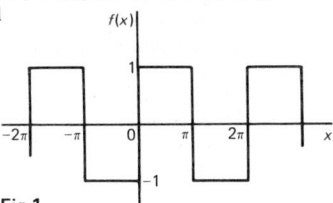

$$f(x) = \begin{cases} -1, & \text{when } -\pi < x < 0 \\ 1, & \text{when } 0 < x < \pi \end{cases}$$

**Fig 1**

3   If a graph of a function has no sudden jumps or breaks it is called a **continuous function**, examples being the graphs of sine and cosine functions. However, other graphs make finite jumps at a point or points in the interval. The square wave shown in *Fig 1* has **finite discontinuities** at $x = \pi$, $2\pi$, $3\pi$, and so on. A great advantage of Fourier series over other series is that it can be applied to functions which are discontinuous as well as those which are continuous.

4   (i) The basis of a Fourier series is that all functions of practical significance which are defined in the interval $-\pi \leqslant x \leqslant \pi$ can be expressed in terms of a convergent **trigonometric series** of the form :
$$f(x) = a_0 + a_1 \cos x + a_2 \cos 2x + a_3 \cos 3x + \ldots\ldots\ldots$$
$$+ b_1 \sin x + b_2 \sin 2x + b_3 \sin 3x + \ldots\ldots\ldots,$$
when $a_0, a_1, a_2, \ldots\ldots b_1, b_2, \ldots\ldots$ are real constants,

i.e., $f(x) = a_0 + \sum_{n=1}^{\infty} (a_n \cos nx + b_n \sin nx),$   (1)

where for the range $-\pi$ to $\pi$:

$$a_0 = \frac{1}{2\pi} \int_{-\pi}^{\pi} f(x)\, dx$$

$$a_n = \frac{1}{\pi} \int_{-\pi}^{\pi} f(x) \cos nx\, dx \quad (n = 1, 2, 3, \ldots\ldots)$$

and $b_n = \frac{1}{\pi} \int_{-\pi}^{\pi} f(x) \sin nx\, dx \quad (n = 1, 2, 3, \ldots\ldots)$

(ii) $a_0$, $a_n$ and $b_n$ are called the **Fourier coefficients** of the series and if these can be determined, the series of equation (1) is called the Fourier series corresponding to $f(x)$.

(iii) An alternative way of writing the series is by using the $a \cos x + b \sin x = c \sin(x+\alpha)$ relationship, i.e.,

$f(x) = a_0 + c_1 \sin(x+\alpha_1) + c_2 \sin(2x+\alpha_2) + \ldots\ldots + c_n \sin(nx+\alpha_n)$,

where $a_0$ is a constant, $c_1 = \sqrt{(a_1{}^2 + b_1{}^2)}, \ldots c_n = \sqrt{(a_n{}^2 + b_n{}^2)}$ are the amplitudes of the various components, and phase angle

$$\alpha_n = \arctan \frac{a_n}{b_n}.$$

(iv) For the series of equation (i):
the term $(a_1 \cos x + b_1 \sin x)$ or $c_1 \sin(x+\alpha_1)$ is called the **first harmonic** or the **fundamental**, the term $(a_2 \cos 2x + b_2 \sin 2x)$ or $c_2 \sin(2x+\alpha_2)$ is called the **second harmonic**, and so on.

5    For an exact representation of a complex wave, an infinite number of terms are, in general, required. In many practical cases, however, it is sufficient to take the first few terms only. (See *Problem 2*.)

6    The **sum of a Fourier series at a point of discontinuity** is given by the arithmetic mean of the two limiting values of $f(x)$ as $x$ approaches the point of discontinuity from the two sides. For example, for the waveform shown in *Fig 2*, the sum of the Fourier series at the points of discontinuity (i.e., at $0, \frac{\pi}{2}, \pi, \ldots\ldots$) is given by:

$$\frac{8+(-3)}{2} = \frac{5}{2} \text{ of } 2\frac{1}{2}$$

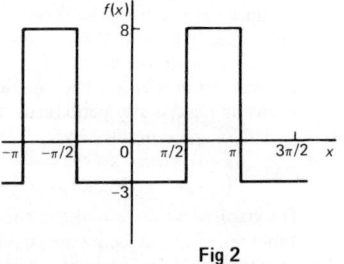

Fig 2

---

## B. WORKED PROBLEMS ON FOURIER SERIES OF PERIODIC FUNCTIONS OF PERIOD $2\pi$

*Problem 1* Obtain a Fourier series for the periodic function $f(x)$ defined as:

$$f(x) = \begin{cases} -k, & \text{when } -\pi < x < 0 \\ +k, & \text{when } 0 < x < \pi \end{cases}$$

The function is periodic outside of this range with period $2\pi$.

142

The square wave function defined is shown in *Fig 3*. Since $f(x)$ is given by two different expressions in the two halves of the range the integration is done in two parts, one from $-\pi$ to 0 and the other from 0 to $\pi$.

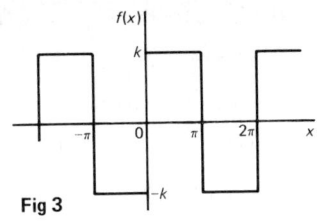

**Fig 3**

From para. 4(i): $a_0 = \dfrac{1}{2\pi} \displaystyle\int_{-\pi}^{\pi} f(x)\, dx = \dfrac{1}{2\pi}\left[\int_{-\pi}^{0} -k\, dx + \int_{0}^{\pi} k\, dx\right]$

$$= \frac{1}{2\pi}\left\{[-kx]_{-\pi}^{0} + [kx]_{0}^{\pi}\right\} = 0$$

($a_0$ is in fact the **mean value** of the waveform over a complete period of $2\pi$ and this could have been deduced on sight from *Fig 3*.)

From para. 4(i): $a_n = \dfrac{1}{\pi}\displaystyle\int_{-\pi}^{\pi} f(x)\cos nx\, dx = \dfrac{1}{\pi}\left\{\int_{-\pi}^{0} -k\cos nx\, dx + \int_{0}^{\pi} k\cos nx\, dx\right\}$

$$= \frac{1}{\pi}\left\{\left[\frac{-k\sin nx}{n}\right]_{-\pi}^{0} + \left[\frac{k\sin nx}{n}\right]_{0}^{\pi}\right\} = 0$$

Hence $a_1, a_2, a_3, \ldots\ldots$ are all zero and no cosine terms will appear in the Fourier series.

From para. 4(i): $b_n = \dfrac{1}{\pi}\displaystyle\int_{-\pi}^{\pi} f(x)\sin nx\, dx = \dfrac{1}{\pi}\left\{\int_{-\pi}^{0} -k\sin nx\, dx + \int_{0}^{\pi} k\sin nx\, dx\right\}$

$$= \frac{1}{\pi}\left\{\left[\frac{k\cos nx}{n}\right]_{-\pi}^{0} + \left[\frac{-k\cos nx}{n}\right]_{0}^{\pi}\right\}$$

When $n$ is odd: $b_n = \dfrac{k}{\pi}\left\{\left[\left(\dfrac{1}{n}\right) - \left(-\dfrac{1}{n}\right)\right] + \left[-\left(-\dfrac{1}{n}\right) - \left(-\dfrac{1}{n}\right)\right]\right\} = \dfrac{k}{\pi}\left\{\dfrac{2}{n} + \dfrac{2}{n}\right\}$,

i.e. $b_n = \dfrac{4k}{n\pi}$

Hence $b_1 = \dfrac{4k}{\pi}$, $b_3 = \dfrac{4k}{3\pi}$, $b_5 = \dfrac{4k}{5\pi}$, and so on.

When $n$ is even: $b_n = \dfrac{k}{\pi}\left\{\left[\dfrac{1}{n} - \dfrac{1}{n}\right] + \left[-\dfrac{1}{n} - \left(-\dfrac{1}{n}\right)\right]\right\} = 0$

Hence the Fourier series for the function shown in *Fig 3* is given by:

$$f(x) = \frac{4k}{\pi}\left(\sin x + \frac{1}{3}\sin 3x + \frac{1}{5}\sin 5x + \ldots\ldots\right)$$

*Problem 2* For the Fourier series of *Problem 1* let $k = \pi$. Show by plotting the first three partial sums of this Fourier series that as the series is added together term by term the result approximates more and more closely to the function it represents.

If $k = \pi$ in the Fourier series of *Problem 1* then:

$$f(x) = 4 \left( \sin x + \frac{1}{3} \sin 3x + \frac{1}{5} \sin 5x + \ldots \ldots \right).$$

$4 \sin x$ is termed the first partial sum of the Fourier series of $f(x)$,

$\left( 4 \sin x + \frac{4}{3} \sin 3x \right)$ is termed the second partial sum of the Fourier series, and

$\left( 4 \sin x + \frac{4}{3} \sin 3x + \frac{4}{5} \sin 5x \right)$ is termed the third partial sum, and so on. Let

$P_1 = 4 \sin x$, $P_2 = \left( 4 \sin x + \frac{4}{3} \sin 3x \right)$ and $P_3 = \left( 4 \sin x + \frac{4}{3} \sin 3x + \frac{4}{5} \sin 5x \right)$.

Graphs of $P_1$, $P_2$ and $P_3$, obtained by drawing up tables of values, and adding waveforms, are shown in *Figs 4(a) to (c)* and they show that the series is convergent, i.e., continually approximating towards a definite limit as more and more partial sums are taken, and in the limit will have the sum $f(x) = \pi$.

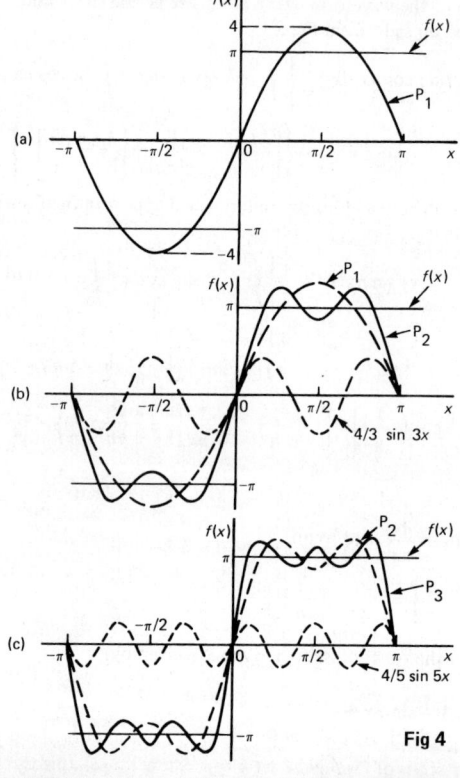

Fig 4

*Problem 3* If in the Fourier series of *Problem 1*, $k = 1$, deduce a series for $\frac{\pi}{4}$ at the point $x = \frac{\pi}{2}$.

If $k = 1$ in the Fourier series of *Problem 1*:

$$f(x) = \frac{4}{\pi} \left( \sin x + \frac{1}{3} \sin 3x + \frac{1}{5} \sin 5x + \ldots \ldots \right)$$

When $x = \frac{\pi}{2}$, $f(x) = 1$, $\sin x = \sin \frac{\pi}{2} = 1$, $\sin 3x = \sin \frac{3\pi}{2} = -1$,

$$\sin 5x = \sin \frac{5\pi}{2} = 1, \text{ and so on.}$$

Hence $1 = \frac{4}{\pi} \left[ 1 + \frac{1}{3}(-1) + \frac{1}{5}(1) + \frac{1}{7}(-1) + \ldots \ldots \right]$,

i.e., $\frac{\pi}{4} = 1 - \frac{1}{3} + \frac{1}{5} - \frac{1}{7} + \ldots \ldots \ldots$ .

*Problem 4* Determine the Fourier series for the full wave rectified sine wave $i = 5 \sin \frac{\theta}{2}$ shown in *Fig 5*.

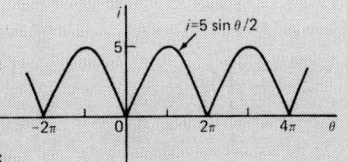

**Fig 5**

$i = 5 \sin \frac{\theta}{2}$ is a periodic function of period $2\pi$.

Thus $i = f(\theta) = a_0 + \sum\limits_{n=1}^{\infty} (a_n \cos n\theta + b_n \sin n\theta)$

In this case it is better to take the range 0 to $2\pi$ instead of $-\pi$ to $+\pi$ since the waveform is continuous between 0 and $2\pi$.

$$a_0 = \frac{1}{2\pi} \int_0^{2\pi} f(\theta)\, d\theta = \frac{1}{2\pi} \int_0^{2\pi} 5 \sin \frac{\theta}{2}\, d\theta = \frac{5}{2\pi} \left[ -2 \cos \frac{\theta}{2} \right]_0^{2\pi}$$

$$= \frac{5}{\pi} \left[ \left( -\cos \frac{2\pi}{2} \right) - (-\cos 0) \right] = \frac{5}{\pi} \left[ (1) - (-1) \right] = \frac{10}{\pi}$$

$$a_n = \frac{1}{\pi} \int_0^{2\pi} 5 \sin \frac{\theta}{2} \cos n\theta\, d\theta = \frac{5}{\pi} \int_0^{2\pi} \frac{1}{2} \left\{ \sin \left( \frac{\theta}{2} + n\theta \right) + \sin \left( \frac{\theta}{2} - n\theta \right) \right\}\, d\theta$$

(see chapter 10)

$$= \frac{5}{2\pi} \left[ \frac{-\cos \left[ \theta(\frac{1}{2} + n) \right]}{(\frac{1}{2} + n)} - \frac{\cos \left[ \theta(\frac{1}{2} - n) \right]}{(\frac{1}{2} - n)} \right]_0^{2\pi}$$

$$= \frac{5}{2\pi} \left\{ \left[ \frac{-\cos \left[ 2\pi(\frac{1}{2} + n) \right]}{(\frac{1}{2} + n)} - \frac{\cos \left[ 2\pi(\frac{1}{2} - n) \right]}{(\frac{1}{2} - n)} \right] - \left[ \frac{-\cos 0}{(\frac{1}{2} + n)} - \frac{\cos 0}{(\frac{1}{2} - n)} \right] \right\}$$

when $n$ is both odd and even, $a_n = \frac{5}{\pi} \left\{ \left[ \frac{1}{(\frac{1}{2} + n)} + \frac{1}{(\frac{1}{2} - n)} \right] - \left[ \frac{-1}{(\frac{1}{2} + n)} - \frac{-1}{(\frac{1}{2} - n)} \right] \right\}$

$$= \frac{5}{2\pi} \left\{ \frac{2}{(\frac{1}{2} + n)} + \frac{2}{(\frac{1}{2} - n)} \right\} = \frac{5}{\pi} \left\{ \frac{1}{(\frac{1}{2} + n)} + \frac{1}{(\frac{1}{2} - n)} \right\}$$

145

Hence $a_1 = \dfrac{5}{\pi}\left[\dfrac{1}{3/2} + \dfrac{1}{-1/2}\right] = \dfrac{5}{\pi}\left[\dfrac{2}{3} - \dfrac{2}{1}\right] = \dfrac{-20}{3\pi}$

$$a_2 = \dfrac{5}{\pi}\left[\dfrac{1}{5/2} + \dfrac{1}{-3/2}\right] = \dfrac{5}{\pi}\left[\dfrac{2}{5} - \dfrac{2}{3}\right] = \dfrac{-20}{(3)(5)\pi}$$

$$a_3 = \dfrac{5}{\pi}\left[\dfrac{1}{7/2} + \dfrac{1}{-5/2}\right] = \dfrac{5}{\pi}\left[\dfrac{2}{7} - \dfrac{2}{5}\right] = \dfrac{-20}{(5)(7)\pi}\ , \text{ and so on.}$$

$$b_n = \dfrac{1}{\pi}\int_0^{2\pi} 5\,\sin\dfrac{\theta}{2}\sin n\theta\ d\theta = \dfrac{5}{\pi}\int_0^{2\pi} -\dfrac{1}{2}\left\{\cos\left[\theta\left(\dfrac{1}{2}+n\right)\right] -\cos\left[\theta\left(\dfrac{1}{2}-n\right)\right]\right\}d\theta$$

$$= \dfrac{5}{2\pi}\left[\dfrac{\sin\left[\theta(\tfrac{1}{2}-n)\right]}{(\tfrac{1}{2}-n)} - \dfrac{\sin\left[\theta(\tfrac{1}{2}+n)\right]}{(\tfrac{1}{2}+n)}\right]_0^{2\pi}$$

$$= \dfrac{5}{2\pi}\left\{\left[\dfrac{\sin 2\pi(\tfrac{1}{2}-n)}{(\tfrac{1}{2}-n)} - \dfrac{\sin 2\pi(\tfrac{1}{2}+n)}{(\tfrac{1}{2}+n)}\right] - \left[\dfrac{\sin 0}{(\tfrac{1}{2}-n)} - \dfrac{\sin 0}{(\tfrac{1}{2}+n)}\right]\right\}$$

When $n$ is both odd and even, $b_n = 0$ since sin 0, sin $\pi$, sin $3\pi$, .... are all zero. Hence the Fourier series for the rectified sine wave $i = 5\sin\dfrac{\theta}{2}$ is given by:

$$i = f(\theta) = \dfrac{10}{\pi} - \dfrac{20}{3\pi}\cos\theta - \dfrac{20}{(3)(5)\pi}\cos 2\theta - \dfrac{20}{(5)(7)\pi}\cos 3\theta - \ldots\ldots$$

i.e. $i = \dfrac{20}{\pi}\left(\dfrac{1}{2} - \dfrac{\cos\theta}{(3)} - \dfrac{\cos 2\theta}{(3)(5)} - \dfrac{\cos 3\theta}{(5)(7)} - \ldots\ldots\ldots\right)$

## C. FURTHER PROBLEMS ON FOURIER SERIES OF PERIODIC FUNCTIONS OF PERIOD $2\pi$

1   Show that the Fourier series for the periodic function of period $2\pi$ defined by

$f(\theta) = \begin{cases} \theta\ , & \text{when } -\pi < \theta < 0 \\ \sin\theta, & \text{when } 0 < \theta < \pi \end{cases}$   is given by:

$$f(\theta) = \dfrac{2}{\pi}\left(\dfrac{1}{2} - \dfrac{\cos 2\theta}{(3)} - \dfrac{\cos 4\theta}{(3)(5)} - \dfrac{\cos 6\theta}{(5)(7)} - \ldots\ldots\right)$$

2   (a) Determine the Fourier series for the periodic function:

$f(x) = \begin{cases} -2, & \text{when } -\pi < x < 0 \\ +2, & \text{when } 0 < x < \pi \end{cases}$   which is periodic outside this range of

   period $2\pi$.

   (b) Deduce a series for $\dfrac{\pi}{4}$ at the point where $x = \dfrac{\pi}{2}$ in the Fourier series of (a).

$$\left[\begin{array}{l} \text{(a)}\quad f(x) = \dfrac{8}{\pi}\left(\sin x + \dfrac{1}{3}\sin 3x + \dfrac{1}{5}\sin 5x + \ldots\ldots\right) \\[2mm] \text{(b)}\quad \dfrac{\pi}{4} = 1 - \dfrac{1}{3} + \dfrac{1}{5} - \dfrac{1}{7} + \ldots\ldots\ldots \end{array}\right]$$

3   For the waveform shown in *Fig 6* determine (a) the Fourier series for the function and (b) the sum of the Fourier series at the points of discontinuity.

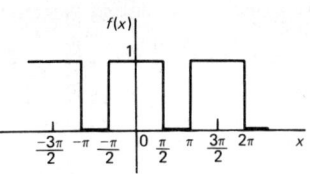

**Fig 6**

$$\left[ \begin{array}{l} \text{(a)} \quad f(x) = \frac{1}{2} + \frac{2}{\pi}\left(\cos x - \frac{1}{3}\cos 3x + \frac{1}{5}\cos 5x - \ldots\ldots\right) \\[2mm] \text{(b)} \quad \frac{1}{2} \end{array} \right]$$

4   For *Problem 3*, draw graphs of the first three partial sums of the Fourier series and show that as the series is added together term by term the result approximates more and more closely to the function it represents.

5   Find the term representing the third harmonic for the period function of period $2\pi$ given by: $f(x) = \begin{cases} 0, \text{ when } -\pi < x < 0 \\ 1, \text{ when } 0 < x < \pi \end{cases}$

$$\left[ \frac{2}{3\pi}\sin 3x \right]$$

6   Determine the Fourier series for the periodic function of period $2\pi$ defined by:

$$f(t) = \begin{cases} 0, \text{ when } -\pi < t < 0 \\ 1, \text{ when } 0 < t < \frac{\pi}{2} \\ 0, \text{ when } \frac{\pi}{2} < t < \pi \end{cases}$$

The function has a period of $2\pi$.

$$\left[ \begin{array}{l} f(t) = \frac{2}{\pi}\left(\cos t - \frac{1}{3}\cos 3t + \frac{1}{5}\cos 5t - \ldots\ldots\ldots \right. \\[2mm] \qquad \left. + \sin 2t + \frac{1}{3}\sin 6t + \frac{1}{5}\sin 10t + \ldots\ldots \right) \end{array} \right]$$

# 18 Fourier series for a non-periodic function over range 2π

## A. MAIN POINTS CONCERNED WITH FOURIER SERIES FOR NON-PERIODIC FUNCTIONS OVER RANGE 2π

1   If a function $f(x)$ is not periodic then it cannot be expanded in a Fourier series for **all** values of $x$. However, it is possible to determine a Fourier series to represent the function over any range of width $2\pi$.

2   Given a non-periodic function, a new function may be constructed by taking the values of $f(x)$ in the given range and then repeating them outside of the given range at intervals of $2\pi$. Since this new function is, by construction, periodic with period $2\pi$, it may then be expanded in a Fourier series for all values of $x$. For example, the function $f(x) = x$ is not a periodic function. However, if a Fourier series for $f(x) = x$ is required then the function is constructed outside of this range so that it is periodic with period $2\pi$ as shown by the broken lines in *Fig 1*.

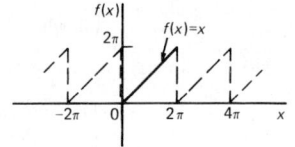

**Fig 1**

3   For non-periodic functions, such as $f(x) = x$, the sum of the Fourier series is equal to $f(x)$ at all points in the given range but it is not equal to $f(x)$ at points outside of the range.

4   For determining a Fourier series of a non-periodic function over a range $2\pi$, exactly the same formulae for the Fourier coefficients are used as in para. 4(i) of chapter 17.

## B. WORKED PROBLEMS ON FOURIER SERIES OF NON-PERIODIC FUNCTIONS OVER A RANGE OF 2π

*Problem 1*   Determine the Fourier series to represent the function $f(x) = 2x$ in the range $-\pi$ to $+\pi$.

The function $f(x) = 2x$ is not periodic. The function is shown in the range $-\pi$ to $\pi$ in *Fig 2* and is then constructed outside of that range so that it is periodic of period $2\pi$ (see broken lines) with the resulting saw-tooth waveform.

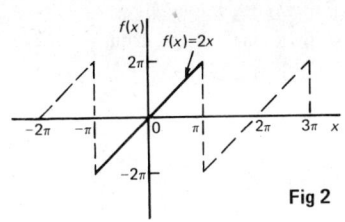

**Fig 2**

For a Fourier series: $f(x) = a_0 + \sum\limits_{n=1}^{\infty} (a_n \cos nx + b_n \sin nx)$

From para. 4(i), chapter 17, $a_0 = \dfrac{1}{2\pi} \int_{-\pi}^{\pi} f(x)\,dx = \dfrac{1}{2\pi} \int_{-\pi}^{\pi} 2x\,dx = \dfrac{2}{2\pi} \left[\dfrac{x^2}{2}\right]_{-\pi}^{\pi} = 0$

$a_n = \dfrac{1}{\pi} \int_{-\pi}^{\pi} f(x) \cos nx\,dx = \dfrac{1}{\pi} \int_{-\pi}^{\pi} 2x \cos nx\,dx$

$\qquad = \dfrac{2}{\pi} \left[\dfrac{x \sin x}{n} - \int \dfrac{\sin nx}{n}\,dx\right]_{-\pi}^{\pi}$, by parts (see chapter 11)

$\qquad = \dfrac{2}{\pi} \left[\dfrac{x \sin nx}{n} + \dfrac{\cos nx}{n^2}\right]_{-\pi}^{\pi}$

$\qquad = \dfrac{2}{\pi} \left[\left(0 + \dfrac{\cos n\pi}{n^2}\right) - \left(0 + \dfrac{\cos n(-\pi)}{n^2}\right)\right] = 0$

$b_n = \dfrac{1}{\pi} \int_{-\pi}^{\pi} f(x) \sin nx\,dx = \dfrac{1}{\pi} \int_{-\pi}^{\pi} 2x \sin nx\,dx$

$\qquad = \dfrac{2}{\pi} \left[\dfrac{-x \cos nx}{n} - \int\left(\dfrac{-\cos nx}{n}\right)\,dx\right]_{-\pi}^{\pi}$, by parts

$\qquad = \dfrac{2}{\pi} \left[\dfrac{-x \cos nx}{n} + \dfrac{\sin nx}{n^2}\right]_{-\pi}^{\pi}$

$\qquad = \dfrac{2}{\pi} \left[\left(\dfrac{-\pi \cos n\pi}{n} + \dfrac{\sin n\pi}{n^2}\right) - \left(\dfrac{-(-\pi) \cos n(-\pi)}{n} + \dfrac{\sin n(-\pi)}{n^2}\right)\right]$

$\qquad = \dfrac{2}{\pi} \left[\dfrac{-\pi \cos n\pi}{n} - \dfrac{\pi \cos n\pi}{n}\right] = \dfrac{-4}{n} \cos n\pi$

When $n$ is odd, $b_n = \dfrac{4}{n}$. Thus $b_1 = 4, b_3 = \dfrac{4}{3}, b_5 = \dfrac{4}{5}$, and so on.

When $n$ is even, $b_n = \dfrac{-4}{n}$. Thus $b_2 = -2, b_4 = -1, b_6 = -\dfrac{2}{3}$, and so on.

Thus $f(x) = 2x = 4 \sin x - 2 \sin 2x + \dfrac{4}{3} \sin 3x - 1 \sin 4x + \dfrac{4}{5} \sin 5x - \dfrac{2}{3} \sin 6x + \ldots$

i.e., $2x = 4(\sin x - \dfrac{1}{2} \sin 2x + \dfrac{1}{3} \sin 3x - \dfrac{1}{4} \sin 4x + \dfrac{1}{5} \sin 5x - \dfrac{1}{6} \sin 6x + \ldots)$    (1)

for values of $f(x)$ between $-\pi$ and $\pi$. For values of $f(x)$ outside the range $-\pi$ to $+\pi$ the sum of the series is not equal to $f(x)$.

*Problem 2* In the Fourier series of *Problem 1*, by letting $x = \pi/2$, deduce a series for $\pi/4$.

When $x = \pi/2, f(x) = \pi$ from *Fig 2*.
Thus, from the Fourier series of equation (1):

$$\pi = 4 \left(\sin \frac{\pi}{2} - \frac{1}{2} \sin \frac{2\pi}{3} + \frac{1}{3} \sin \frac{3\pi}{2} - \frac{1}{4} \sin \frac{4\pi}{2} + \frac{1}{5} \sin \frac{5\pi}{2} - \frac{1}{6} \sin \frac{6\pi}{2} + \ldots \ldots \right)$$

i.e. $\dfrac{\pi}{4} = 1 - \dfrac{1}{3} + \dfrac{1}{5} - \dfrac{1}{7} + \ldots \ldots$

*Problem 3* Obtain a Fourier series for the function defined by:
$$f(x) = \begin{cases} x, \text{ when } 0 < x < \pi \\ 0, \text{ when } \pi < x < 2\pi \end{cases}$$

The defined function is shown in *Fig 3* between 0 and $2\pi$. The function is constructed outside of this range so that it is periodic of period $2\pi$, as shown by the broken lines in *Fig 3*. For a Fourier series:

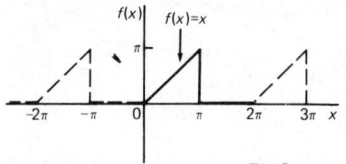

Fig 3

$$f(x) = a_0 + \sum_{n=1}^{\infty} (a_n \cos nx + b_n \sin nx).$$

It is more convenient in this case to take the limits from 0 to $2\pi$ instead of from $-\pi$ to $+\pi$. The value of the Fourier coefficients are unaltered by this change of limits.

Hence $a_0 = \dfrac{1}{2\pi} \displaystyle\int_0^{2\pi} f(x) \, dx = \dfrac{1}{2\pi} \left[ \displaystyle\int_0^{\pi} x \, dx + \displaystyle\int_{\pi}^{2\pi} 0 \, dx \right] = \dfrac{1}{2\pi} \left[ \dfrac{x^2}{2} \right]_0^{\pi} = \dfrac{1}{2\pi} \left( \dfrac{\pi^2}{2} \right) = \dfrac{\pi}{4}$

$a_n = \dfrac{1}{\pi} \displaystyle\int_0^{2\pi} f(x) \cos nx \, dx = \dfrac{1}{\pi} \left[ \displaystyle\int_0^{\pi} x \cos nx \, dx + \displaystyle\int_{\pi}^{2\pi} 0 \, dx \right]$

$\qquad = \dfrac{1}{\pi} \left[ \dfrac{x \sin nx}{n} + \dfrac{\cos nx}{n^2} \right]_0^{\pi}$ , (from *Problem 1*),

$= \dfrac{1}{\pi} \left\{ \left[ \dfrac{\pi \sin n\pi}{n} + \dfrac{\cos n\pi}{n^2} \right] - \left[ 0 + \dfrac{\cos 0}{n^2} \right] \right\} = \dfrac{1}{\pi n^2} \left( \cos n\pi - 1 \right)$

When $n$ is even, $a_n = 0$

When $n$ is odd, $a_n = \dfrac{-2}{\pi n^2}$ . Hence $a_1 = \dfrac{-2}{\pi}, a_3 = \dfrac{-2}{3^2 \pi}, a_5 = \dfrac{-2}{5^2 \pi}$ , and so on.

$b_n = \dfrac{1}{\pi} \displaystyle\int_0^{2\pi} f(x) \sin nx \, dx = \dfrac{1}{\pi} \left[ \displaystyle\int_0^{\pi} x \sin nx \, dx - \displaystyle\int_{\pi}^{2\pi} 0 \, dx \right]$

$= \dfrac{1}{\pi} \left[ \dfrac{-x \cos nx}{n} + \dfrac{\sin nx}{n^2} \right]_0^{\pi}$ (from *Problem 1*)

150

$$= \frac{1}{\pi}\left\{\left[\frac{-\pi\cos n\pi}{n} + \frac{\sin n\pi}{n^2}\right] - \left[0 + \frac{\sin 0}{n^2}\right]\right\}$$

$$= \frac{1}{\pi}\left[\frac{-\pi\cos n\pi}{n}\right] = \frac{-\cos n\pi}{n}$$

Hence $b_1 = -\cos\pi = 1$, $b_2 = -\frac{1}{2}$, $b_3 = \frac{1}{3}$, and so on.

Thus the Fourier series is: $f(x) = \frac{\pi}{4} - \frac{2}{\pi}\cos x - \frac{2}{3^2\pi}\cos 3x - \frac{2}{5^2\pi}\cos 5x - \ldots$

$$+ \sin x - \frac{1}{2}\sin 2x + \frac{1}{3}\sin 3x - \ldots\ldots$$

i.e., $f(x) = \frac{\pi}{4} - \frac{2}{\pi}\left(\cos x + \frac{\cos 3x}{3^2} + \frac{\cos 5x}{5^2} + \ldots\right) + \left(\sin x - \frac{1}{2}\sin 2x + \frac{1}{3}\sin 3x - \ldots\right)$

*Problem 4* For the Fourier series of *Problem 3*:
(a) what is the sum of the series at the point of discontinuity (i.e. at $x = \pi$),
(b) what is the amplitude and phase angle of the third harmonic, and
(c) let $x = 0$, and deduce a series of $\pi^2/8$.

(a) The sum of the Fourier series at the point of discontinuity is given by the arithmetic mean of the two limiting values of $f(x)$ as $x$ approaches the point of discontinuity from the two sides.

Hence sum of the series at $x = \pi = \frac{\pi-0}{2} = \frac{\pi}{2}$

(b) The third harmonic term of the Fourier series is $\left(-\frac{2}{3^2\pi}\cos 3x + \frac{1}{3}\sin 3x\right)$

This may also be written in the form $c\sin(3x+\alpha)$,

where amplitude, $c = \sqrt{\left[\left(\frac{-2}{3^2\pi}\right)^2 + \left(\frac{1}{3}\right)^2\right]} = 0.341$

and phase angle, $\alpha = \arctan\left(\frac{\frac{-2}{3^2\pi}}{\frac{1}{3}}\right) = 168°\ 1'$ or 2.93 radians.

Hence the third harmonic is given by $0.341\sin(3x+2.93)$.

(c) When $x = 0$, $f(x) = 0$  (see *Fig 3*).
Hence, from the Fourier series: $0 = \frac{\pi}{4} - \frac{2}{\pi}\left(\cos 0 + \frac{1}{3^2}\cos 0 + \frac{1}{5^2}\cos 0 + \ldots\right) + (0)$

i.e., $-\frac{\pi}{4} = -\frac{2}{\pi}\left(1 + \frac{1}{3^2} + \frac{1}{5^2} + \frac{1}{7^2} + \ldots\right)$

Hence $\frac{\pi^2}{8} = 1 + \frac{1}{3^2} + \frac{1}{5^2} + \frac{1}{7^2} + \ldots\ldots$

*Problem 5* Deduce the Fourier series for the function $f(\theta) = \theta^2$ in the range 0 to $2\pi$.

$f(\theta) = \theta^2$ is shown in *Fig 4* in the range 0 to $2\pi$. The function is not periodic but is constructed outside of this range so that it is periodic of period $2\pi$, as shown by the broken lines.

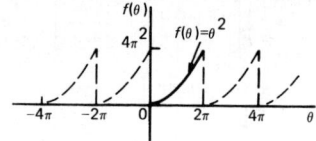

**Fig 4**

For a Fourier series: $f(x) = a_0 + \sum\limits_{n=1}^{\infty} (a_n \cos nx + b_n \sin nx)$

$$a_0 = \frac{1}{2\pi} \int_0^{2\pi} f(\theta)\, d\theta = \frac{1}{2\pi} \int_0^{2\pi} \theta^2 \, d\theta = \frac{1}{2\pi} \left[\frac{\theta^3}{3}\right]_0^{2\pi} = \frac{4\pi^2}{3}$$

$$a_n = \frac{1}{\pi} \int_0^{2\pi} f(\theta) \cos n\theta \, d\theta = \frac{1}{\pi} \int_0^{2\pi} \theta^2 \cos n\theta \, d\theta$$

$$= \frac{1}{\pi} \left[\frac{\theta^2 \sin n\theta}{n} + \frac{2\theta \cos n\theta}{n^2} - \frac{2 \sin n\theta}{n^3}\right]_0^{2\pi} \text{ by parts}$$

$$= \frac{1}{\pi} \left[\left(0 + \frac{4\pi \cos 2\pi n}{n^2} - 0\right) - (0)\right]$$

$$= \frac{4}{n^2} \cos 2\pi n = \frac{4}{n^2} \text{ when } n = 1, 2, 3, \ldots .$$

Hence $a_1 = \frac{4}{1^2}, a_2 = \frac{4}{2^2}, a_3 = \frac{4}{3^2}$, and so on.

$$b_n = \frac{1}{\pi} \int_0^{2\pi} f(\theta) \sin n\theta \, d\theta = \frac{1}{\pi} \int_0^{2\pi} \theta^2 \sin n\theta \, d\theta$$

$$= \frac{1}{\pi} \left[\frac{-\theta^2 \cos n\theta}{n} + \frac{2\theta \sin n\theta}{n^2} + \frac{2 \cos n\theta}{n^3}\right]_0^{2\pi} \text{ by parts}$$

$$= \frac{1}{\pi} \left[\left(\frac{-4\pi^2 \cos 2\pi n}{n} + 0 + \frac{2 \cos 2\pi n}{n^3}\right) - \left(0 + 0 - \frac{2 \cos 0}{n^3}\right)\right]$$

$$= \frac{1}{\pi} \left[\frac{-4\pi^2}{n} + \frac{2}{n^3} - \frac{2}{n^3}\right] = \frac{-4\pi}{n}$$

Hence $b_1 = \frac{-4\pi}{1}, b_2 = \frac{-4\pi}{2}, b_3 = \frac{-4\pi}{3}$, and so on.

Thus $f(\theta) = \theta^2 = \frac{4\pi^2}{3} + \sum\limits_{n=1}^{\infty} \left(\frac{4}{n^2} \cos n\theta - \frac{4\pi}{n} \sin n\theta\right)$

i.e., $\theta^2 = \frac{4\pi^2}{3} + 4(\cos \theta + \frac{1}{2^2} \cos 2\theta + \frac{1}{3^2} \cos 3\theta + \ldots .)$

$$- 4\pi(\sin \theta + \frac{1}{2} \sin 2\theta + \frac{1}{3} \sin 3\theta + \ldots \ldots)$$

for values of $\theta$ between 0 and $2\pi$.

*Problem 6* In the Fourier series of *Problem 5*, let $\theta = \pi$ and determine a series for $\pi^2/12$.

When $\theta = \pi$, $f(\theta) = \pi^2$.

Hence $\pi^2 = \dfrac{4\pi^2}{3} + 4(\cos \pi + \dfrac{1}{4}\cos 2\pi + \dfrac{1}{9}\cos 3\pi + \dfrac{1}{16}\cos 4\pi + \ldots\ldots)$

$\qquad\qquad - 4\pi(\sin \pi + \dfrac{1}{2}\sin 2\pi + \dfrac{1}{3}\sin 3\pi + \ldots\ldots\ldots)$

i.e., $\pi^2 - \dfrac{4\pi^2}{3} = 4(-1 + \dfrac{1}{4} - \dfrac{1}{9} + \dfrac{1}{16} - \ldots\ldots) - 4\pi(0)$

$\qquad\dfrac{-\pi^2}{3} = 4(-1 + \dfrac{1}{4} - \dfrac{1}{9} + \dfrac{1}{16} - \ldots\ldots)$

$\qquad\dfrac{\pi^2}{3} = 4(1 - \dfrac{1}{4} + \dfrac{1}{9} - \dfrac{1}{16} + \ldots\ldots)$

Hence $\quad\dfrac{\pi^2}{12} = 1 - \dfrac{1}{4} + \dfrac{1}{9} - \dfrac{1}{16} + \ldots\ldots$

or $\qquad\dfrac{\pi^2}{12} = 1 - \dfrac{1}{2^2} + \dfrac{1}{3^2} - \dfrac{1}{4^2} + \ldots\ldots\ldots$

## C. FURTHER PROBLEMS ON FOURIER SERIES OF NON-PERIODIC FUNCTIONS OVER A RANGE OF $2\pi$

1   Show that the Fourier series for the function $f(x) = x$ over the range $x = 0$ to $x = 2\pi$ is given by:

$$f(x) = \pi - 2(\sin x + \tfrac{1}{2}\sin 2x + \tfrac{1}{3}\sin 3x + \tfrac{1}{4}\sin 4x + \ldots\ldots)$$

2   Determine the Fourier series for the function defined by:

$$f(t) = \begin{cases} 1-t, & \text{when } -\pi < t < 0 \\ 1+t, & \text{when } 0 < t < \pi \end{cases}$$

Draw a graph of the function within and outside of the given range.

$$\left[ f(t) = \frac{\pi}{2} + 1 - \frac{4}{\pi}\left(\cos t + \frac{\cos 3t}{3^2} + \frac{\cos 5t}{5^2} + \ldots\ldots\right) \right]$$

3   Find the Fourier series for the function $f(x) = x + \pi$ within the range $-\pi < x < \pi$.

$$\left[ f(x) = \pi + 2(\sin x - \tfrac{1}{2}\sin 2x + \tfrac{1}{3}\sin 3x - \ldots\ldots) \right]$$

4   Determine the Fourier series up to and including the third harmonic for the function defined by:

$$f(x) = \begin{cases} x, & \text{when } 0 < x < \pi \\ 2\pi - x, & \text{when } \pi < x < 2\pi \end{cases}$$

Sketch a graph of the function within and outside of the given range, assuming the period is $2\pi$.

$$\left[ f(x) = \frac{\pi}{2} - \frac{4}{\pi}\left(\cos x + \frac{\cos 3x}{3^2} + \frac{\cos 5x}{5^2} + \ldots\ldots\right) \right]$$

5   Show that the Fourier series for the function $f(x) = \sin^2 x$ for the range $-\pi \leqslant x \leqslant \pi$ is given by $f(x) = \dfrac{1}{2} - \dfrac{1}{2}\cos 2x$.

6   Expand the function $f(\theta) = \theta^2$ in a Fourier series in the range $-\pi < \theta < \pi$. Sketch the function within and outside of the given range.

$$\left[ f(\theta) = \frac{\pi^2}{3} - 4\left(\cos\theta - \frac{1}{2^2}\cos 2\theta + \frac{1}{3^2}\cos 3\theta - \ldots\ldots\right) \right]$$

7   For the Fourier series obtained in *Problem 6*, let $x = \pi$ and deduce the series for

$$\sum_{n=1}^{\infty} \frac{1}{n^2} \,.$$

$$\left[ 1 + \frac{1}{2^2} + \frac{1}{3^2} + \frac{1}{4^2} + \frac{1}{5^2} + \ldots\ldots\ldots = \frac{\pi^2}{6} \right]$$

8   Show that the Fourier series for the triangular waveform shown in *Fig 5* is given by:

$$y = \frac{8}{\pi^2}\left( \sin\theta - \frac{1}{3^2}\sin 3\theta + \frac{1}{5^2}\sin 5\theta \right.$$

$$\left. - \frac{1}{7^2}\sin 7\theta + \ldots \right)$$

in the range 0 to $2\pi$.

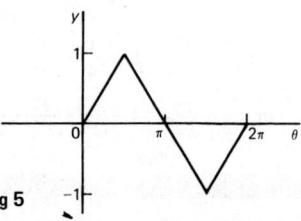

Fig 5

9   Given the function $f(t) = t+\pi$ in the range $-\pi < t < \pi$ determine the Fourier series.

$$\left[ f(t) = \pi + 2\left( \sin t - \frac{1}{2}\sin 2t + \frac{1}{3}\sin 3t - \ldots\ldots \right) \right]$$

10  Sketch the waveform defined by: $f(x) = \begin{cases} 1 + \dfrac{2x}{\pi}, & \text{when } -\pi < x < 0 \\ 1 - \dfrac{2x}{\pi}, & \text{when } 0 < x < \pi \end{cases}$

Determine the Fourier series in this range.

$$\left[ f(x) = \frac{8}{\pi^2}\left( \cos x + \frac{1}{3^2}\cos 3x + \frac{1}{5^2}\cos 5x + \frac{1}{7^2}\cos 7x + \ldots\ldots \right) \right]$$

11  For the Fourier series of *Problem 10*, deduce a series for $\dfrac{\pi^2}{8}$.

$$\left[ \frac{\pi^2}{8} = 1 + \frac{1}{3^2} + \frac{1}{5^2} + \frac{1}{7^2} + \frac{1}{9^2} + \ldots\ldots \right]$$

# 19 Even and odd functions and half-range series

## A. MAIN POINTS CONCERNED WITH EVEN AND ODD FUNCTIONS AND HALF RANGE FOURIER SERIES

1 **Even functions**

A function $y = f(x)$ is said to be **even** if $f(-x) = f(x)$ for all values of $x$. Graphs of even functions are always **symmetrical about the $y$-axis** (i.e., is a mirror image). Two examples of even functions are $y = x^2$ and $y = \cos x$ as shown in *Fig 1*.

2 **Odd functions**

A function $y = f(x)$ is said to be **odd** if $f(-x) = -f(x)$ for all values of $x$. Graphs of odd functions are always **symmetrical about the origin**. Two examples of odd functions are $y = x^3$ and $y = \sin x$ as shown in *Fig 2*.

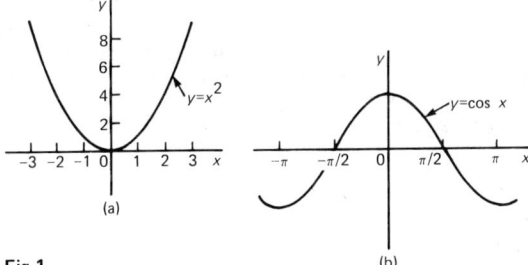

**Fig 1**

(a)

(b)

**Fig 2**    (a)

(b)

155

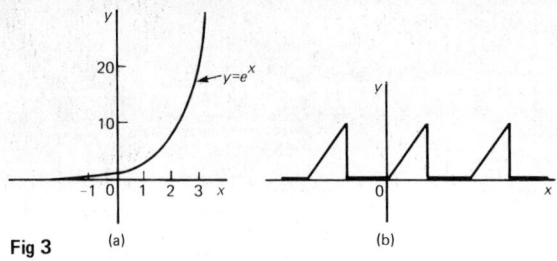

**Fig 3**    (a)    (b)

3    Many functions are neither even nor odd, two such examples being shown in *Fig 3*. (See *Problems 1 and 2*)

4    **Fourier cosine series**

The Fourier series of an **even** periodic function $f(x)$ having period $2\pi$ contains **cosine terms only** (i.e. contains no sine terms) and may contain a constant term.

Hence    $f(x) = a_0 + \sum_{n=1}^{\infty} a_n \cos nx$

where    $a_0 = \dfrac{1}{2\pi} \int_{-\pi}^{\pi} f(x) = \dfrac{1}{\pi} \int_{0}^{\pi} f(x)\, dx$    (due to symmetry)

and    $a_n = \dfrac{1}{\pi} \int_{-\pi}^{\pi} f(x) \cos nx\, dx = \dfrac{2}{\pi} \int_{0}^{\pi} f(x) \cos nx\, dx$

5    **Fourier sine series**

The Fourier series of an **odd** periodic function $f(x)$ having period $2\pi$ contains **sine terms only** (i.e. contains no constant term and no cosine terms).

Hence $f(x) = \sum_{n=1}^{\infty} b_n \sin nx$

where $b_n = \dfrac{1}{\pi} \int_{-\pi}^{\pi} f(x) \sin nx\, dx = \dfrac{2}{\pi} \int_{0}^{\pi} f(x) \sin nx\, dx$

(See *Problems 3 to 8*)

**Half range Fourier series**

6    When a function is defined over the range say 0 to $\pi$ instead of from 0 to $2\pi$ it may be expanded in a series of sine terms only or of cosine terms only. The series produced is called a **half range Fourier series**.

7    If a **half range cosine series** is required for the function $f(x) = x$ in the range 0 to $\pi$ then an **even** periodic function is required. In *Fig 4*, $f(x) = x$ is shown plotted from $x = 0$ to $x = \pi$. Since an even function is symmetrical about the $f(x)$ axis the line AB is constructed as shown. If the triangular waveform produced is assumed to be periodic of period $2\pi$ outside of this range then the waveform is as shown in *Fig 4*. When a half range cosine series is required then the Fourier coefficients $a_0$ and $a_n$ are calculated as in para. 4.

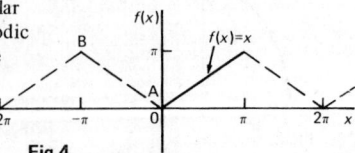

**Fig 4**

156

8  If a **half range sine series** is required for the function $f(x) = x$ in the range 0 to $\pi$ then an **odd** periodic function is required. In *Fig 5*, $f(x) = x$ is shown plotted from $x = 0$ to $x = \pi$. Since an odd function is symmetrical about the origin the line CD is constructed as shown. If the sawtooth waveform produced is assumed to be periodic of period $2\pi$ outside of this range, then the waveform is as shown in *Fig 5*. When a half range sine series is required then the Fourier coefficient $b_n$ is calculated as in para. 5. (See *Problems 9 to 11*)

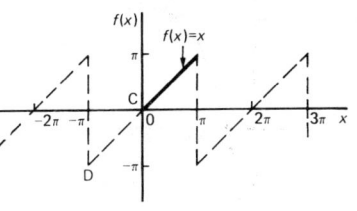

**Fig 5**

---

B. **WORKED PROBLEMS ON EVEN AND ODD FUNCTIONS AND HALF RANGE FOURIER SERIES**

*Problem 1* Sketch the following functions and state whether they are even or odd functions:

(a) $y = \tan x$; (b) $f(x) = \begin{cases} 1, & \text{when } 0 < x < \dfrac{\pi}{2} \\ -1, & \text{when } \dfrac{\pi}{2} < x < \dfrac{3\pi}{2}, \text{ and is periodic of period } 2\pi. \\ 1, & \text{when } \dfrac{3\pi}{2} < x < 2\pi \end{cases}$

(a) A graph of $y = \tan x$ is shown in *Fig 6(a)* and is symmetrical about the origin and is thus an **odd function** (i.e., $\tan(-x) = -\tan x$).

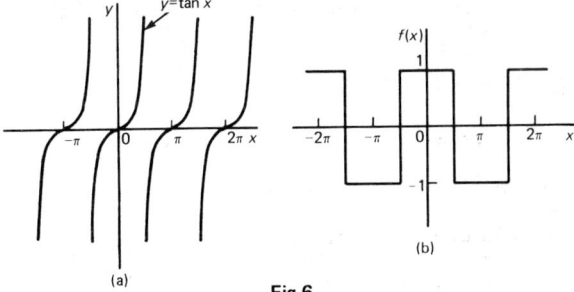

**Fig 6**

(b) A graph of $f(x)$ is shown in *Fig 6(b)* and is symmetrical about the $f(x)$ axis hence the function is an **even** one. $(f(-x) = f(x))$.

*Problem 2* Sketch the following graphs and state whether the functions are even, odd or neither even nor odd:
(a) $y = \ln x$, (b) $y = x \sin x$, (c) $f(x) = x$ in the range $-\pi$ to $\pi$ and is periodic of period $2\pi$.

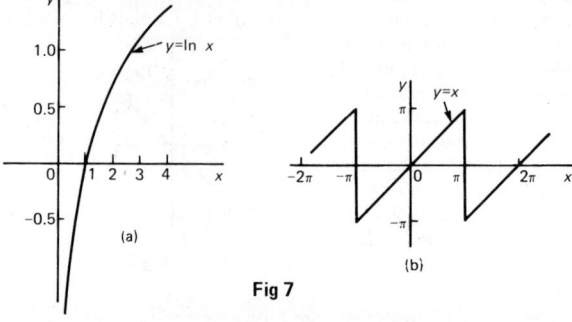

**Fig 7**

(a) A graph of $y = \ln x$ is shown in *Fig 7(a)* and the curve is neither symmetrical about the $y$ axis nor symmetrical about the origin and is thus **neither even nor odd**.

(b) A graph of $y = x \sin x$ is shown in *Fig 7(b)* (obtained by drawing up a table of values) and is symmetrical about the $y$ axis and is thus an **even function**.

(c) A graph of $y = x$ in the range $-\pi$ to $\pi$ is shown in *Fig 7(c)* and is symmetrical about the origin and is thus an **odd function**.

*Problem 3* Deduce (a) that an even function has a Fourier series containing a constant term and cosine terms only, and (b) that an odd function has a Fourier series containing sine terms only.

Let $g(x)$ be an even function and $h(x)$ be an odd function. Then the product $g(x) . h(x)$ is odd since $g(-x)h(-x) = g(x)[-h(x)] = -g(x) . h(x)$.

Similarly it may be shown that a product of two even functions and a product of two odd functions both result in a function which is even. A Fourier series of a function $f(x)$ is defined by:

$$f(x) = a_0 + \sum_{n=1}^{\infty} (a_n \cos nx + b_n \sin nx)$$

(a) If $f(x)$ is an even function:

(i) $a_0 = \int_{-\pi}^{\pi} f(x)dx = 2\int_{0}^{\pi} f(x)dx$ due to symmetry about the $f(x)$ axis

(see *Fig 1*). Hence $a_0$ has a certain value (which may be zero).

(ii) The product $f(x) \sin nx$ is odd since $\sin nx$ is odd. Hence the Fourier

coefficient $b_n = \int_{-\pi}^{\pi} f(x) \sin nx \, dx = 0$ due to symmetry about the

origin.

(iii) The product $f(x) \cos nx$ is even since $\cos nx$ is even. Hence the Fourier

coefficient $a_n = \int_{-\pi}^{\pi} f(x) \cos nx \, dx$ has a certain value.

Thus for an even function a Fourier series contains a constant term and cosine terms only.

158

**(b) If $f(x)$ is an odd function:**

(i) $a_0 = \displaystyle\int_{-\pi}^{\pi} f(x)\, dx = 0$ due to symmetry about the origin (see *Fig 2*).

(ii) The product $f(x) \sin nx$ is even hence the Fourier coefficient

$b_n = \displaystyle\int_{-\pi}^{\pi} f(x) \sin nx\, dx$ has a certain value.

(iii) The product $f(x) \cos nx$ is odd since $\cos nx$ is even. Hence the Fourier

coefficient $a_n = \displaystyle\int_{-\pi}^{\pi} f(x) \cos nx\, dx = 0$ due to symmetry about the origin

**Thus for an odd function a Fourier series contains sine terms only.**

*Problem 4* Determine the Fourier series for the periodic function defined by:

$$f(x) = \begin{cases} -2, & \text{when } -\pi < x < -\dfrac{\pi}{2} \\[2mm] 2, & \text{when } -\dfrac{\pi}{2} < x < \dfrac{\pi}{2} \\[2mm] -2, & \text{when } \dfrac{\pi}{2} < x < \pi \end{cases}$$

The function has a period of $2\pi$.

The square wave shown in *Fig 8* is an even function since it is symmetrical about the $f(x)$ axis.

Hence from para. 4, the Fourier series is given by $f(x) = a_0 + \displaystyle\sum_{n=1}^{\infty} a_n \cos nx$

(i.e. the series contains no sine terms).

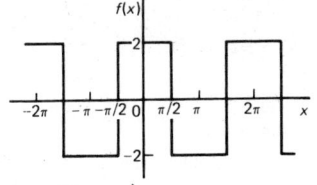

**Fig 8**

From para. 4, $a_0 = \dfrac{1}{\pi}\displaystyle\int_0^{\pi} f(x)\, dx = \dfrac{1}{\pi}\left\{\int_0^{\frac{\pi}{2}} 2\, dx + \int_{\frac{\pi}{2}}^{\pi} -2\, dx\right\}$

$= \dfrac{1}{\pi}\left\{[2x]_0^{\frac{\pi}{2}} + [-2x]_{\frac{\pi}{2}}^{\pi}\right\} = \dfrac{1}{\pi}\{(\pi) + [(-2\pi)-(-\pi)]\} = 0$

$a_n = \dfrac{2}{\pi}\displaystyle\int_0^{\pi} f(x) \cos nx\, dx = \dfrac{2}{\pi}\left\{\int_0^{\frac{\pi}{2}} 2 \cos nx\, dx + \int_{\frac{\pi}{2}}^{\pi} -2 \cos nx\, dx\right\}$

$= \dfrac{4}{\pi}\left\{\left[\dfrac{\sin nx}{n}\right]_0^{\frac{\pi}{2}} + \left[\dfrac{-\sin nx}{n}\right]_{\frac{\pi}{2}}^{\pi}\right\} = \dfrac{4}{\pi}\left\{\left(\dfrac{\sin\frac{\pi}{2} n}{n} - 0\right) + \left(0 - \dfrac{-\sin\frac{\pi}{2} n}{n}\right)\right\}$

$= \dfrac{4}{\pi}\left(\dfrac{2\sin\frac{\pi}{2} n}{n}\right) = \dfrac{8}{\pi n}\left(\sin\dfrac{n\pi}{2}\right).$

When $n$ is even, $a_n = 0$

When $n$ is odd, $a_n = \dfrac{8}{\pi n}$ for $n = 1, 5, 9 \ldots\ldots$ and $a_n = \dfrac{-8}{\pi n}$ for $n = 3, 7, 11, \ldots$

159

Hence, $a_1 = \dfrac{8}{\pi}$, $a_3 = \dfrac{-8}{3\pi}$, $a_5 = \dfrac{8}{5\pi}$, and so on.

Hence the Fourier series for the waveform of *Fig 8* is given by:

$$f(x) = \frac{8}{\pi}\left(\cos x - \frac{1}{3}\cos 3x + \frac{1}{5}\cos 5x - \frac{1}{7}\cos 7x + \ldots \ldots\right)$$

*Problem 5* In the Fourier series of *Problem 4* let $x=0$ and deduce a series for $\pi/4$.

When $x = 0$, $f(x) = 2$ (from *Fig 8*)

Thus, from the Fourier series, $2 = \dfrac{8}{\pi}\left(\cos 0 - \dfrac{1}{3}\cos 0 + \dfrac{1}{5}\cos 0 - \dfrac{1}{7}\cos 0 + \ldots \ldots\right)$

Hence $\quad \dfrac{2\pi}{8} = 1 - \dfrac{1}{3} + \dfrac{1}{5} - \dfrac{1}{7} + \ldots \ldots$

i.e. $\quad \dfrac{\pi}{4} = 1 - \dfrac{1}{3} + \dfrac{1}{5} - \dfrac{1}{7} + \ldots \ldots$

*Problem 6* Obtain the Fourier series for the square wave shown in *Fig 9*.

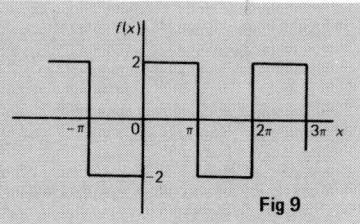

**Fig 9**

The square wave shown in *Fig 9* is an odd function since it is symmetrical about the origin.

Hence, from para. 5, the Fourier series is given by: $f(x) = \displaystyle\sum_{n=1}^{\infty} b_n \sin nx$.

The function is defined by: $f(x) = \begin{cases} -2, \text{ when } -\pi < x < 0 \\ 2, \text{ when } 0 < x < \pi \end{cases}$

From para. 5, $b_n = \dfrac{2}{\pi}\displaystyle\int_0^{\pi} f(x)\sin nx\, dx = \dfrac{2}{\pi}\displaystyle\int_0^{\pi} 2\sin nx\, dx = \dfrac{4}{\pi}\left[\dfrac{-\cos nx}{n}\right]_0^{\pi}$

$$= \frac{4}{\pi}\left[\left(\frac{-\cos n\pi}{n}\right) - \left(-\frac{1}{n}\right)\right] = \frac{4}{\pi n}(1 - \cos n\pi).$$

When $n$ is even, $b_n = 0$. When $n$ is odd, $b_n = \dfrac{4}{\pi n}(1--1) = \dfrac{8}{\pi n}$.

Hence $b_1 = \dfrac{8}{\pi}$, $b_3 = \dfrac{8}{3\pi}$, $b_5 = \dfrac{8}{5\pi}$, and so on.

Hence the Fourier series is:

$$f(x) = \frac{8}{\pi}\left(\sin x + \frac{1}{3}\sin 3x + \frac{1}{5}\sin 5x + \frac{1}{7}\sin 7x + \ldots \ldots\right)$$

*Problem 7* Determine the Fourier series for the function $f(\theta) = \theta^2$ in the range $-\pi < \theta < \pi$. The function has a period of $2\pi$.

A graph of $f(\theta) = \theta^2$ is shown in *Fig 10* in the range $-\pi$ to $\pi$ with period $2\pi$. The function is symmetrical about the $f(\theta)$ axis and is thus an even function. Thus a Fourier cosine series will result of the form:

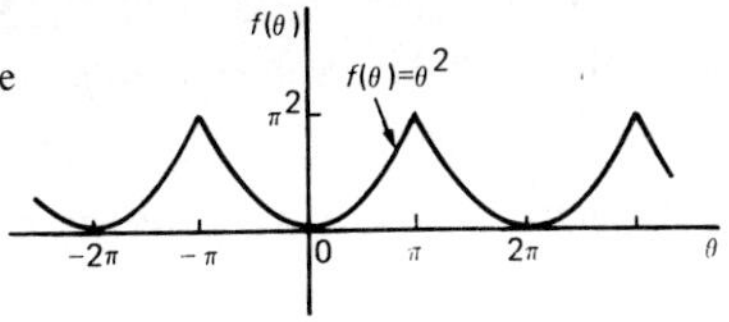

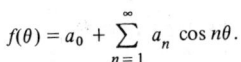

$$f(\theta) = a_0 + \sum_{n=1}^{\infty} a_n \cos n\theta.$$

**Fig 10**

From para. 4, $a_0 = \dfrac{1}{\pi} \displaystyle\int_0^\pi f(\theta)\, d\theta = \dfrac{1}{\pi} \displaystyle\int_0^\pi \theta^2\, d\theta = \dfrac{1}{\pi}\left[\dfrac{\theta^3}{3}\right]_0^\pi = \dfrac{\pi^2}{3}$

and $\quad a_n = \dfrac{2}{\pi} \displaystyle\int_0^\pi f(\theta) \cos n\theta\, d\theta = \dfrac{2}{\pi} \displaystyle\int_0^\pi \theta^2 \cos n\theta\, d\theta$

$$= \frac{2}{\pi}\left[\frac{\theta^2 \sin n\theta}{n} + \frac{2\theta \cos n\theta}{n^2} - \frac{2 \sin n\theta}{n^3}\right]_0^\pi \quad \text{by parts,}$$

$$= \frac{2}{\pi}\left[\left(0 + \frac{2\pi \cos n\pi}{n^2} - 0\right) - (0)\right] = \frac{4}{n^2} \cos n\pi$$

When $n$ is odd, $a_n = \dfrac{-4}{n^2}$. Hence $a_1 = \dfrac{-4}{1^2}$, $a_3 = \dfrac{-4}{3^2}$, $a_5 = \dfrac{-4}{5^2}$, and so on.

When $n$ is even, $a_n = \dfrac{4}{n^2}$. Hence $a_2 = \dfrac{4}{2^2}$, $a_4 = \dfrac{4}{4^2}$, and so on.

Hence the Fourier series is:

$$f(\theta) = \theta^2 = \frac{\pi^2}{3} - 4\left(\cos\theta - \frac{1}{2^2}\cos 2\theta + \frac{1}{3^2}\cos 3\theta - \frac{1}{4^2}\cos 4\theta + \frac{1}{5^2}\cos 5\theta - \ldots\right)$$

*Problem 8* For the Fourier series of *Problem 7*, let $\theta = \pi$ and show that $\displaystyle\sum_{n=1}^{\infty} \frac{1}{n^2} = \frac{\pi^2}{6}$.

When $\theta = \pi$, $f(\theta) = \pi^2$ (see *Fig 10*). Hence from the Fourier series:

$$\pi^2 = \frac{\pi^2}{3} - 4\left(\cos\pi - \frac{1}{2^2}\cos 2\pi + \frac{1}{3^2}\cos 3\pi - \frac{1}{4^2}\cos 4\pi + \frac{1}{5^2}\cos 5\pi - \ldots\right)$$

i.e. $\pi^2 - \dfrac{\pi^2}{3} = -4\left(-1 - \dfrac{1}{2^2} - \dfrac{1}{3^2} - \dfrac{1}{4^2} - \dfrac{1}{5^2} - \ldots\right)$

$\quad\dfrac{2\pi^2}{3} = 4\left(1 + \dfrac{1}{2^2} + \dfrac{1}{3^2} + \dfrac{1}{4^2} + \dfrac{1}{5^2} + \ldots\right)$

i.e. $\dfrac{2\pi^2}{(3)(4)} = 1 + \dfrac{1}{2^2} + \dfrac{1}{3^2} + \dfrac{1}{4^2} + \dfrac{1}{5^2} + \ldots$

i.e. $\dfrac{\pi^2}{6} = \dfrac{1}{1^2} + \dfrac{1}{2^2} + \dfrac{1}{3^2} + \dfrac{1}{4^2} + \dfrac{1}{5^2} + \ldots$

Hence $\displaystyle\sum_{n=1}^{\infty} \frac{1}{n^2} = \frac{\pi^2}{6}$.

*Problem 9* Determine the half range Fourier cosine series to represent the function $f(x) = 3x$ in the range $0 \leqslant x \leqslant \pi$.

161

From para. 7, for a half range cosine series: $f(x) = a_0 + \sum_{n=1}^{\infty} a_n \cos nx$.

When $f(x) = 3x$, $a_0 = \frac{1}{\pi} \int_0^{\pi} f(x)\, dx = \frac{1}{\pi} \int_0^{\pi} 3x\, dx = \frac{3}{\pi} \left[ \frac{x^2}{2} \right]_0^{\pi} = \frac{3\pi}{2}$

$a_n = \frac{2}{\pi} \int_0^{\pi} f(x) \cos nx\, dx = \frac{2}{\pi} \int_0^{\pi} 3x \cos nx\, dx = \frac{6}{\pi} \left[ \frac{x \sin nx}{n} + \frac{\cos nx}{n^2} \right]_0^{\pi}$ by parts

$$= \frac{6}{\pi} \left[ \left( \frac{\pi \sin n\pi}{n} + \frac{\cos n\pi}{n^2} \right) - \left( 0 + \frac{\cos 0}{n^2} \right) \right]$$

$$= \frac{6}{\pi} \left( 0 + \frac{\cos n\pi}{n^2} - \frac{\cos 0}{n^2} \right) = \frac{6}{\pi n^2} (\cos n\pi - 1)$$

When $n$ is even, $a_n = 0$.

When $n$ is odd, $a_n = \frac{6}{\pi n^2} (-1-1) = \frac{-12}{\pi n^2}$.

Hence $a_1 = \frac{-12}{\pi}$, $a_3 = \frac{-12}{\pi 3^2}$, $a_5 = \frac{-12}{\pi 5^2}$, and so on.

Hence the half range Fourier cosine series is given by:

$$f(x) = 3x = \frac{3\pi}{2} - \frac{12}{\pi} \left( \cos x + \frac{1}{3^2} \cos 3x + \frac{1}{5^2} \cos 5x + \ldots \right)$$

*Problem 10* Find the half-range Fourier sine series to represent the function $f(x) = 3x$ in the range $0 \leqslant x \leqslant \pi$.

From para. 8, for a half range sine series: $f(x) = \sum_{n=1}^{\infty} b_n \sin nx$.

When $f(x) = 3x$, $b_n = \frac{2}{\pi} \int_0^{\pi} f(x) \sin nx\, dx = \frac{2}{\pi} \int_0^{\pi} 3x \sin nx\, dx$

$$= \frac{6}{\pi} \left[ \frac{-x \cos nx}{n} + \frac{\sin nx}{n^2} \right]_0^{\pi}, \text{ by parts,}$$

$$= \frac{6}{\pi} \left[ \left( \frac{-\pi \cos n\pi}{n} + \frac{\sin n\pi}{n^2} \right) - (0+0) \right] = -\frac{6}{n} \cos n\pi.$$

When $n$ is odd, $b_n = \frac{6}{n}$. Hence $b_1 = \frac{6}{1}$, $b_3 = \frac{6}{3}$, $b_5 = \frac{6}{5}$, and so on.

When $n$ is even, $b_n = -\frac{6}{n}$. Hence $b_2 = -\frac{6}{2}$, $b_4 = -\frac{6}{4}$, $b_6 = -\frac{6}{6}$, and so on.

Hence the half range Fourier sine series is given by:

$$f(x) = 3x = 6 \left( \sin x - \frac{1}{2} \sin 2x + \frac{1}{3} \sin 3x - \frac{1}{4} \sin 4x + \frac{1}{5} \sin 5x - \ldots \right)$$

*Problem 11* Expand $f(x) = \cos x$ as a half range Fourier sine series in the range $0 \leqslant x \leqslant \pi$, and sketch the function within and outside of the given range.

When a half range sine series is required then an odd function is implied, i.e. a function symmetrical about the origin. A graph of $y = \cos x$ is shown in *Fig 11* in the range 0 to $\pi$. For $\cos x$ to be symmetrical about the origin the function is as shown by the broken lines in *Fig 11* outside of the given range.

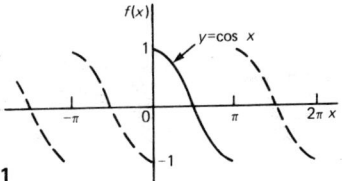

**Fig 11**

From para. 8, for a half range Fourier sine series: $f(x) = \sum\limits_{n=1}^{\infty} b_n \sin nx\, dx$.

$$b_n = \frac{2}{\pi}\int_0^\pi f(x) \sin nx\, dx = \frac{2}{\pi}\int_0^\pi \cos x \sin nx\, dx$$

$$= \frac{2}{\pi}\int_0^\pi \frac{1}{2}\left[\sin(x+nx) - \sin(x-nx)\right] dx$$

$$= \frac{1}{\pi}\left[\frac{-\cos[x(1+n)]}{(1+n)} + \frac{\cos[x(1-n)]}{(1-n)}\right]_0^\pi$$

$$= \frac{1}{\pi}\left[\left(\frac{-\cos[\pi(1+n)]}{(1+n)} + \frac{\cos[\pi(1-n)]}{(1-n)}\right) - \left(\frac{-\cos 0}{(1+n)} + \frac{\cos 0}{(1-n)}\right)\right]$$

When $n$ is odd, $b_n = \frac{1}{\pi}\left[\left(\frac{-1}{(1+n)} + \frac{1}{(1-n)}\right) - \left(\frac{-1}{(1+n)} + \frac{1}{(1-n)}\right)\right] = 0$

When $n$ is even, $b_n = \frac{1}{\pi}\left[\left(\frac{1}{(1+n)} - \frac{1}{(1-n)}\right) - \left(\frac{-1}{(1+n)} + \frac{1}{(1-n)}\right)\right]$

$$= \frac{1}{\pi}\left(\frac{2}{(1+n)} - \frac{2}{(1-n)}\right) = \frac{1}{\pi}\left(\frac{-4n}{1-n^2}\right) = \frac{4n}{\pi(n^2-1)}$$

Hence $b_2 = \frac{8}{3\pi}$, $b_4 = \frac{16}{15\pi}$, $b_6 = \frac{24}{35\pi}$, and so on.

Hence the half range Fourier sine series for $f(x)$ in the range 0 to $\pi$ is given by:

$$f(x) = \frac{8}{3\pi}\sin 2x + \frac{16}{15\pi}\sin 4x + \frac{24}{35\pi}\sin 6x + \ldots\ldots$$

or $f(x) = \frac{8}{\pi}\left(\frac{1}{3}\sin 2x + \frac{2}{(3)(5)}\sin 4x + \frac{3}{(5)(7)}\sin 6x + \ldots\ldots\right)$

## C. FURTHER PROBLEMS ON EVEN AND ODD FUNCTIONS AND HALF RANGE FOURIER SERIES

In *Problems 1 and 2* determine whether the given functions are even, odd or neither even nor odd.

163

1 (a) $x^4$ (b) $\tan 3x$ (c) $2e^{3t}$ (d) $\sin^2 x$

[(a) even (b) odd (c) neither (d) even]

2 (a) $5t^3$ (b) $e^x + e^{-x}$ (c) $\dfrac{\cos\theta}{\theta}$ (e) $e^{x^2}$ [(a) odd (b) even (c) odd (d) even]

3 State whether the following functions which are periodic of period $2\pi$ are even or odd:

(a) $f(\theta) = \begin{cases} \theta, \text{ when } -\pi < \theta < 0 \\ -\theta, \text{ when } 0 < \theta < \pi \end{cases}$, (b) $f(x) = \begin{cases} x, \text{ when } -\dfrac{\pi}{2} < x < \dfrac{\pi}{2} \\ 0, \text{ when } \dfrac{\pi}{2} < x < \dfrac{3\pi}{2} \end{cases}$

[(a) even (b) odd]

4 Determine the Fourier series for the function defined by:

$$f(x) = \begin{cases} -1, & -\pi < x < -\dfrac{\pi}{2} \\ 1, & -\dfrac{\pi}{2} < x < \dfrac{\pi}{2} \\ -1, & \dfrac{\pi}{2} < x < \pi \end{cases},$$

which is periodic outside of this range of period $2\pi$.

$$\left[ f(x) = \frac{4}{\pi}\left(\cos x - \frac{1}{3}\cos 3x + \frac{1}{5}\cos 5x - \frac{1}{7}\cos 7x + \ldots\ldots\right) \right]$$

5 Find the Fourier series for the function $f(x) = \dfrac{1}{2}$ in the range $x = 0$ to $x = \pi$.

$$\left[ f(x) = \frac{2}{\pi}\left(\sin x + \frac{1}{3}\sin 3x + \frac{1}{5}\sin 5x + \frac{1}{7}\sin 7x + \ldots\ldots\right) \right]$$

6 For the Fourier series of *Problem 5* plot graphs of the partial sums of the series and show that the result approximates more and more closely to the function it represents.

7 Obtain the Fourier series of the function defined by: $f(t) = \begin{cases} t+\pi, & -\pi < t < 0 \\ t-\pi, & 0 < t < \pi \end{cases}$,

which is periodic of period $2\pi$. Sketch the given function.

$$\left[ f(t) = -2\left(\sin t + \frac{1}{2}\sin 2t + \frac{1}{3}\sin 3t + \frac{1}{4}\sin 4t + \ldots.\right) \right]$$

8 Determine the Fourier series defined by $f(x) = \begin{cases} 1-x, & -\pi < x < 0 \\ 1+x, & 0 < x < \pi \end{cases}$,

which is periodic of period $2\pi$.

$$\left[ f(x) = \frac{\pi}{2} + 1 - \frac{4}{\pi}\left(\cos x + \frac{1}{3^2}\cos 3x + \frac{1}{5^2}\cos 5x + \ldots\ldots\right) \right]$$

9 In the Fourier series of *Problem 8*, let $x = 0$ and deduce a series for $\pi^2/8$.

$$\left[ \frac{\pi^2}{8} = 1 + \frac{1}{3^2} + \frac{1}{5^2} + \frac{1}{7^2} + \ldots. \right]$$

10 Determine the half range sine series for the function defined by:

$$f(x) = \begin{cases} x, & 0 < x < \dfrac{\pi}{2} \\ 0, & \dfrac{\pi}{2} < x < \pi \end{cases}.$$

$$\left[ f(x) = \frac{2}{\pi}\left(\sin x + \frac{\pi}{4}\sin 2x - \frac{1}{9}\sin 3x - \frac{\pi}{8}\sin 4x + \ldots.\right) \right]$$

11 Obtain (a) the half range cosine series and (b) the half range sine series for the

function $f(t) = \begin{cases} 0, & 0 < t < \frac{\pi}{2} \\ 1, & \frac{\pi}{2} < t < \pi \end{cases}$ .

$$\left[ \begin{array}{l} \text{(a) } f(t) = \frac{1}{2} - \frac{2}{\pi} \left( \cos t - \frac{1}{3} \cos 3t + \frac{1}{5} \cos 5t - \ldots \right) \\ \text{(b) } f(t) = \frac{2}{\pi} \left( \sin t - \sin 2t + \frac{1}{3} \sin 3t + \frac{1}{5} \sin 5t - \frac{1}{6} \sin 6t + \ldots \right) \end{array} \right]$$

12 Find (a) the half range Fourier sine series and (b) the half range Fourier cosine series for the function $f(x) = \sin^2 x$ in the range $0 \leqslant x \leqslant \pi$. Sketch the function within and outside of the given range.

$$\left[ \begin{array}{l} \text{(a) } f(x) = \frac{8}{\pi} \left( \frac{\sin x}{(1)(3)} - \frac{\sin 3x}{(1)(3)(5)} - \frac{\sin 5x}{(3)(5)(7)} - \frac{\sin 7x}{(5)(7)(9)} - \ldots \right) \\ \text{(b) } f(x) = \frac{1}{2}(1 - \cos 2x) \end{array} \right]$$

13 If $f(x) = e^x$ show that the Fourier sine series in the range 0 to $\pi$ is given by:

$$f(x) = \frac{2}{\pi} \sum_{n=1}^{\infty} \frac{n}{1+n^2} \{1 + (-1)^{n-1} e^\pi\} \sin nx.$$

14 Given the function $f(x) = \sin x$ in the range 0 to $\pi$ show that the amplitude of the fourth harmonic of the half range Fourier cosine series is $4/15\pi$.

15 Determine the half range Fourier cosine series in the range $x = 0$ to $x = \pi$ for the

function defined by: $f(x) = \begin{cases} x, & 0 < x < \frac{\pi}{2} \\ (\pi - x), & \frac{\pi}{2} < x < \pi \end{cases}$ .

$$\left[ f(x) = \frac{\pi}{4} - \frac{2}{\pi} \left( \cos 2x + \frac{\cos 6x}{3^2} + \frac{\cos 10x}{5^2} + \ldots \right) \right]$$

# 20 Fourier series over any range

## A. MAIN POINTS CONCERNED WITH FOURIER SERIES OVER ANY RANGE

1    A periodic function $f(x)$ of period $l$ repeats itself when $x$ increases by $l$, i.e. $f(x+l) = f(x)$. The change from functions dealt with previously having period $2\pi$ to functions having period $l$ is not difficult since it may be achieved by a change of variable.

2    To find a Fourier series for a function $f(x)$ in the range $-\frac{l}{2} \leqslant x \leqslant \frac{l}{2}$ a new variable $u$

is introduced such that $f(x)$, as a function of $u$, has period $2\pi$. If $u = \frac{2\pi x}{l}$ then,

when $x = -\frac{l}{2}$, $u = -\pi$ and when $x = \frac{l}{2}$, $u = +\pi$. Also, let $f(x) = f\left(\frac{lu}{2\pi}\right) = F(u)$.

The Fourier series for $F(u)$ is given by: $F(u) = a_0 + \sum_{n=1}^{\infty} (a_n \cos nu + b_n \sin nu)$,

where $a_0 = \frac{1}{2\pi}\int_{-\pi}^{\pi} F(u)\, du$, $a_n = \frac{1}{\pi}\int_{-\pi}^{\pi} F(u) \cos nu\, du$ and $b_n = \frac{1}{\pi}\int_{-\pi}^{\pi} F(u) \sin nu\, du$

3    It is however more usual to change the formula of para. 2 to terms of $x$. Since

$u = \frac{2\pi x}{l}$, then $du = \frac{2\pi}{l}\, dx$, and the limits of integration are $-\frac{l}{2}$ to $+\frac{l}{2}$ instead of

from $-\pi$ to $+\pi$. Hence the Fourier series expressed in terms of $x$ is given by:

$$f(x) = a_0 + \sum_{n=1}^{\infty} \left[ a_n \cos\left(\frac{2\pi nx}{l}\right) + b_n \sin\left(\frac{2\pi nx}{l}\right) \right],$$

where, in the range $-\frac{l}{2}$ to $+\frac{l}{2}$:

$$a_0 = \frac{1}{l}\int_{-\frac{l}{2}}^{\frac{l}{2}} f(x)\, dx, \qquad a_n = \frac{2}{l}\int_{-\frac{l}{2}}^{\frac{l}{2}} f(x) \cos\left(\frac{2\pi nx}{l}\right) dx$$

and $b_n = \frac{2}{l}\int_{-\frac{l}{2}}^{\frac{l}{2}} f(x) \sin\left(\frac{2\pi nx}{l}\right) dx$

The limits of integration may be replaced by any interval of length $l$, such as from 0 to $l$. (See *Problems 1 to 3*)

**Half range series**

4  By making the substitution $u = \frac{\pi x}{l}$ (see para. 2), the range $x = 0$ to $x = l$ corresponds to the range $u = 0$ to $u = \pi$. Hence a function may be expanded in a series of either cosine terms or sine terms only, i.e. a **half range Fourier series**.

5  A **half range cosine series** in the range 0 to $l$ can be expanded as:

$$f(x) = a_0 + \sum_{n=1}^{\infty} a_n \cos\left(\frac{n\pi x}{l}\right)$$

where $a_0 = \frac{1}{l} \int_0^l f(x)\, dx$ and $a_n = \frac{2}{l} \int_0^l f(x) \cos\left(\frac{n\pi x}{l}\right)\, dx$

6  A **half range sine series** in the range 0 to $l$ can be expanded as:

$$f(x) = \sum_{n=1}^{\infty} b_n \sin\left(\frac{n\pi x}{l}\right) ,$$

where $b_n = \frac{2}{l} \int_0^l f(x) \sin\left(\frac{n\pi x}{l}\right)\, dx$

(See *Problems 4 and 5*)

# B. WORKED PROBLEMS ON FOURIER SERIES OVER ANY RANGE

*Problem 1* The voltage from a square wave generator is of the form:
$$v(t) = \begin{cases} 0, & -4 < t < 0 \\ 10, & 0 < t < 4 \end{cases} \quad \text{and has a period of 8 ms.}$$

Find the Fourier series for this periodic function.

The square wave is shown in *Fig 1*. From para. 3, the Fourier series is of the form:

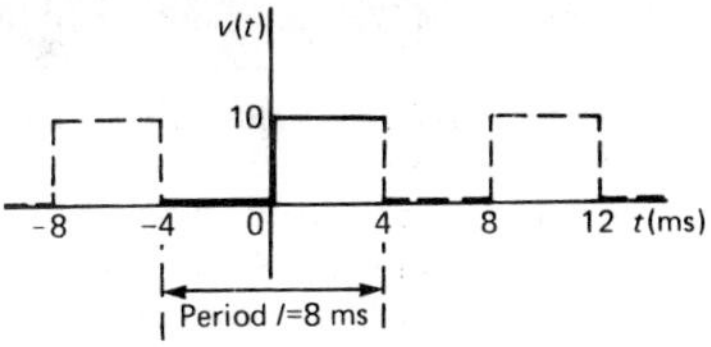

**Fig 1**

$$v(t) = a_0 + \sum_{n=1}^{\infty} \left[ a_n \cos\left(\frac{2\pi n t}{l}\right) + b_n \sin\left(\frac{2\pi n t}{l}\right) \right]$$

$$a_0 = \frac{1}{l} \int_{-\frac{l}{2}}^{\frac{l}{2}} v(t)\, dt = \frac{1}{8} \int_{-4}^{4} v(t)\, dt = \frac{1}{8}\left\{ \int_{-4}^{0} 0\, dt + \int_{0}^{4} 10\, dt \right\} = \frac{1}{8}\left[ 10t \right]_0^4 = 5$$

$$a_n = \frac{2}{l} \int_{-\frac{l}{2}}^{\frac{l}{2}} v(t) \cos\left(\frac{2\pi n t}{l}\right)\, dt = \frac{2}{8} \int_{-4}^{4} v(t) \cos\left(\frac{2\pi n t}{8}\right)\, dt$$

$$= \frac{1}{4}\left\{ \int_{-4}^{0} 0 \cos\left(\frac{\pi n t}{4}\right) dt + \int_{0}^{4} 10 \cos\left(\frac{\pi n t}{4}\right) dt \right\}$$

167

$$= \frac{1}{4}\left[\frac{10 \sin\left(\frac{\pi nt}{4}\right)}{\left(\frac{\pi n}{4}\right)}\right]_0^4 = \frac{10}{\pi n}\ [\sin \pi n - \sin 0] = 0$$

$$\text{for } n = 1, 2, 3, \ldots.$$

$$b_n = \frac{2}{l}\int_{-\frac{l}{2}}^{\frac{l}{2}} v(t) \sin\left(\frac{2\pi nt}{l}\right)dt = \frac{2}{8}\int_{-4}^{4} v(t) \sin\left(\frac{2\pi nt}{8}\right)dt$$

$$= \frac{1}{4}\left\{\int_{-4}^{0} 0 \sin\left(\frac{\pi nt}{4}\right)dt + \int_{0}^{4} 10 \sin\left(\frac{\pi nt}{4}\right)dt\right\}$$

$$= \frac{1}{4}\left[\frac{-10 \cos\frac{\pi nt}{4}}{\frac{\pi n}{4}}\right]_0^4 = \frac{-10}{\pi n}\ [\cos \pi n - \cos 0]$$

When $n$ is even, $b_n = 0$.

When $n$ is odd, $b_1 = \frac{-10}{\pi}\ (-1-1) = \frac{20}{\pi}$, $b_3 = \frac{-10}{3\pi}\ (-1-1) = \frac{20}{3\pi}$, $b_5 = \frac{20}{5\pi}$, and so on.

Thus the Fourier series for the function $v(t)$ is given by:

$$v(t) = 5 + \frac{20}{\pi}\left[\sin\frac{\pi t}{4} + \frac{1}{3}\sin\left(\frac{3\pi t}{4}\right) + \frac{1}{5}\sin\left(\frac{5\pi t}{4}\right) + \ldots\ldots\right]$$

*Problem 2*  Obtain the Fourier series for the function defined by:
$$f(x) = \begin{cases} 0, \text{ when } -2 < x < -1 \\ 5, \text{ when } -1 < x < 1 \\ 0, \text{ when } 1 < x < 2 \end{cases}$$
The function is periodic outside of this range of period 4.

The function $f(x)$ is shown in *Fig 2* where period, $l = 4$. Since the function is symmetrical about the $f(x)$ axis it is an even function and the Fourier series contains no sine terms (i.e. $b_n = 0$).

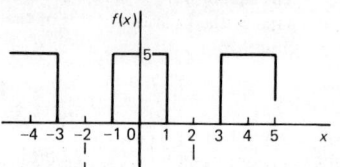

Fig 2

Thus, from para. 3, $f(x) = a_0 + \sum_{n=1}^{\infty} a_n \cos\left(\frac{2\pi nx}{l}\right)$.

$$a_0 = \frac{1}{l}\int_{-\frac{l}{2}}^{\frac{l}{2}} f(x)\ dx = \frac{1}{4}\int_{-2}^{2} f(x)\ dx = \frac{1}{4}\left\{\int_{-2}^{-1} 0\ dx + \int_{-1}^{1} 5\ dx + \int_{1}^{2} 0\ dx\right\}$$

$$= \frac{1}{4}\left[5x\right]_{-1}^{1} = \frac{1}{4}[(5)-(-5)] = \frac{10}{4} = \frac{5}{2}$$

$$a_n = \frac{2}{l}\int_{-\frac{l}{2}}^{\frac{l}{2}} f(x) \cos\left(\frac{2\pi nx}{l}\right)dx = \frac{2}{4}\int_{-2}^{2} f(x) \cos\left(\frac{2\pi nx}{4}\right)dx$$

168

$$= \frac{1}{2} \left\{ \int_{-2}^{-1} 0 \cos\left(\frac{\pi n x}{2}\right) dx + \int_{-1}^{1} 5 \cos\left(\frac{\pi n x}{2}\right) dx + \int_{1}^{2} 0 \cos\left(\frac{\pi n x}{2}\right) dx \right\}$$

$$= \frac{5}{2} \left[ \frac{\sin\frac{\pi n x}{2}}{\frac{\pi n}{2}} \right]_{-1}^{1} = \frac{5}{\pi n} \left[ \sin\left(\frac{\pi n}{2}\right) - \sin\left(\frac{-\pi n}{2}\right) \right]$$

When $n$ is even, $a_n = 0$.

When $n$ is odd, $a_1 = \frac{5}{\pi}(1--1) = \frac{10}{\pi}$, $a_3 = \frac{5}{3\pi}(-1-1) = \frac{-10}{3\pi}$,

$$a_5 = \frac{5}{5\pi}(1--1) = \frac{10}{5\pi}, \text{ and so on.}$$

Hence the Fourier series for the function $f(x)$ is given by:

$$f(x) = \frac{5}{2} + \frac{10}{\pi} \left[ \cos\left(\frac{\pi x}{2}\right) - \frac{1}{3}\cos\left(\frac{3\pi x}{2}\right) + \frac{1}{5}\cos\left(\frac{5\pi x}{2}\right) - \frac{1}{7}\cos\left(\frac{7\pi x}{2}\right) + \dots \right]$$

*Problem 3* Determine the Fourier series for the function $f(t) = t$ in the range $t = 0$ to $t = 3$.

The function $f(t) = t$ in the interval 0 to 3 is shown in *Fig 3*. Although the function is not periodic it may be constructed outside of this range so that it is periodic of period 3, as shown by the broken lines in *Fig 3*. From para. 3, the Fourier series is given by:

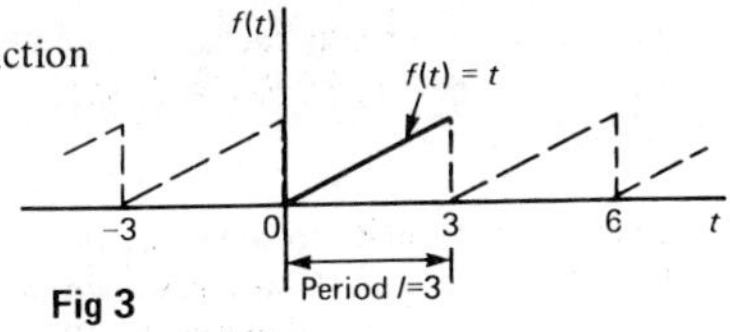

**Fig 3**

$$f(t) = a_0 + \sum_{n=1}^{\infty} \left[ a_n \cos\left(\frac{2\pi n t}{l}\right) + b_n \sin\left(\frac{2\pi n t}{l}\right) \right]$$

$$a_0 = \frac{1}{l} \int_{-\frac{l}{2}}^{\frac{l}{2}} f(t) dt = \frac{1}{l} \int_{0}^{l} f(t) dt = \frac{1}{3} \int_{0}^{3} t \, dt = \frac{1}{3} \left[ \frac{t^2}{2} \right]_{0}^{3} = \frac{3}{2}$$

$$a_n = \frac{2}{l} \int_{-\frac{l}{2}}^{\frac{l}{2}} f(t) \cos\left(\frac{2\pi n t}{l}\right) dt = \frac{2}{l} \int_{0}^{l} t \cos\left(\frac{2\pi n t}{l}\right) dt = \frac{2}{3} \int_{0}^{3} t \cos\left(\frac{2\pi n t}{3}\right) dt$$

$$= \frac{2}{3} \left[ \frac{t \sin\left(\frac{2\pi n t}{3}\right)}{\left(\frac{2\pi n}{3}\right)} + \frac{\cos\left(\frac{2\pi n t}{3}\right)}{\left(\frac{2\pi n}{3}\right)^2} \right]_{0}^{3} \quad \text{by parts}$$

$$= \frac{2}{3} \left[ \left\{ \frac{3 \sin 2\pi n}{\left(\frac{2\pi n}{3}\right)} + \frac{\cos 2\pi n}{\left(\frac{2\pi n}{3}\right)^2} \right\} - \left\{ 0 - \frac{\cos 0}{\left(\frac{2\pi n}{3}\right)^2} \right\} \right] = 0$$

169

$$b_n = \frac{2}{l}\int_{-\frac{l}{2}}^{\frac{l}{2}} f(t)\sin\left(\frac{2\pi nt}{l}\right) dt = \frac{2}{l}\int_0^l t\sin\left(\frac{2\pi nt}{l}\right) dt = \frac{2}{3}\int_0^3 t\sin\left(\frac{2\pi nt}{3}\right) dt$$

$$= \frac{2}{3}\left[\frac{-t\cos\left(\frac{2\pi nt}{3}\right)}{\left(\frac{2\pi n}{3}\right)} + \frac{\sin\left(\frac{2\pi nt}{3}\right)}{\left(\frac{2\pi n}{3}\right)^2}\right]_0^3 \quad \text{by parts}$$

$$= \frac{2}{3}\left[\left\{\frac{-3\cos 2\pi n}{\left(\frac{2\pi n}{3}\right)} + \frac{\sin 2\pi n}{\left(\frac{2\pi n}{3}\right)^2}\right\} - \left\{0 + \frac{\sin 0}{\left(\frac{2\pi n}{3}\right)^2}\right\}\right]$$

$$= \frac{2}{3}\left[\frac{-3\cos 2\pi n}{\left(\frac{2\pi n}{3}\right)}\right] = \frac{-3}{\pi n}\cos 2\pi n = \frac{-3}{\pi n}$$

Hence $b_1 = \frac{-3}{\pi}$, $b_2 = \frac{-3}{2\pi}$, $b_3 = \frac{-3}{3\pi}$, and so on.

Thus the Fourier series for the function $f(t)$ in the range 0 to 3 is given by:

$$f(t) = \frac{3}{2} - \frac{3}{\pi}\left[\sin\left(\frac{2\pi t}{3}\right) + \frac{1}{2}\sin\left(\frac{4\pi t}{3}\right) + \frac{1}{3}\sin\left(\frac{6\pi t}{3}\right) + \dots\dots\right]$$

*Problem 4* Determine the half range Fourier cosine series for the function $f(x) = x$ in the range $0 \leqslant x \leqslant 2$. Sketch the function within and outside of the given range.

A half range Fourier cosine series indicates an even function. Thus the graph of $f(x) = x$ in the range 0 to 2 is shown in *Fig 4* and is extended outside of this range so as to be symmetrical about the $f(x)$ axis as shown by the broken lines.

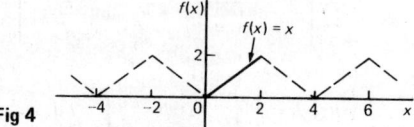

**Fig 4**

From para. 5, for a half range cosine series: $f(x) = a_0 + \sum_{n=1}^{\infty} a_n \cos\left(\frac{n\pi x}{l}\right)$

$$a_0 = \frac{1}{l}\int_0^l f(x)\, dx = \frac{1}{2}\int_0^2 x\, dx = \frac{1}{2}\left[\frac{x^2}{2}\right]_0^2 = 1$$

$$a_n = \frac{2}{l}\int_0^l f(x)\cos\left(\frac{n\pi x}{l}\right)\, dx = \frac{2}{2}\int_0^2 x\cos\left(\frac{n\pi x}{2}\right) dx = \left[\frac{x\sin\left(\frac{n\pi x}{2}\right)}{\left(\frac{n\pi}{2}\right)} + \frac{\cos\left(\frac{n\pi x}{2}\right)}{\left(\frac{n\pi}{2}\right)^2}\right]_0^2$$

$$= \left[\left(\frac{2\sin n\pi}{\left(\frac{n\pi}{2}\right)} + \frac{\cos n\pi}{\left(\frac{n\pi}{2}\right)^2}\right) - \left(0 + \frac{\cos 0}{\left(\frac{n\pi}{2}\right)^2}\right)\right]$$

$$= \left[ \frac{\cos n\pi}{\left(\frac{n\pi}{2}\right)^2} - \frac{1}{\left(\frac{n\pi}{2}\right)^2} \right] = \left(\frac{2}{\pi n}\right)^2 (\cos n\pi - 1)$$

When $n$ is even, $a_n = 0$.

$b_1 = \frac{-8}{\pi^2}$, $b_3 = \frac{-8}{\pi^2 3^2}$, $b_5 = \frac{-8}{\pi^2 5^2}$, and so on.

Hence the half range Fourier cosine series for $f(x)$ in the range 0 to 2 is given by:

$$f(x) = 1 - \frac{8}{\pi^2} \left[ \cos\left(\frac{\pi x}{2}\right) + \frac{1}{3^2} \cos\left(\frac{3\pi x}{2}\right) + \frac{1}{5^2} \cos\left(\frac{5\pi x}{2}\right) + \ldots \right]$$

*Problem 5* Find the half range Fourier sine series for the function $f(x) = x$ in the range $0 \leqslant x \leqslant 2$. Sketch the function within and outside of the given range.

A half range Fourier sine series indicates an odd function. Thus the graphs of $f(x) = x$ in the range 0 to 2 is shown in *Fig 5* and is extended outside of this range so as to be symmetrical about the origin, as shown by the broken lines.

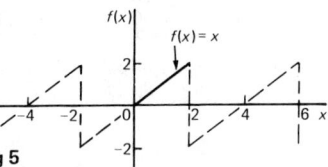

**Fig 5**

From para. 6, for a half range sine series: $f(x) = \sum_{n=1}^{\infty} b_n \sin\left(\frac{n\pi x}{l}\right)$

$$b_n = \frac{2}{l} \int_0^l f(x) \sin\left(\frac{n\pi x}{l}\right) dx = \frac{2}{2} \int_0^2 x \sin\left(\frac{n\pi x}{2}\right) dx = \left[ \frac{-x \cos\left(\frac{n\pi x}{2}\right)}{\left(\frac{n\pi}{2}\right)} - \frac{\sin\left(\frac{n\pi x}{2}\right)}{\left(\frac{n\pi}{2}\right)^2} \right]_0^2$$

$$= \left[ \left( \frac{-2 \cos n\pi}{\left(\frac{n\pi}{2}\right)} - \frac{\sin n\pi}{\left(\frac{n\pi}{2}\right)^2} \right) - \left( 0 - \frac{\sin 0}{\left(\frac{n\pi}{2}\right)^2} \right) \right] = \frac{-2 \cos n\pi}{\frac{n\pi}{2}} = \frac{-4}{n\pi} \cos n\pi.$$

Hence $b_1 = \frac{-4}{\pi}(-1) = \frac{4}{\pi}$, $b_2 = \frac{-4}{2\pi}(1) = \frac{-4}{2\pi}$, $b_3 = \frac{-4}{3\pi}(-1) = \frac{4}{3\pi}$, and so on.

Thus the half range Fourier sine series in the range 0 to 2 is given by:

$$f(x) = \frac{4}{\pi} \left[ \sin\left(\frac{\pi x}{2}\right) - \frac{1}{2}\sin\left(\frac{2\pi x}{2}\right) + \frac{1}{3}\sin\left(\frac{3\pi x}{2}\right) - \frac{1}{4}\sin\left(\frac{4\pi x}{2}\right) + \ldots \right]$$

## C. FURTHER PROBLEMS ON FOURIER SERIES OVER ANY RANGE

1   Determine the Fourier series for the half wave rectified sinusoidal voltage $V \sin \omega t$ defined by:

$$f(t) = \begin{cases} V \sin \omega t, & 0 < t < \frac{\pi}{\omega} \\ 0, & \frac{\pi}{\omega} < t < \frac{2\pi}{\omega} \end{cases}$$

which is periodic of period $\frac{2\pi}{\omega}$.

$$\left[ f(t) = \frac{V}{\pi} + \frac{V}{2} \sin \omega t - \frac{2V}{\pi} \left( \frac{\cos 2\omega t}{(1)(3)} + \frac{\cos 4\omega t}{(3)(5)} + \frac{\cos 6\omega t}{(5)(7)} + \ldots \right) \right]$$

2   The voltage from a square wave generator is of the form:
$$v(t) = \begin{cases} 0, & -10 < t < 0 \\ 5, & 0 < t < 10 \end{cases}$$
and is periodic of period 20. Show that the Fourier series for the function is given by:

$$v(t) = \frac{5}{2} + \frac{10}{\pi}\left[ \sin\left(\frac{\pi t}{10}\right) + \frac{1}{3}\sin\left(\frac{3\pi t}{10}\right) + \frac{1}{5}\sin\left(\frac{5\pi t}{10}\right) + \ldots\ldots \right]$$

3   Find the Fourier series for $f(x) = x$ in the range $x = 0$ to $x = 5$.

$$\left[ f(x) = \frac{5}{2} - \frac{5}{\pi}\left[ \sin\left(\frac{2\pi x}{5}\right) + \frac{1}{2}\sin\left(\frac{4\pi x}{5}\right) + \frac{1}{3}\sin\left(\frac{6\pi x}{5}\right) + \ldots\ldots \right]\right]$$

4   A periodic function of period $2\pi$ is defined by:
$$f(x) = \begin{cases} -3, & -2 < x < 0 \\ +3, & 0 < x < 2 \end{cases}$$
Sketch the function and obtain the Fourier series for the function.

$$\left[ f(x) = \frac{6}{\pi}\left( \sin\left(\frac{\pi x}{3}\right) + \frac{1}{3}\sin\left(\frac{3\pi x}{3}\right) + \frac{1}{5}\sin\left(\frac{5\pi x}{3}\right) + \ldots\ldots \right)\right]$$

5   Determine the half range Fourier cosine series for the function $f(x) = x$ in the range $0 \leqslant x \leqslant 3$. Sketch the function within and outside of the given range.

$$\left[ f(x) = \frac{3}{2} + \frac{12}{\pi^2}\left( \cos\left(\frac{\pi x}{3}\right) + \frac{1}{3^2}\cos\left(\frac{3\pi x}{3}\right) + \frac{1}{5^2}\cos\left(\frac{5\pi x}{3}\right) + \ldots\ldots \right)\right]$$

6   Find the half range Fourier sine series for the function $f(x) = x$ in the range $0 \leqslant x \leqslant 3$. Sketch the function within and outside of the given range.

$$\left[ f(x) = \frac{6}{\pi}\left( \sin\left(\frac{\pi x}{3}\right) - \frac{1}{2}\sin\left(\frac{2\pi x}{3}\right) + \frac{1}{3}\sin\left(\frac{3\pi x}{3}\right) - \frac{1}{4}\sin\left(\frac{4\pi x}{3}\right) + \ldots\ldots \right)\right]$$

7   Determine the half range Fourier sine series for the function defined by:
$$f(t) = \begin{cases} t, & 0 < t < 1 \\ (2-t), & 1 < t < 2 \end{cases}$$

$$\left[ f(t) = \frac{8}{\pi^2}\left( \sin\left(\frac{\pi t}{2}\right) - \frac{1}{3^2}\sin\left(\frac{3\pi t}{2}\right) + \frac{1}{5^2}\sin\left(\frac{5\pi t}{2}\right) - \ldots\ldots \right)\right]$$

8   Show that the half range Fourier cosine series for the function $f(\theta) = \theta^2$ in the range 0 to 4 is given by:

$$f(\theta) = \frac{16}{3} - \frac{64}{\pi^2}\left( \cos\left(\frac{\pi\theta}{4}\right) - \frac{1}{2^2}\cos\left(\frac{2\pi\theta}{4}\right) + \frac{1}{3^2}\cos\left(\frac{3\pi\theta}{4}\right) - \ldots\ldots \right)$$

Sketch the function within and outside of the given range.

# 21 A numerical method of harmonic analysis

## A. MAIN POINTS CONCERNED WITH A NUMERICAL METHOD OF HARMONIC ANALYSIS

1 Many practical waveforms can be represented by simple mathematical expressions, and, by using Fourier series, the magnitude of their harmonic components determined. For waveforms not in this category, analysis may be achieved by numerical methods.

2 **Harmonic analysis** is the process of resolving a periodic, non-sinusoidal quantity into a series of sinusoidal components of ascending order of frequency.

3 The Fourier coefficients $a_0$, $a_n$ and $b_n$ used in chapters 17 to 20 all require functions to be integrated, i.e.,

$$a_0 = \frac{1}{2\pi} \int_{-\pi}^{\pi} f(x)\,dx = \frac{1}{2\pi} \int_{0}^{2\pi} f(x)\,dx = \text{mean value of } f(x) \text{ in the range } -\pi \text{ to } \pi$$
$$\text{or } 0 \text{ to } 2\pi.$$

$$a_n = \frac{1}{\pi} \int_{-\pi}^{\pi} f(x) \cos nx\,dx = \frac{1}{\pi} \int_{0}^{2\pi} f(x) \cos nx\,dx = \text{twice the mean value of}$$
$$f(x) \cos nx \text{ in the range } 0 \text{ to } 2\pi$$

$$b_n = \frac{1}{\pi} \int_{-\pi}^{\pi} f(x) \sin nx\,dx = \frac{1}{\pi} \int_{0}^{2\pi} f(x) \sin nx\,dx = \text{twice the mean value of}$$
$$f(x) \sin nx \text{ in the range } 0 \text{ to } 2\pi$$

However, irregular waveforms are not usually defined by mathematical expressions and thus Fourier coefficients cannot be determined by using calculus. In these cases, approximate methods, such as the **trapezoidal rule** can be used to evaluate the Fourier coefficients.

4 Most practical waveforms to be analysed are periodic. Let the period of a waveform be $2\pi$ and be divided into $p$ equal parts as shown in *Fig 1*. The width of each interval is thus $2\pi/p$. Let the ordinates be labelled $y_0, y_1, y_2, \ldots\ldots, y_p$ (note that $y_0 = y_p$). The trapezoidal rule states:

Area $\simeq$ (width of interval) $\left[\frac{1}{2}(\text{first} + \text{last ordinate}) + \text{sum of remaining ordinates}\right]$

$$\simeq \frac{2\pi}{p}\left[\frac{1}{2}(y_0 + y_p) + y_1 + y_2 + y_3 + \ldots\ldots\ldots\right]$$

Since $y_0 = y_p$, then $\frac{1}{2}(y_0 + y_p) = y_0 = y_p$. Hence area $\simeq \frac{2\pi}{p} \sum_{k=1}^{p} y_k$

173

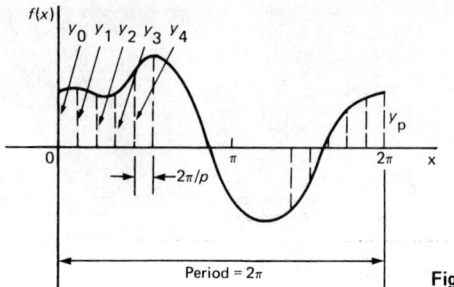

Fig 1

Mean value $= \dfrac{\text{area}}{\text{length of base}} \simeq \dfrac{1}{2\pi}\left(\dfrac{2\pi}{p}\right)\displaystyle\sum_{k=1}^{p} y_k \simeq \dfrac{1}{p}\displaystyle\sum_{k=1}^{p} y_k.$

However, $a_0 = $ mean value of $f(x)$ in the range 0 to $2\pi$.

Thus $a_0 \simeq \dfrac{1}{p}\displaystyle\sum_{k=1}^{p} y_k$  (1)

Similarly, $a_n = $ twice the mean value of $f(x)\cos nx$ in the range 0 to $2\pi$,

thus $a_n \simeq \dfrac{2}{p}\displaystyle\sum_{k=1}^{p} y_k \cos nx_k$  (2)

and $b_n = $ twice the mean value of $f(x)\sin nx$ in the range 0 to $2\pi$,

thus $b_n \simeq \dfrac{2}{p}\displaystyle\sum_{k=1}^{p} y_k \sin nx_k$  (3)

(See *Problem 1*)

5  It is sometimes possible to predict the harmonic content of a waveform on inspection of particular waveform characteristics.

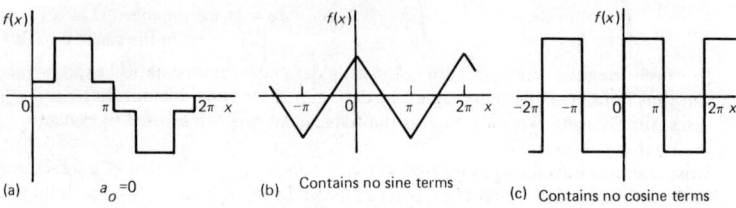

(a)  $a_o = 0$

(b)  Contains no sine terms

(c)  Contains no cosine terms

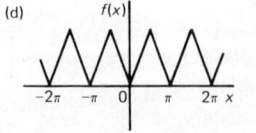

(d)

(e)

Contains only even harmonics  Fig 2  Contains only odd harmonics

174

(i) If a periodic waveform is such that the area above the horizontal axis is equal to the area below then the mean value is zero. Hence $a_0 = 0$ (see *Fig 2(a)*).

(ii) An **even function** is symmetrical about the vertical axis and contains **no sine terms** (see *Fig 2(b)*).

(iii) An **odd function** is symmetrical about the origin and contains no **cosine terms** (see *Fig 2(c)*).

(iv) $f(x) = f(x+\pi)$ represents a waveform which repeats after half a cycle and **only even harmonics** are present (see *Fig 2(d)*).

(v) $f(x) = -f(x+\pi)$ represents a waveform for which the positive and negative cycles are identical in shape and **only odd harmonics** are present (see *Fig 2(e)*). (See *Problems 2 and 3*)

## B. WORKED PROBLEMS ON A NUMERICAL METHOD OF HARMONIC ANALYSIS

*Problem 1* The values of the voltage $v$ volts at different moments in a cycle are given by:

| $\theta$ degrees | 30 | 60 | 90 | 120 | 150 | 180 | 210 | 240 | 270 | 300 | 330 | 360 |
|---|---|---|---|---|---|---|---|---|---|---|---|---|
| $v$ (volts) | 62 | 35 | −38 | −64 | −63 | −52 | −28 | 24 | 80 | 96 | 90 | 70 |

Draw the graph of voltage $v$ against angle $\theta$ and analyse the voltage into its first three constituent harmonics, each coefficient correct to 2 decimal places.

The graph of voltage $v$ against angle $\theta$ is shown in *Fig 3*. The range 0 to $2\pi$ is divided into 12 equal intervals giving an interval width of $2\pi/12$, i.e. $\pi/6$ or $30°$. The values of the ordinates $y_1, y_2, y_3, \ldots\ldots$ are 62, 35, −38, $\ldots\ldots$ from the given table of

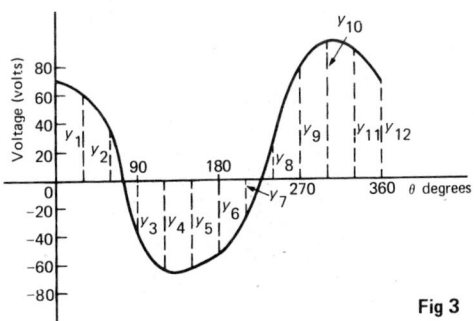

**Fig 3**

values. If a larger number of intervals are used, results having a greater accuracy are achieved. The data is tabulated in the proforma shown in *Table 1*.

From equation (1), para. 4, $a_0 \simeq \dfrac{1}{p}\displaystyle\sum_{k=1}^{p} y_k = \dfrac{1}{12}(212) = 17.67$ (since $p = 12$)

From equation (2), para. 4, $a_n \simeq \dfrac{2}{p}\displaystyle\sum_{k=1}^{p} y_k \cos nx_k$

175

TABLE 1

| Ordinates | $\theta°$ | $V$ | $\cos\theta$ | $V\cos\theta$ | $\sin\theta$ | $V\sin\theta$ | $\cos 2\theta$ | $V\cos 2\theta$ | $\sin 2\theta$ | $V\sin 2\theta$ | $\cos 3\theta$ | $V\cos 3\theta$ | $\sin 3\theta$ | $V\sin 3\theta$ |
|---|---|---|---|---|---|---|---|---|---|---|---|---|---|---|
| $y_1$ | 30 | 62 | 0.866 | 53.69 | 0.5 | 31 | 0.5 | 31 | 0.866 | 53.69 | 0 | 0 | 1 | 62 |
| $y_2$ | 60 | 35 | 0.5 | 17.5 | 0.866 | 30.31 | -0.5 | -17.5 | 0.866 | 30.31 | -1 | -35 | 0 | 0 |
| $y_3$ | 90 | -38 | 0 | 0 | 1 | -38 | -1 | 38 | 0 | 0 | 0 | 0 | -1 | 38 |
| $y_4$ | 120 | -64 | -0.5 | 32 | 0.866 | -55.42 | -0.5 | 32 | -0.866 | 55.42 | 1 | -64 | 0 | 0 |
| $y_5$ | 150 | -63 | -0.866 | 54.56 | 0.5 | -31.5 | 0.5 | -31.5 | -0.866 | 54.56 | 0 | 0 | 1 | -63 |
| $y_6$ | 180 | -52 | -1 | 52 | 0 | 0 | 1 | -52 | 0 | 0 | -1 | 52 | 0 | 0 |
| $y_7$ | 210 | -28 | -0.866 | 24.25 | -0.5 | 14 | 0.5 | -14 | 0.866 | -24.25 | 0 | 0 | -1 | 28 |
| $y_8$ | 240 | 24 | -0.5 | -12 | -0.866 | -20.78 | -0.5 | -12 | 0.866 | 20.78 | 1 | 24 | 0 | 0 |
| $y_9$ | 270 | 80 | 0 | 0 | -1 | -80 | -1 | -80 | 0 | 0 | 0 | 0 | 1 | 80 |
| $y_{10}$ | 300 | 96 | 0.5 | 48 | -0.866 | -83.14 | -0.5 | -48 | -0.866 | -83.14 | -1 | -96 | 0 | 0 |
| $y_{11}$ | 330 | 90 | 0.866 | 77.94 | -0.5 | -45 | 0.5 | 45 | -0.866 | -77.94 | 0 | 0 | -1 | -90 |
| $y_{12}$ | 360 | 70 | 1 | 70 | 0 | 0 | 1 | 70 | 0 | 0 | 1 | 70 | 0 | 0 |
| $\sum_{k=1}^{12} y_k = 212$ | | | $\sum_{k=1}^{12} y_k\cos\theta_k$ $= 417.94$ | | $\sum_{k=1}^{12} y_k\sin\theta_k$ $= -278.53$ | | $\sum_{k=1}^{12} y_k\cos 2\theta_k$ $= -39$ | | $\sum_{k=1}^{12} y_k\sin 2\theta_k$ $= 29.43$ | | $\sum_{k=1}^{12} y_k\cos 3\theta_k$ $= -49$ | | $\sum_{k=1}^{12} y_k\sin 3\theta_k$ $= 55$ | |

Hence $a_1 \simeq \frac{2}{12}(417.94) = 69.66$; $a_2 \simeq \frac{2}{12}(-39) = -6.50$;

and $a_3 \simeq \frac{2}{12}(-49) = -8.17$

From equation (3), para. 4, $b_n \simeq \frac{2}{p} \sum_{k=1}^{p} y_k \sin nx_k$

Hence $b_1 \simeq \frac{2}{12}(-278.53) = -46.42$; $b_2 \simeq \frac{2}{12}(29.43) = 4.91$;

and $b_3 \simeq \frac{2}{12}(55) = 9.17$

Substituting these values into the Fourier series: $f(x) = a_0 + \sum_{n=1}^{\infty}(a_n \cos nx + b_n \sin nx)$

gives: $v = 17.67 + 69.66 \cos \theta - 6.50 \cos 2\theta - 8.17 \cos 3\theta + \ldots \ldots$
$-46.42 \sin \theta + 4.91 \sin 2\theta + 9.17 \sin 3\theta + \ldots \ldots$

*Problem 2* Without calculating Fourier coefficients state which harmonics will be present in the waveforms shown in *Fig 4*.

(a) The waveform shown in *Fig 4(a)* is symmetrical about the origin and is thus an odd function. An odd function contains no cosine terms. Also, the waveform has the characteristic $f(x) = -f(x+\pi)$, i.e., the positive and negative half cycles are identical in shape. Only odd harmonics can be present in such a waveform. Thus the waveform shown in *Fig 4(a)* contains **only odd sine terms**. Since the area above the x-axis is equal to the area below, $a_0 = 0$.

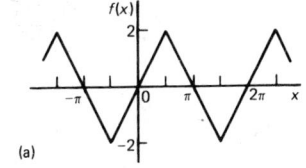

**Fig 4**

(b) The waveform shown in *Fig 4(b)* is symmetrical about the $f(x)$ axis and is thus an even function. An even function contains no sine terms. Also, the waveform has the characteristic $f(x) = f(x+\pi)$, i.e. the waveform repeats itself after half a cycle. Only even harmonics can be present in such a waveform. Thus the waveform shown in *Fig 4(b)* contains **only even cosine terms** (together with a constant term, $a_0$).

*Problem 3* An alternating current $i$ amperes is shown in *Fig 5*. Analyse the waveform into its constituent harmonics as far as and including the fifth harmonic, correct to 2 decimal places, by taking $30°$ intervals.

With reference to *Fig 5*, the following characteristics are noted:
(i) The mean value is zero since the area above the $\theta$ axis is equal to the area below it. Thus the constant term, or d.c. component, $a_0 = 0$.
(ii) Since the waveform is symmetrical about the origin the function $i$ is odd, which means that there are no cosine terms present in the Fourier series.
(iii) The waveform is of the form $f(\theta) = -f(\theta+\pi)$ which means that only odd harmonics are present.

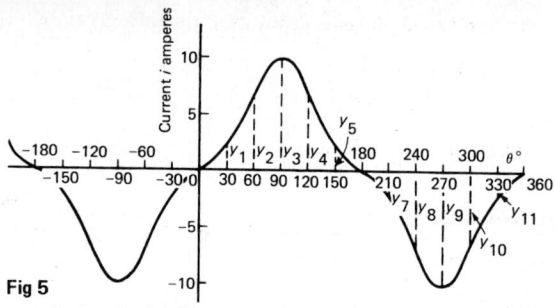

Fig 5

Investigating waveform characteristics has thus saved unnecessary calculations and in this case the Fourier series has only odd sine terms present, i.e.,

$i = b_1 \sin \theta + b_3 \sin 3\theta + b_5 \sin 5\theta + \ldots\ldots\ldots$

A proforma, similar to *Table 1*, but without the 'cosine terms' columns and without the 'even sine terms' columns is shown in *Table 2* up to, and including, the fifth

TABLE 2

| Ordinate | $\theta°$ | $i$ | $\sin \theta$ | $i \sin \theta$ | $\sin 3\theta$ | $i \sin 3\theta$ | $\sin 5\theta$ | $i \sin 5\theta$ |
|---|---|---|---|---|---|---|---|---|
| $y_1$ | 30 | 2 | 0.5 | 1 | 1 | 2 | 0.5 | 1 |
| $y_2$ | 60 | 7 | 0.866 | 6.06 | 0 | 0 | −0.866 | −6.06 |
| $y_3$ | 90 | 10 | 1 | 10 | −1 | −10 | 1 | 10 |
| $y_4$ | 120 | 7 | 0.866 | 6.06 | 0 | 0 | −0.866 | −6.06 |
| $y_5$ | 150 | 2 | 0.5 | 1 | 1 | 2 | 0.5 | 1 |
| $y_6$ | 180 | 0 | 0 | 0 | 0 | 0 | 0 | 0 |
| $y_7$ | 210 | −2 | −0.5 | 1 | −1 | 2 | −0.5 | 1 |
| $y_8$ | 240 | −7 | −0.866 | 6.06 | 0 | 0 | 0.866 | −6.06 |
| $y_9$ | 270 | −10 | −1 | 10 | 1 | −10 | −1 | 10 |
| $y_{10}$ | 300 | −7 | −0.866 | 6.06 | 0 | 0 | 0.866 | −6.06 |
| $y_{11}$ | 330 | −2 | −0.5 | 1 | −1 | 2 | −0.5 | 1 |
| $y_{12}$ | 360 | 0 | 0 | 0 | 0 | 0 | 0 | 0 |
| | | | $\sum_{k=1}^{12} i_k \sin \theta_k$ | | $\sum_{k=1}^{12} i_k \sin 3\theta_k$ | | $\sum_{k=1}^{12} i_k \sin 5\theta_k$ | |
| | | | = 48.24 | | = −12 | | = −0.24 | |

harmonic, from which the Fourier coefficients $b_1$, $b_3$ and $b_5$ can be determined. Twelve co-ordinates are chosen and labelled $y_1, y_2, y_3, \ldots\ldots\ldots y_{12}$ as shown in *Fig 5*.

From equation (3), para. 4, $b_n = \dfrac{2}{p} \sum\limits_{k=1}^{p} i_k \sin n\theta_k$, where $p = 12$.

Hence $b_1 \simeq \dfrac{2}{12}(48.24) = 8.04$; $b_3 \simeq \dfrac{2}{12}(-12) = -2.00$;

and $b_5 \simeq \dfrac{2}{12}(-0.24) = -0.04$

Thus the Fourier series for current $i$ is given by: $i = 8.04 \sin \theta - 2.00 \sin 3\theta - 0.04 \sin 5\theta$.

## C. FURTHER PROBLEMS ON A NUMERICAL METHOD OF HARMONIC ANALYSIS

Determine the Fourier series to represent the periodic functions given by the tables of values in *Problems 1 to 3*, up to and including the third harmonics and each coefficient correct to 2 decimal places. Use 12 ordinates in each case.

1

| Angle $\theta°$ | 30 | 60 | 90 | 120 | 150 | 180 | 210 | 240 | 270 | 300 | 330 | 360 |
|---|---|---|---|---|---|---|---|---|---|---|---|---|
| Displacement $y$ | 40 | 43 | 38 | 30 | 23 | 17 | 11 | 9 | 10 | 13 | 21 | 32 |

$[y = 23.92+7.81 \cos \theta+14.61 \sin \theta+0.17 \cos 2\theta+2.31 \sin 2\theta-0.33 \cos 3\theta+0.50 \sin 3\theta]$

2

| Angle $\theta°$ | 0 | 30 | 60 | 90 | 120 | 150 | 180 | 210 | 240 | 270 | 300 | 330 |
|---|---|---|---|---|---|---|---|---|---|---|---|---|
| Voltage $v$ | −5.0 | −1.5 | 6.0 | 12.5 | 16.0 | 16.5 | 15.0 | 12.5 | 6.5 | −4.0 | −7.0 | −7.5 |

$[v = 5.00-10.78 \cos \theta+6.83 \sin \theta-1.96 \cos 2\theta+0.80 \sin 2\theta+0.58 \cos 3\theta-1.08 \sin 3\theta]$

3

| Angle $\theta°$ | 30 | 60 | 90 | 120 | 150 | 180 | 210 | 240 | 270 | 300 | 330 | 360 |
|---|---|---|---|---|---|---|---|---|---|---|---|---|
| Current $i$ | 0 | −1.4 | −1.8 | −1.9 | −1.8 | −1.3 | 0 | 2.2 | 3.8 | 3.9 | 3.5 | 2.5 |

$[i = 0.64+1.58 \cos \theta-2.73 \sin \theta-0.23 \cos 2\theta-0.42 \sin 2\theta+0.27 \cos 3\theta+0.05 \sin 3\theta]$

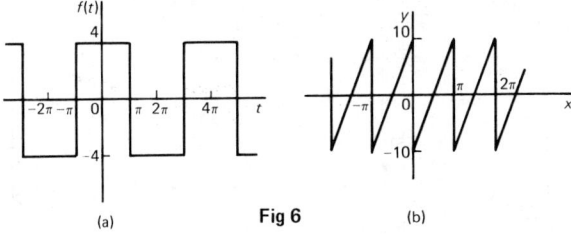

(a)　　**Fig 6**　　(b)

4   Without performing calculations, state which harmonics will be present in the waveforms shown in *Fig 6*.  $\begin{bmatrix}\text{(a) only odd cosine terms present}\\ \text{(b) only even sine terms present}\end{bmatrix}$

5   Analyse the periodic waveform of displacement $y$ against angle $\theta$ in *Fig 7(a)* into its constituent harmonics as far as and including the third harmonic, by taking 30° intervals.
$\begin{bmatrix}y = 9.33+13.57 \cos \theta-23.87 \sin \theta+1.25 \cos 2\theta\\ \qquad\qquad -0.72 \sin 2\theta+1.00 \cos 3\theta+1.00 \sin 3\theta\end{bmatrix}$

6   For the waveform of current shown in *Fig 7(b)* state why only a d.c. component and even cosine terms will appear in the Fourier series and determine the series, using $\pi/6$ rad. intervals, up to and including the sixth harmonic.
$[I = 3.83-4.50 \cos 2\theta+1.17 \cos 4\theta-1.00 \cos 6\theta]$

7   Determine the Fourier series as far as the third harmonic to represent the periodic function $y$ given by the waveform in *Fig 8*. Take 12 intervals when analysing the waveform.
$[y = 83.37 \sin \theta-29.28 \cos \theta+13.33 \sin 3\theta+17.00 \cos 3\theta]$

179

y
40
30
20
10
0
−10
−20
90°
180°
270°
360°
θ°
(a)
Current I amperes
10
5
0
π/2
π
3π/2
2π
θ rads
(b)

Fig 7

y
100
80
60
40
20
0
−20
−40
−60
−80
−100
−90°
90°
180°
270°
360°
θ°

Fig 8

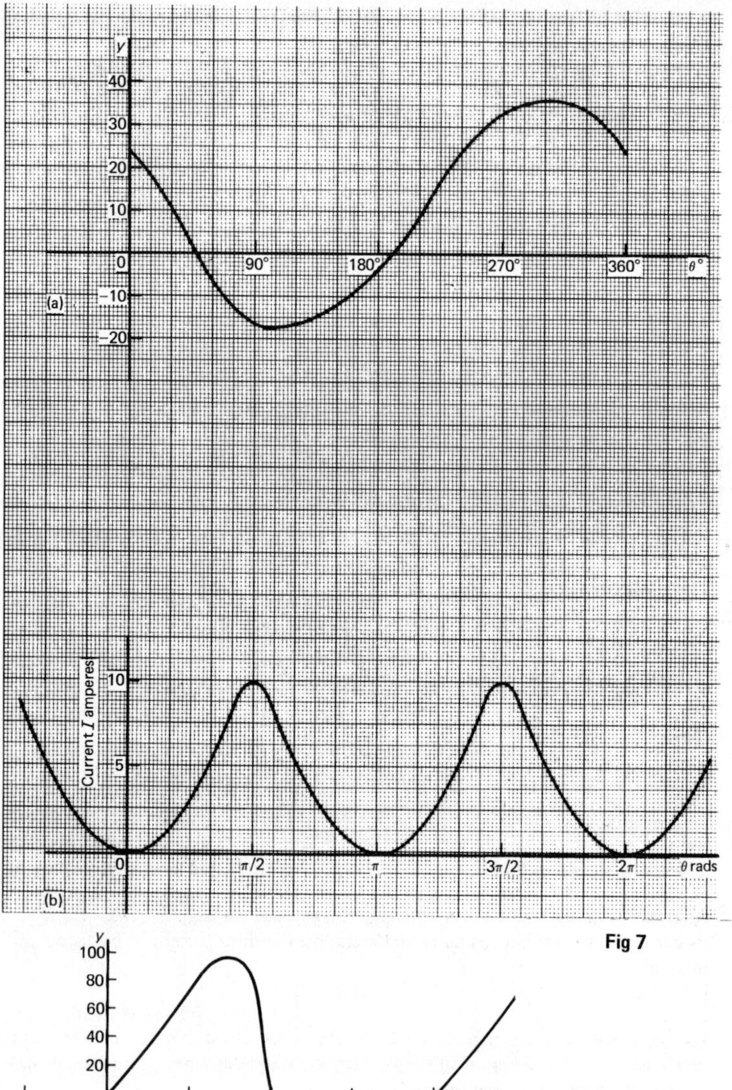

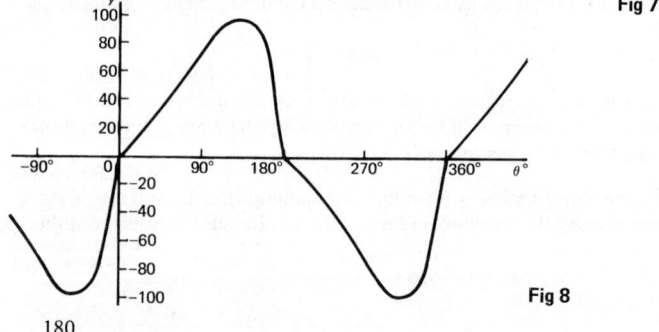

# 22 Introduction to Laplace transforms

## A. MAIN POINTS CONCERNED WITH LAPLACE TRANSFORMS

1   The solution of most electrical circuit problems can be reduced ultimately to the solution of differential equations. The use of **Laplace transforms** provides an alternative method to those discussed in chapter 12 to 16 for solving linear differential equations.

2   **Definition**
The Laplace transform of the function $f(t)$ is defined by the integral $\int_0^\infty e^{-st} f(t)dt$, where $s$ is a parameter assumed to be a real number.

3   **Common notations used for the Laplace transform**
There are various commonly used notations for the Laplace transform of $f(t)$ and these include:

(i)   $\mathcal{L}\{f(t)\}$ or $L\{f(t)\}$

(ii)  $\mathcal{L}(f)$ or $Lf$

(iii) $\bar{f}(s)$ or $\bar{f}(s)$

Also, the letter $p$ is sometimes used instead of $s$ as the parameter. The notation adopted in this book will be $f(t)$ for the original function and $\mathcal{L}\{f(t)\}$ for its Laplace transform.

Hence, from para. 2:  $\mathcal{L}\{f(t)\} = \int_0^\infty e^{-st} f(t)\, dt$  (1)

4   **Linearity property of the Laplace transform**

From equation (1),  $\mathcal{L}\{kf(t)\} = \int_0^\infty e^{-st} kf(t)dt = k \int_0^\infty e^{-st} f(t)dt$

i.e.,        $\mathcal{L}\{kf(t)\} = k\mathcal{L}\{f(t)\}$,  (2)

where $k$ is any constant.

Similarly, $\mathcal{L}\{af(t)+bg(t)\} = \int_0^\infty e^{-st}(af(t)+bg(t))dt$

$$= a \int_0^\infty e^{-st} f(t)dt + b \int_0^\infty e^{-st} g(t)dt$$

i.e. $\mathcal{L}\{af(t)+bg(t)\} = a\mathcal{L}\{f(t)\} + b\mathcal{L}\{g(t)\}$,  (3)

where $a$ and $b$ are any real constants.

The Laplace transform is termed a **linear operator** because of the properties shown in equations (2) and (3).

## 5 Laplace transforms of elementary functions

Using the definition of the Laplace transform in equation (1) a number of elementary functions may be transformed. For example:

(a) $f(t) = 1$

From equation (1), $\mathcal{L}\{1\} = \int_0^\infty e^{-st}(1)dt = \left[\dfrac{e^{-st}}{-s}\right]_0^\infty = -\dfrac{1}{s}[e^{-s(\infty)} - e^0]$

$$= -\frac{1}{s}[0-1] = \frac{1}{s} \text{ (provided } s > 0)$$

(b) $f(t) = k$. From equation (2), $\mathcal{L}\{k\} = k\mathcal{L}\{1\}$

Hence $\mathcal{L}\{k\} = k\left(\dfrac{1}{s}\right) = \dfrac{k}{s}$, from (a) above.

(c) $f(t) = e^{at}$ (where $a$ is a real constant $\neq 0$)

From equation (1), $\mathcal{L}\{e^{at}\} = \int_0^\infty e^{-st}(e^{at})\,dt = \int_0^\infty e^{-(s-a)t}dt$,

from the laws of indices,

$$= \left[\frac{e^{-(s-a)t}}{-(s-a)}\right]_0^\infty = \frac{1}{-(s-a)}(0-1)$$

$$= \frac{1}{s-a} \text{ (provided } (s-a) > 0, \text{ i.e. } s > a)$$

(d) $f(t) = \cos at$ (where $a$ is a real constant)

From equation (1), $\mathcal{L}\{\cos at\} = \int_0^\infty e^{-st}\cos at\,dt$

$$= \left[\frac{e^{-st}}{s^2+a^2}(a\sin at - s\cos at)\right]_0^\infty \begin{array}{l}\text{by integration by parts twice} \\ \text{(see page 100)}\end{array}$$

$$= \frac{s}{s^2+a^2} \text{ (provided } s > 0)$$

(e) $f(t) = t$

From equation (1), $\mathcal{L}\{t\} = \int_0^\infty e^{-st}t\,dt = \left[\dfrac{te^{-st}}{-s} - \int\dfrac{e^{-st}}{-s}\,dt\right]_0^\infty$

$$= \left[\frac{te^{-st}}{s} - \frac{e^{-st}}{s^2}\right]_0^\infty \text{ by integration by parts}$$

$$= \left[\frac{\infty e^{-s(\infty)}}{s} - \frac{e^{-s(\infty)}}{s^2}\right] - \left[0 - \frac{e^0}{s^2}\right]$$

$$= (0-0) - (0 - \frac{1}{s^2}), \text{ since } (\infty \times 0) = 0,$$

$$= \frac{1}{s^2}, \text{ (provided } s > 0)$$

(f) $f(t) = t^n$ (where $n = 0, 1, 2, 3, \ldots\ldots$).

By a similar method to (e) it may be shown that $\mathcal{L}\{t^2\} = \dfrac{2}{s^3}$

TABLE 1. Elementary standard Laplace transforms

| Function $f(t)$ | Laplace transforms $\mathcal{L}\{f(t)\} = \displaystyle\int_0^\infty e^{-st} f(t)\, dt$ |
|---|---|
| (i)    1 | $\dfrac{1}{s}$ |
| (ii)    $k$ | $\dfrac{k}{s}$ |
| (iii)    $e^{at}$ | $\dfrac{1}{s-a}$ |
| (iv)    $\sin at$ | $\dfrac{a}{s^2 + a^2}$ |
| (v)    $\cos at$ | $\dfrac{s}{s^2 + a^2}$ |
| (vi)    $t$ | $\dfrac{1}{s^2}$ |
| (vii)    $t^2$ | $\dfrac{2!}{s^3}$ |
| (viii)    $t^n \ (n = 1, 2, 3, \ldots)$ | $\dfrac{n!}{s^{n+1}}$ |
| (ix)    $\cosh at$ | $\dfrac{s}{s^2 - a^2}$ |
| (x)    $\sinh at$ | $\dfrac{a}{s^2 - a^2}$ |

and $\mathcal{L}\{t^3\} = \dfrac{(3)(2)}{s^4} = \dfrac{3!}{s^4}$ . These results can be extended to $n$ being any positive integer.

Thus $\mathcal{L}\{t^n\} = \dfrac{n!}{s^{n+1}}$ (provided $s > 0$).

(g) $f(t) = \sinh at$. From chapter 2, $\sinh at = \dfrac{1}{2}(e^{at} - e^{-at})$.

Hence $\mathcal{L}\{\sinh at\} = \mathcal{L}\left\{\dfrac{1}{2}e^{at} - \dfrac{1}{2}e^{-at}\right\} = \dfrac{1}{2}\mathcal{L}\{e^{at}\} - \dfrac{1}{2}\mathcal{L}\{e^{-at}\}$,

from equations (2) and (3)

$= \dfrac{1}{2}\left[\dfrac{1}{s-a}\right] - \dfrac{1}{2}\left[\dfrac{1}{s+a}\right]$ , from (c) above,

$= \dfrac{1}{2}\left[\dfrac{1}{s-a} - \dfrac{1}{s+a}\right] = \dfrac{a}{s^2 - a^2}$ (provided $s > a$)

6   A list of elementary standard Laplace transforms are summarised in *Table 1*.

## B. WORKED PROBLEMS ON STANDARD LAPLACE TRANSFORMS

*Problem 1* Determine the Laplace transforms of (a) $1+2t-\frac{1}{3}t^4$; (b) $5e^{2t}-3e^{-t}$

(a) $\mathcal{L}\{1+2t-\frac{1}{3}t^4\} = \mathcal{L}\{1\}+2\mathcal{L}\{t\}-\frac{1}{3}\mathcal{L}\{t^4\}$, from equations (2) and (3), para. 4

$$= \frac{1}{s}+2\left(\frac{1}{s^2}\right) -\frac{1}{3}\left(\frac{4!}{s^{4+1}}\right) \text{, from (i), (vi) and (viii) of } Table\ 1$$

$$= \frac{1}{s}+\frac{2}{s^2}-\frac{1}{3}\left(\frac{4.3.2.1}{s^5}\right) = \frac{1}{s}+\frac{2}{s^2}-\frac{8}{s^5}$$

(b) $\mathcal{L}\{5e^{2t}-3e^{-t}\} = 5\mathcal{L}\{e^{2t}\}-3\mathcal{L}\{e^{-t}\}$, from equations (2) and (3), para. 4

$$= 5\left(\frac{1}{s-2}\right) -3\left(\frac{1}{s--1}\right) \text{, from (iii) of } Table\ 1,$$

$$= \frac{5}{s-2} -\frac{3}{s+1} = \frac{5(s+1)-3(s-2)}{(s-2)(s+1)} = \frac{2s+11}{(s-2)(s+1)}$$

*Problem 2* Find the Laplace transforms of (a) $6 \sin 3t-4 \cos 5t$; (b) $2 \cosh 2\theta-\sinh 3\theta$.

(a) $\mathcal{L}\{6 \sin 3t-4 \cos 5t\} = 6\mathcal{L}\{\sin 3t\}-4\mathcal{L}\{\cos 5t\}$ from equations (2) and (3), para. 4,

$$= 6\left(\frac{3}{s^2+3^2}\right) -4\left(\frac{s}{s^2+5^2}\right) = \frac{18}{s^2+9} - \frac{4s}{s^2+25}$$

(b) $\mathcal{L}\{2 \cosh 2\theta-\sinh 3\theta\} = 2\mathcal{L}\{\cosh 2\theta\}-\mathcal{L}\{\sinh 3\theta\}$

$$= 2\left(\frac{s}{s^2-2^2}\right) -\left(\frac{3}{s^2-3^2}\right)\text{, from (ix) and (x) of } Table\ 1.$$

$$= \frac{2s}{s^2-4} - \frac{3}{s^2-9}$$

*Problem 3* Prove that (a) $\mathcal{L}\{\sin at\}=\frac{a}{s^2+a^2}$; (b) $\mathcal{L}\{t^2\}=\frac{2}{s^3}$; (c) $\mathcal{L}\{\cosh at\}=\frac{s}{s^2-a^2}$

(a) From equation (1), $\mathcal{L}\{\sin at\} = \int_0^\infty e^{-st} \sin at\ dt$

$$= \left[\frac{e^{-st}}{s^2+a^2} (-s \sin at-a \cos at)\right]_0^\infty,$$

by integration by parts

$$= \frac{1}{s^2+a^2} [e^{-s(\infty)}(-s \sin a(\infty)-a \cos a(\infty))-e^0 (-s \sin 0-a \cos 0)]$$

$$= \frac{1}{s^2+a^2} [(0)-1(0-a)] = \frac{a}{s^2+a^2} \text{ (provided } s>0)$$

(b) From equation (1),

$$\mathcal{L}\{t^2\} = \int_0^\infty e^{-st} t^2 \, dt = \left[ \frac{t^2 e^{-st}}{-s} - \frac{2t e^{-st}}{s^2} - \frac{2e^{-st}}{s^3} \right]_0^\infty, \text{ by integration by parts twice,}$$

$$= \left[ (0-0-0) - \left( 0-0-\frac{2}{s^3} \right) \right] = \frac{2}{s^3} \text{ (provided } s > 0)$$

(c) From equation (1),

$$\mathcal{L}\{\cosh at\} = \mathcal{L}\left\{ \frac{1}{2}(e^{at} + e^{-at}) \right\}, \text{ from chapter 2,}$$

$$= \frac{1}{2}\mathcal{L}\{e^{at}\} + \frac{1}{2}\mathcal{L}\{e^{-at}\}, \text{ from equations (2) and (3),}$$

$$= \frac{1}{2}\left( \frac{1}{s-a} \right) + \frac{1}{2}\left( \frac{1}{s--a} \right), \text{ from (iii) of } Table\ 1,$$

$$= \frac{1}{2}\left[ \frac{1}{s-a} + \frac{1}{s+a} \right] = \frac{1}{2}\left[ \frac{(s+a)+(s-a)}{(s-a)(s+a)} \right] = \frac{s}{s^2-a^2} \text{ (provided } s > a)$$

*Problem 4* Determine the Laplace transforms of (a) $\sin^2 t$; (b) $\cosh^2 3x$.

(a) Since $\cos 2t = 1 - 2\sin^2 t$ then $\sin^2 t = \frac{1}{2}(1 - \cos 2t)$

Hence $\mathcal{L}\{\sin^2 t\} = \mathcal{L}\left\{ \frac{1}{2}(1 - \cos 2t) \right\} = \frac{1}{2}\mathcal{L}\{1\} - \frac{1}{2}\mathcal{L}\{\cos 2t\}$

$$= \frac{1}{2}\left( \frac{1}{s} \right) - \frac{1}{2}\left( \frac{s}{s^2+2^2} \right) \text{ from (i) and (v) of } Table\ 1$$

$$= \frac{(s^2+4)-s^2}{2s(s^2+4)} = \frac{4}{2s(s^2+4)} = \frac{2}{s(s^2+4)}$$

(b) Since $\cosh 2x = 2\cosh^2 x - 1$ then $\cosh^2 x = \frac{1}{2}(1 + \cosh 2x)$ from chapter 2.

Hence $\cosh^2 3x = \frac{1}{2}(1 + \cosh 6x)$

Thus $\mathcal{L}\{\cosh^2 3x\} = \mathcal{L}\left\{ \frac{1}{2}(1 + \cosh 6x) \right\} = \frac{1}{2}\mathcal{L}\{1\} + \frac{1}{2}\mathcal{L}\{\cosh 6x\}$

$$= \frac{1}{2}\left( \frac{1}{s} \right) + \frac{1}{2}\left( \frac{s}{s^2-6^2} \right) = \frac{1}{2s} + \frac{s}{2(s^2-36)}$$

$$= \frac{(s^2-36)+s^2}{2s(s^2-36)} = \frac{2s^2-36}{2s(s^2-36)} = \frac{s^2-18}{s(s^2-36)}$$

*Problem 5* Find the Laplace transform of $3\sin(wt+\alpha)$, where $w$ and $\alpha$ are constants.

Using the compound angle formula for $\sin(A+B)$, $\sin(wt+\alpha)$ may be expanded to $(\sin wt \cos \alpha + \cos wt \sin \alpha)$.

Hence $\mathcal{L}\{3\sin(wt+\alpha)\} = \mathcal{L}\{3(\sin wt \cos \alpha + \cos wt \sin \alpha)\}$

$$= 3\cos\alpha\,\mathcal{L}\{\sin wt\} + 3\sin\alpha\,\mathcal{L}\{\cos wt\}, \text{ since } \alpha \text{ is a constant}$$

$$= 3\cos\alpha\left( \frac{w}{s^2+w^2} \right) + 3\sin\alpha\left( \frac{s}{s^2+w^2} \right) \text{ from (iv) and (v) of } Table\ 1$$

$$= \frac{3}{(s^2+w^2)}(w\cos\alpha + s\sin\alpha)$$

## C. FURTHER PROBLEMS ON STANDARD LAPLACE TRANSFORMS

Determine the Laplace transforms in *Problems 1 to 10*

1  (a) $2t-3$;  (b) $5t^2+4t-3$

$$\left[(a)\ \frac{2}{s^2} - \frac{3}{s}\ ;\ (b)\ \frac{10}{s^3} + \frac{4}{s^2} - \frac{3}{s}\right]$$

2  (a) $\frac{t^3}{24} -3t+2$; (b) $\frac{x^5}{15} -2x^4 + \frac{x^2}{2}$

$$\left[(a)\ \frac{1}{4s^4} - \frac{3}{s^2} + \frac{2}{s};\ (b)\ \frac{8}{s^6} - \frac{48}{s^5} + \frac{1}{s^3}\right]$$

3  (a) $5e^{3t}$; (b) $2e^{-2t}$

$$\left[(a)\ \frac{5}{s-3}\ ;\ (b)\ \frac{2}{s+2}\right]$$

4  (a) $4 \sin 3t$; (b) $3 \cos 2t$

$$\left[(a)\ \frac{12}{s^2+9};\ (b)\ \frac{3s}{s^2+4}\right]$$

5  (a) $7 \cosh 2x$;  (b) $\frac{1}{3}\sinh 3t$

$$\left[(a)\ \frac{7s}{s^2-4};\ (b)\ \frac{1}{s^2-9}\right]$$

6  (a) $2 \cos^2 t$;  (b) $3 \sin^2 2x$

$$\left[(a)\ \frac{2(s^2+2)}{s(s^2+4)};\ (b)\ \frac{24}{s(s^2+16)}\right]$$

7  (a) $\cosh^2 t$;  (b) $2 \sinh^2 2\theta$

$$\left[(a)\ \frac{s^2-2}{s(s^2-4)}\ ;\ (b)\ \frac{16}{s(s^2-16)}\right]$$

8  $4 \sin (at+b)$, where $a$ and $b$ are constants.

$$\left[\frac{4}{s^2+a^2}\ (a \cos b+s \sin b)\right]$$

9  $3 \cos (wt-\alpha)$, where $w$ and $\alpha$ are constants.

$$\left[\frac{3}{s^2+w^2}\ (s \cos \alpha+w \sin \alpha)\right]$$

10  $2 \sin (t + \frac{\pi}{3})$

$$\left[\frac{1+\sqrt{3}s}{(s^2+1)}\right]$$

11  Show that $\mathcal{L}\{\cos^2 3t-\sin^2 3t\} = \frac{s}{s^2+36}$

12  Show that $\mathcal{L}\left\{\frac{1}{a} \sin a\theta - \frac{1}{b} \sin b\theta \right\} = \frac{b^2-a^2}{(s^2+a^2)(s^2+b^2)}$, where $a$ and $b$ are constants.

# 23 Properties of Laplace transforms

## A. MAIN POINTS CONCERNED WITH PROPERTIES OF LAPLACE TRANSFORMS

1 **Laplace transform of $e^{at}f(t)$**

From chapter 22, the definition of the Laplace transform of $f(t)$ is:

$$\mathcal{L}\{f(t)\} = \int_0^\infty e^{-st}f(t)\,dt \tag{1}$$

Thus $\mathcal{L}\{e^{at}f(t)\} = \int_0^\infty e^{-st}(e^{at}f(t))dt = \int_0^\infty e^{-(s-a)t}f(t)\,dt \tag{2}$

(where $a$ is a real constant)

Hence the substitution of $(s-a)$ for $s$ in the transform shown in equation (1) corresponds to the multiplication of the original function $f(t)$ by $e^{at}$. This is known as a shift theorem.

2 From equation (2) of para. 1, **Laplace transforms of the form $e^{at}f(t)$** may be deduced. For example:

(i) $\mathcal{L}\{e^{at}t^n\}$

Since $\mathcal{L}\{t^n\} = \dfrac{n!}{s^{n+1}}$ from (viii) of *Table 1* of chapter 22, page 183.

thus $\mathcal{L}\{e^{at}t^n\} = \dfrac{n!}{(s-a)^{n+1}}$ from equation (2) above, (provided $s > a$)

(ii) $\mathcal{L}\{e^{at}\sin \omega t\}$

Since $\mathcal{L}\{\sin \omega t\} = \dfrac{\omega}{s^2+\omega^2}$ from (iv) of *Table 1* of chapter 22, page 183.

thus $\mathcal{L}\{e^{at}\sin \omega t\} = \dfrac{\omega}{(s-a)^2+\omega^2}$ from equation (2), (provided $s > a$)

(iii) $\mathcal{L}\{e^{at}\cosh \omega t\}$

Since $\mathcal{L}\{\cosh \omega t\} = \dfrac{s}{s^2-\omega^2}$ from (ix) of *Table 1* of chapter 22, page 183.

thus $\mathcal{L}\{e^{at}\cosh \omega t\} = \dfrac{s-a}{(s-a)^2-\omega^2}$ from equation (2) (provided $s > a$)

A summary of Laplace transforms of the form $e^{at}f(t)$ is shown in *Table 1*.
(See *Problems 1 to 4*)

# TABLE 1. Laplace transforms of the form $e^{at} f(t)$

| Function $e^{at} f(t)$ ($a$ is a real constant) | Laplace transform $\mathcal{L}\{e^{at} f(t)\}$ |
|---|---|
| (i) $\quad e^{at} t^n$ | $\dfrac{n!}{(s-a)^{n+1}}$ |
| (ii) $\quad e^{at} \sin \omega t$ | $\dfrac{\omega}{(s-a)^2 + \omega^2}$ |
| (iii) $\quad e^{at} \cos \omega t$ | $\dfrac{s-a}{(s-a)^2 + \omega^2}$ |
| (iv) $\quad e^{at} \sinh \omega t$ | $\dfrac{\omega}{(s-a)^2 - \omega^2}$ |
| (v) $\quad e^{at} \cosh \omega t$ | $\dfrac{s-a}{(s-a)^2 - \omega^2}$ |

## 3 Laplace transforms of derivatives

### (a) First derivative

Let the first derivative of $f(t)$ be $f'(t)$ then, from equation (1),

$$\mathcal{L}\{f'(t)\} = \int_0^\infty e^{-st} f'(t)\, dt$$

From chapter 11, when integrating by parts $\int u \dfrac{dv}{dt}\, dt = uv - \int v \dfrac{du}{dt}\, dt$.

When evaluating $\int_0^\infty e^{-st} f'(t)\, dt$, let $u = e^{-st}$ and $\dfrac{dv}{dt} = f'(t)$,

from which, $\dfrac{du}{dt} = -se^{-st}$ and $v = \int f'(t) = f(t)$.

Hence $\int_0^\infty e^{-st} f'(t)\, dt = [e^{-st} f(t)]_0^\infty - \int_0^\infty f(t)(-se^{-st})\, dt$

$$= [0 - f(0)] + s \int_0^\infty e^{-st} f(t)\, dt$$

$= -f(0) + s\mathcal{L}\{f(t)\}$, assuming $e^{-st} f(t) \to 0$ as $t \to 0$, and $f(0)$ is the value of $f(t)$ at $t = 0$.

Hence $\mathcal{L}\{f'(t)\} = s\mathcal{L}\{f(t)\} - f(0)$

or $\mathcal{L}\left\{\dfrac{dy}{dx}\right\} = s\mathcal{L}\{y\} - y(0),$

$$\tag{3}$$

where $y(0)$ is the value of $y$ at $x = 0$.

### (b) Second derivative

Let the second derivative of $f(t)$ be $f''(t)$, then from equation (1),

$$\mathcal{L}\{f''(t)\} = \int_0^\infty e^{-st} f''(t)\, dt$$

Integrating by parts gives:

$$\int_0^\infty e^{-st} f''(t)\, dt = \left[ e^{-st} f'(t) \right]_0^\infty + s \int_0^\infty e^{-st} f'(t)\, dt$$

$$= [0 - f'(0)] + s\mathcal{L}\{f't\},$$

assuming $e^{-st} f'(t) \to 0$ as $t \to 0$, and $f'(0)$ is the value of $f'(t)$ at $t = 0$.
Hence $\mathcal{L}\{f''(t)\} = -f'(0) + s[s\mathcal{L}\{f(t)\} - f(0)]$, from equation (3),

$$\left.\begin{array}{l} \text{i.e.,} \quad \mathcal{L}\{\,f''(t)\} = s^2\, \mathcal{L}\{f(t)\} - sf(0) - f'(0) \\[2mm] \text{or} \quad \mathcal{L}\left\{\dfrac{dy^2}{dx^2}\right\} = s^2\, \mathcal{L}\{y\} - sy(0) - y'(0) \end{array}\right\} \tag{4}$$

where $y'(0)$ is the value of $\dfrac{dy}{dx}$ at $x = 0$.

Equations (3) and (4) are important and are used in the solution of differential equations, (chapter 24). (See *Problems 5 and 6*)

4   There are several Laplace transform theorems used to simplify and interpret the solution of certain problems. Two such theorems are the initial value theorem and the final value theorem.

(a) **The initial value theorem** states:

$$\underset{t \to 0}{\text{limit}}\ [f(t)] = \underset{s \to \infty}{\text{limit}}\ [s\mathcal{L}\{f(t)\}]$$

For example, if $f(t) = 3e^{4t}$ then $\mathcal{L}\{3e^{4t}\} = \dfrac{3}{s-4}$, from (iii) of *Table 1* of chapter 22, page 183.

By the initial value theorem, $\underset{t \to 0}{\text{limit}}\ [3e^{4t}] = \underset{s \to \infty}{\text{limit}}\ \left[s\left(\dfrac{3}{s-4}\right)\right]$

i.e.   $3e^0 = \infty\left(\dfrac{3\infty}{\infty-4}\right)$

i.e.   $3 = 3$, which illustrates the theorem.

(See *Problems 7 and 8*)

(b) **The final value theorem** states:

$$\underset{t \to \infty}{\text{limit}}\ [f(t)] = \underset{s \to 0}{\text{limit}}\ [s\mathcal{L}\{f(t)\}]$$

For example, if $f(t) = 3e^{4t}$ then:
$\underset{t \to \infty}{\text{limit}}\ [3e^{4t}] = \underset{s \to 0}{\text{limit}}\ \left[s\left(\dfrac{3}{s-4}\right)\right]$

i.e.   $3e^\infty = 0\left(\dfrac{3}{0-4}\right)$

i.e.   $0 = 0$, which illustrates the theorem. (See *Problem 9*)

The initial and final value theorems are used in pulse circuit applications where the response of the circuit for small periods of time, or the behaviour immediately after the switch is closed are of interest. The final value theorem is particularly useful in investigating the stability of systems (such as in automatic aircraft-landing systems) and is concerned with the steady state response for large values of time $t$, i.e. after all transient effects have died away.

## B. WORKED PROBLEMS ON THE PROPERTIES OF LAPLACE TRANSFORMS

*Problem 1*  Find the Laplace transforms of (a) $2t^4 e^{3t}$; (b) $4e^{3t} \cos 5t$

(a)  From (i) of *Table 1*, $\mathcal{L}\{2t^4 e^{3t}\} = 2\mathcal{L}\{t^4 e^{3t}\}$

$$= 2\left(\frac{4!}{(s-3)^{4+1}}\right) = \frac{2(4)(3)(2)}{(s-3)^5} = \frac{48}{(s-3)^5}$$

(b)  From (iii) of *Table 1*, $\mathcal{L}\{4e^{3t} \cos 5t\} = 4\mathcal{L}\{e^{3t} \cos 5t\}$

$$= 4\left(\frac{s-3}{(s-3)^2 + 5^2}\right) = \frac{4(s-3)}{s^2 - 6s + 9 + 25} = \frac{4(s-3)}{s^2 - 6s + 34}$$

*Problem 2*  Find the Laplace transforms of (a) $e^{-2t} \sin 3t$; (b) $3e^{\theta} \cosh 4\theta$.

(a)  From (ii) of *Table 1*, $\mathcal{L}\{e^{-2t} \sin 3t\} = \frac{3}{(s--2)^2 + 3^2} + \frac{3}{(s+2)^2 + 9}$

$$= \frac{3}{s^2 + 4s + 4 + 9} = \frac{3}{s^2 + 4s + 13}$$

(b)  From (v) of *Table 1*, $\mathcal{L}\{3e^{\theta} \cosh 4\theta\} = 3\mathcal{L}\{e^{\theta} \cosh 4\theta\} = \frac{3(s-1)}{(s-1)^2 - 4^2}$

$$= \frac{3(s-1)}{s^2 - 2s + 1 - 16} = \frac{3(s-1)}{s^2 - 2s - 15}$$

*Problem 3*  Determine the Laplace transforms of (a) $5e^{-3t} \sinh 2t$;
(b) $2e^{3t}(4 \cos 2t - 5 \sin 2t)$

(a)  From (iv) of *Table 1*, $\mathcal{L}\{5e^{-3t} \sinh 2t\} - 5\mathcal{L}\{e^{-3t} \sinh 2t\}$

$$= 5\left(\frac{2}{(s--3)^2 - 2^2}\right) = \frac{10}{(s+3)^2 - 2^2} = \frac{10}{s^2 + 6s + 9 - 4} = \frac{10}{s^2 + 6s + 5}$$

(b)  $\mathcal{L}\{2e^{3t}(4 \cos 2t - 5 \sin 2t)\} = 8\mathcal{L}\{e^{3t} \cos 2t\} - 10\mathcal{L}\{e^{3t} \sin 2t\}$

$$= \frac{8(s-3)}{(s-3)^2 + 2^2} - \frac{10(2)}{(s-3)^2 + 2^2}, \text{ from (iii) and (ii) of}$$
$$\text{Table 1}$$

$$= \frac{8(s-3) - 10(2)}{(s-3)^2 + 2^2} = \frac{8s - 44}{s^2 - 6s + 13}$$

*Problem 4*  Show that $\mathcal{L}\{3e^{-\frac{1}{2}x} \sin^2 x\} = \frac{48}{(2s+1)(4s^2 + 4s + 17)}$

Since $\cos 2x = 1 - 2 \sin^2 x$, $\sin^2 x = \frac{1}{2}(1 - \cos 2x)$

Hence $\mathcal{L}\{3e^{-\frac{1}{2}x} \sin^2 x\} = \mathcal{L}\{3e^{-\frac{1}{2}x} \frac{1}{2}(1 - \cos 2x)\}$

$$= \frac{3}{2}\mathcal{L}\{e^{-\frac{1}{2}x}\} - \frac{3}{2}\mathcal{L}\{e^{-\frac{1}{2}x} \cos 2x\}$$

$$= \frac{3}{2}\left(\frac{1}{s--\frac{1}{2}}\right) - \frac{3}{2}\left(\frac{(s--\frac{1}{2})}{(s--\frac{1}{2})^2 + 2^2}\right),$$

from (iii) of *Table 1* of chapter 22 (page 183) and (iii) of *Table 1* of chapter 23, page 188.

$$= \frac{3}{2(s+\frac{1}{2})} - \frac{3(s+\frac{1}{2})}{2[(s+\frac{1}{2})^2 + 2^2]} = \frac{3}{2s+1} - \frac{3(s+\frac{1}{2})}{2(s^2+s+\frac{1}{4}+4)}$$

$$= \frac{3}{2s+1} - \frac{3(s+\frac{1}{2})}{2s^2+2s+8\frac{1}{2}} = \frac{3}{2s+1} - \frac{6s+3}{4s^2+4s+17}$$

$$= \frac{3(4s^2+4s+17)-(6s+3)(2s+1)}{(2s+1)(4s^2+4s+17)}$$

$$= \frac{12s^2+12s+51-12s^2-6s-6s-3}{(2s+1)(4s^2+4s+17)} = \frac{48}{(2s+1)(4s^2+4s+17)}$$

*Problem 5* Use the Laplace transform of the first derivative to derive:

(a) $\mathcal{L}\{k\} = \frac{k}{s}$;

(b) $\mathcal{L}\{2t\} = \frac{2}{s^2}$;

(c) $\mathcal{L}\{e^{-at}\} = \frac{1}{s+a}$

From equation (3), para. 3, $\mathcal{L}\{f'(t)\} = s\mathcal{L}\{f(t)\} - f(0)$

(a) Let $f(t) = k$, then $f'(t) = 0$ and $f(0) = k$
Substituting into equation (3) gives: $\mathcal{L}\{0\} = s\mathcal{L}\{k\} - k$

i.e. $k = s\mathcal{L}\{k\}$. Hence $\mathcal{L}\{k\} = \frac{k}{s}$

(b) Let $f(t) = 2t$ then $f'(t) = 2$ and $f(0) = 0$.
Substituting into equation (3) gives: $\mathcal{L}\{2\} = s\mathcal{L}\{2t\} - 0$

i.e. $\frac{2}{s} = s\mathcal{L}\{2t\}$. Hence $\mathcal{L}\{2t\} = \frac{2}{s^2}$

(c) Let $f(t) = e^{-at}$ then $f'(t) = -ae^{-at}$ and $f(0) = 1$
Substituting into equation (3) gives: $\mathcal{L}\{-ae^{-at}\} = s\mathcal{L}\{e^{-at}\} - 1$
$$-a\mathcal{L}\{e^{-at}\} = s\mathcal{L}\{e^{-at}\} - 1$$
$$1 = s\mathcal{L}\{e^{-at}\} + a\mathcal{L}\{e^{-at}\}$$
$1 = (s+a)\mathcal{L}\{e^{-at}\}$. Hence $\mathcal{L}\{e^{-at}\} = \frac{1}{s+a}$

*Problem 6* Use the Laplace transform of the second derivative to derive

$\mathcal{L}\{\cos at\} = \frac{s}{s^2+a^2}$.

From equation (4), para. 3, $\mathcal{L}\{f''(t)\} = s^2 \mathcal{L}\{f(t)\} - sf(0) - f'(0)$
Let $f(t) = \cos at$, then $f'(t) = -a \sin at$ and $f''(t) = -a^2 \cos at$, $f(0) = 1$ and $f'(0) = 0$.
Substituting into equation (4) gives: $\mathcal{L}\{-a^2 \cos at\} = s^2 \mathcal{L}\{\cos at\} - s(1) - 0$
i.e. $-a^2 \mathcal{L}\{\cos at\} = s^2 \mathcal{L}\{\cos at\} - s$

$s = (s^2+a^2)\mathcal{L}\{\cos at\}$ from which, $\mathcal{L}\{\cos at\} = \frac{s}{s^2+a^2}$

*Problem 7* Verify the initial value theorem for the voltage function $(5+2\cos 3t)$ volts, and state its initial value.

Let $f(t) = 5 + 2 \cos 3t$

$\mathcal{L}\{f(t)\} = \mathcal{L}\{5 + 2 \cos 3t\} = \dfrac{5}{s} + \dfrac{2s}{s^2 + 9}$ from (ii) and (v) of *Table 1*

of chapter 22, page 183.

By the initial value theorem (see para. 4), $\underset{t \to 0}{\text{limit}} \, [f(t)] = \underset{s \to \infty}{\text{limit}} \, [s\mathcal{L}\{f(t)\}]$

i.e., $\underset{t \to 0}{\text{limit}} \, [5 + 2 \cos 3t] = \underset{s \to \infty}{\text{limit}} \left[ s\left( \dfrac{5}{s} + \dfrac{2s}{s^2 + 9} \right) \right] = \underset{s \to \infty}{\text{limit}} \left[ 5 + \dfrac{2s^2}{s^2 + 9} \right]$

i.e. $5 + 2(1) = 5 + \dfrac{2\infty^2}{\infty^2 + 9} = 5 + 2$

i.e. **7 = 7**, which verifies the theorem in this case. The initial value of the voltage is thus **7 V**.

Let $f(t) = (2t-3)^2 = 4t^2 - 12t + 9$

Let $\mathcal{L}\{f(t)\} = \mathcal{L}\{4t^2 - 12t + 9\} = 4\left(\dfrac{2}{s^3}\right) - \dfrac{12}{s^2} + \dfrac{9}{s}$ from (vii), (vi) and (ii) of
*Table 1* of chapter 22, page 183.

By the initial value theorem, $\underset{t \to 0}{\text{limit}} \, [(2t-3)^2] = \underset{s \to \infty}{\text{limit}} \left[ s\left( \dfrac{8}{s^3} - \dfrac{12}{s^2} + \dfrac{9}{s} \right) \right]$

$= \underset{s \to \infty}{\text{limit}} \left[ \dfrac{8}{s^2} - \dfrac{12}{s} + 9 \right]$

i.e. $(0-3)^2 = \dfrac{8}{\infty^2} - \dfrac{12}{\infty} + 9$

i.e. **9 = 9**, which verifies the theorem in this case.
The initial value of the given function is thus 9.

Let $f(t) = 2 + 3e^{-2t} \sin 4t$

$\mathcal{L}\{f(t)\} = \mathcal{L}\{2 + 3e^{-2t} \sin 4t\} = \dfrac{2}{s} + 3\left( \dfrac{(s--2)}{(s--2)^2 + 4^2} \right)$

$= \dfrac{2}{s} + \dfrac{3(s+2)}{(s+2)^2 + 16}$, from (ii) of *Table 1* of chapter 22, page 183
and (iii) of *Table 1* on page 188.

By the final value theorem, $\underset{t \to \infty}{\text{limit}} \, [f(t)] = \underset{s \to \infty}{\text{limit}} \, [s\mathcal{L}\{f(t)\}]$

i.e. $\underset{t \to \infty}{\text{limit}} \, [2 + 3e^{-2t} \sin 4t] = \underset{s \to 0}{\text{limit}} \left[ s\left( \dfrac{2}{s} + \dfrac{3(s+2)}{(s+2)^2 + 16} \right) \right]$

$= \underset{s \to 0}{\text{limit}} \left[ 2 + \dfrac{3s(s+2)}{(s+2)^2 + 16} \right]$

i.e. $2 + 0 = 2 + 0$

i.e. **2 = 2**, which verifies the theorem in this case.
The final value of the displacement is thus **2 cm**.

## C. FURTHER PROBLEMS ON THE PROPERTIES OF LAPLACE TRANSFORMS

Determine the Laplace transforms of the functions given in *Problems 1 to 8.*

1  (a) $2te^{2t}$; (b) $t^2 e^t$  $\qquad\left[\text{(a) } \dfrac{2}{(s-2)^2}; \text{ (b) } \dfrac{2}{(s-1)^3}\right]$

2  (a) $4t^3 e^{-2t}$; (b) $\dfrac{1}{2}t^4 e^{-3t}$  $\qquad\left[\text{(a) } \dfrac{24}{(s+2)^4}; \text{ (b) } \dfrac{12}{(s+3)^5}\right]$

3  (a) $e^t \cos t$; (b) $3e^{2t} \sin 2t$  $\qquad\left[\text{(a) } \dfrac{s-1}{s^2-2s+2}; \text{ (b) } \dfrac{6}{s^2-4s+8}\right]$

4  (a) $5e^{-2t} \cos 3t$; (b) $4e^{-5t} \sin t$  $\qquad\left[\text{(a) } \dfrac{5(s+2)}{s^2+4s+13}; \text{ (b) } \dfrac{4}{s^2+10s+26}\right]$

5  (a) $2e^t \sin^2 t$; (b) $\dfrac{1}{2}e^{3t} \cos^2 t$  $\qquad\left[\text{(a) } \dfrac{1}{s-1} - \dfrac{s-1}{s^2-2s+5}; \text{ (b) } \dfrac{1}{4}\left(\dfrac{1}{s-3} + \dfrac{s-3}{s^2-6s+13}\right)\right]$

6  (a) $e^t \sinh t$; (b) $3e^{2t} \cosh 4t$  $\qquad\left[\text{(a) } \dfrac{1}{s(s-2)}; \text{ (b) } \dfrac{3(s-2)}{s^2-4s-12}\right]$

7  (a) $2e^{-t} \sinh 3t$; (b) $\dfrac{1}{4}e^{-3t} \cosh 2t$  $\qquad\left[\text{(a) } \dfrac{6}{s^2+2s-8}; \text{ (b) } \dfrac{s+3}{4(s^2+6s+5)}\right]$

8  (a) $2e^t (\cos 3t-3 \sin 3t)$; (b) $3e^{-2t} (\sinh 2t-2 \cosh 2t)$

$\qquad\left[\text{(a) } \dfrac{2(s-10)}{s^2-2s+10}; \text{ (b) } \dfrac{-6(s+1)}{s(s+4)}\right]$

9  Show that $\mathcal{L}\{e^t (\cos 2t-\sin 2t)^2\} = \dfrac{s^2-6s+21}{(s-1)(s^2-2s+17)}$.

10  Prove that $\mathcal{L}\{e^{6t} \sinh^2 3t\} = \dfrac{18}{s(s-12)(s-6)}$.

11  Derive the Laplace transform of the first derivative from the definition of a Laplace transform. Hence derive the transform $\mathcal{L}\{1\} = 1/s$.

12  Use the Laplace transform of the first derivative to derive the transforms:

(a) $\mathcal{L}\{e^{at}\} = \dfrac{1}{s-a}$; (b) $\mathcal{L}\{3t^2\} = \dfrac{6}{s^3}$.

13  Derive the Laplace transform of the second derivative from the definition of a Laplace transform. Hence derive the transform $\mathcal{L}\{\sin at\} = \dfrac{a}{s^2+a^2}$

14  Use the Laplace transform of the second derivative to derive the transforms:

(a) $\mathcal{L}\{\sinh at\} = \dfrac{a}{s^2-a^2}$; (b) $\mathcal{L}\{\cosh at\} = \dfrac{s}{s^2-a^2}$

15  State the initial value theorem. Verify the theorem for the functions (a) $3-4 \sin t$; (b) $(t-4)^2$ and state their initial values.  [(a) 3; (b) 16]

16  Verify the initial value theorem for the voltage functions: (a) $4+2 \cos t$; (b) $t-\cos 3t$ and state their initial values.  [(a) 6; (b) $-1$]

17  State the final value theorem and state a practical application where it is of use. Verify the theorem for the function $4+e^{-2t} (\sin t+\cos t)$ representing a displacement and state its final value.  [4]

18  Verify the final value theorem for the function $3t^2 e^{4t}$ and determine its steady state value.  [0]

# 24 Inverse Laplace transforms and solution of differential equations

## A. MAIN POINTS CONCERNED WITH INVERSE LAPLACE TRANSFORMS AND THE SOLUTION OF DIFFERENTIAL EQUATIONS

1   If the Laplace transform of a function $f(t)$ is $F(s)$, i.e., $\mathcal{L}\{f(t)\} = F(s)$, then $f(t)$ is called the **inverse Laplace transform** of $F(s)$ and is written as $f(t) = \mathcal{L}^{-1}\{F(s)\}$.

   For example, since $\mathcal{L}\{1\} = \dfrac{1}{s}$ then $\mathcal{L}^{-1}\{\dfrac{1}{s}\} = 1$

   Similarly, since $\mathcal{L}\{\sin at\} = \dfrac{a}{s^2 + a^2}$  then $\mathcal{L}^{-1}\left\{\dfrac{a}{s^2 + a^2}\right\} = \sin at$, and so on.

2   Tables of Laplace transforms, such as the tables in chapters 22 and 23 (see pages 183 and 188) may be used to find inverse Laplace transforms. However, a summary of inverse Laplace transforms is shown in *Table 1*. (See *Problems 1 to 6*)

3   Sometimes the function whose inverse is required is not recognisable as a standard type, such as those listed in *Table 1*. In such cases it may be possible, by using **partial fractions**, to resolve the function into simpler fractions which may be inverted on sight. For example, the function,

$$F(s) = \frac{2s-3}{s(s-3)}$$

   cannot be inverted on sight from *Table 1*. However, by using partial fractions, $\dfrac{2s-3}{s(s-3)} \equiv \dfrac{1}{s} + \dfrac{1}{s-3}$ which may be inverted as $1 + e^{3t}$ from (i) and (iii) of *Table 1*.

   Partial fractions are discussed in *Mathematics 3 Checkbook* (chapter 5), and a summary of the forms of partial fractions is given in *Table 3* in chapter 10 of this book (page 17). (See *Problems 7 to 11*)

4   An alternative method of solving differential equations to that used in chapters 12 to 16 is possible by using Laplace transforms.

   **Procedure to solve differential equations by using Laplace transforms**

   (i)   Take the Laplace transform of both sides of the differential equation by applying the formulae for the Laplace transforms of derivatives (i.e. equations (3) and (4) of chapter 23) and, where necessary, using a list of standard Laplace transforms, such as the tables in chapters 22 and 23.

   (ii)  Put in the given initial conditions, i.e., $y(0)$ and $y'(0)$.

   (iii) Rearrange the equation to make $\mathcal{L}\{y\}$ the subject.

## TABLE 1. Inverse Laplace transforms

| | $F(s) = \mathcal{L}\{f(t)\}$ | $\mathcal{L}^{-1}\{F(s)\} = f(t)$ |
|---|---|---|
| (i) | $\dfrac{1}{s}$ | 1 |
| (ii) | $\dfrac{k}{s}$ | $k$ |
| (iii) | $\dfrac{1}{s-a}$ | $e^{at}$ |
| (iv) | $\dfrac{a}{s^2+a^2}$ | $\sin at$ |
| (v) | $\dfrac{s}{s^2+a^2}$ | $\cos at$ |
| (vi) | $\dfrac{1}{s^2}$ | $t$ |
| (vii) | $\dfrac{2!}{s^3}$ | $t^2$ |
| (viii) | $\dfrac{n!}{s^{n+1}}$ | $t^n$ |
| (ix) | $\dfrac{a}{s^2-a^2}$ | $\sinh at$ |
| (x) | $\dfrac{s}{s^2-a^2}$ | $\cosh at$ |
| (xi) | $\dfrac{n!}{(s-a)^{n+1}}$ | $e^{at}\, t^n$ |
| (xii) | $\dfrac{\omega}{(s-a)^2+\omega^2}$ | $e^{at} \sin \omega t$ |
| (xiii) | $\dfrac{s-a}{(s-a)^2+\omega^2}$ | $e^{at} \cos \omega t$ |
| (xiv) | $\dfrac{\omega}{(s-a)^2-\omega^2}$ | $e^{at} \sinh \omega t$ |
| (xv) | $\dfrac{s-a}{(s-a)^2-\omega^2}$ | $e^{at} \cosh \omega t$ |

(iv) Determine $y$ by using, where necessary, partial fractions, and taking the inverse of each term by using *Table 1*. (See *Problems 12 to 16*)

## B. WORKED PROBLEMS ON INVERSE LAPLACE TRANSFORMS AND THE SOLUTION OF DIFFERENTIAL EQUATIONS

*Problem 1* Find the following inverse Laplace transforms:

(a) $\mathcal{L}^{-1}\left\{\dfrac{1}{s^2+9}\right\}$; (b) $\mathcal{L}^{-1}\left\{\dfrac{5}{3s-1}\right\}$

(a) From (iv) of *Table 1*, $\mathcal{L}^{-1}\left\{\dfrac{a}{s^2+a^2}\right\} = a\mathcal{L}^{-1}\left\{\dfrac{1}{s^2+a^2}\right\} = \sin at$,

thus $\mathcal{L}^{-1}\left\{\dfrac{1}{s^2+a^2}\right\} = \dfrac{1}{a}\sin at$.

Hence $\mathcal{L}^{-1}\left\{\dfrac{1}{s^2+9}\right\} = \mathcal{L}^{-1}\left\{\dfrac{1}{s^2+3^2}\right\} = \dfrac{1}{3}\sin 3t$

(b) $\mathcal{L}^{-1}\left\{\dfrac{5}{3s-1}\right\} = \mathcal{L}^{-1}\left\{\dfrac{5}{3(s-\frac{1}{3})}\right\} = \dfrac{5}{3}\mathcal{L}^{-1}\left\{\dfrac{1}{s-\frac{1}{3}}\right\} = \dfrac{5}{3}e^{\frac{1}{3}t}$ from (iii) of *Table 1*.

*Problem 2* Find the following inverse Laplace transforms:

(a) $\mathcal{L}^{-1}\left\{\dfrac{6}{s^3}\right\}$ ; (b) $\mathcal{L}^{-1}\left\{\dfrac{3}{s^4}\right\}$.

(a) From (vii) of *Table 1*, $\mathcal{L}^{-1}\left\{\dfrac{2}{s^3}\right\} = t^2$

Hence $\mathcal{L}^{-1}\left\{\dfrac{6}{s^3}\right\} = 3\mathcal{L}^{-1}\left\{\dfrac{2}{s^3}\right\} = 3t^2$

(b) From (viii) of *Table 1*, if $s$ is to have a power of 4 then $n = 3$.

Thus $\mathcal{L}^{-1}\left\{\dfrac{3!}{s^4}\right\} = t^3$, i.e., $\mathcal{L}^{-1}\left\{\dfrac{6}{s^4}\right\} = t^3$

Hence $\mathcal{L}^{-1}\left\{\dfrac{3}{s^4}\right\} = \dfrac{1}{2}\mathcal{L}^{-1}\left\{\dfrac{6}{s^4}\right\} = \dfrac{1}{2}t^3$.

*Problem 3* Determine (a) $\mathcal{L}^{-1}\left\{\dfrac{7s}{s^2+4}\right\}$; (b) $\mathcal{L}^{-1}\left\{\dfrac{4s}{s^2-16}\right\}$

(a) $\mathcal{L}^{-1}\left\{\dfrac{7s}{s^2+4}\right\} = 7\mathcal{L}^{-1}\left\{\dfrac{s}{s^2+2^2}\right\} = 7\cos 2t$, from (v) of *Table 1*.

(b) $\mathcal{L}^{-1}\left\{\dfrac{4s}{s^2-16}\right\} = 4\mathcal{L}^{-1}\left\{\dfrac{s}{s^2-4^2}\right\} = 4\cosh 4t$, from (x) of *Table 1*.

*Problem 4* Find (a) $\mathcal{L}^{-1}\left\{\dfrac{3}{s^2-7}\right\}$; (b) $\mathcal{L}^{-1}\left\{\dfrac{2}{(s-3)^5}\right\}$

(a) From (ix) of *Table 1*, $\mathcal{L}^{-1}\left\{\dfrac{a}{s^2-a^2}\right\} = \sinh at$,

Thus $\mathcal{L}^{-1}\left\{\dfrac{1}{s^2-a^2}\right\} = \dfrac{1}{a}\sinh at$.

Thus $\mathcal{L}^{-1}\left\{\dfrac{3}{s^2-7}\right\} = 3\mathcal{L}^{-1}\left\{\dfrac{1}{s^2-(\sqrt{7})^2}\right\} = \dfrac{3}{\sqrt{7}}\sinh\sqrt{7}t$

(b) From (xi) of *Table 1*, $\mathcal{L}^{-1}\left\{\dfrac{n!}{(s-a)^{n+1}}\right\} = e^{at}t^n$,

Thus $\mathcal{L}^{-1}\left\{\dfrac{1}{(s-a)^{n+1}}\right\} = \dfrac{1}{n!}e^{at}t^n$, and comparing with $\mathcal{L}^{-1}\left\{\dfrac{2}{(s-3)^5}\right\}$ shows that $n = 4$ and $a = 3$.

Hence $\mathcal{L}^{-1}\left\{\dfrac{2}{(s-3)^5}\right\} = 2\mathcal{L}^{-1}\left\{\dfrac{1}{(s-3)^5}\right\} = 2\left(\dfrac{1}{4!}\,e^{3t}t^4\right) = \dfrac{1}{12}e^{3t}t^4$

*Problem 5* Determine (a) $\mathcal{L}^{-1}\left\{\dfrac{3}{s^2-4s+13}\right\}$; (b) $\mathcal{L}^{-1}\left\{\dfrac{2(s+1)}{s^2+2s+10}\right\}$

(a) $\mathcal{L}^{-1}\left\{\dfrac{3}{s^2-4s+13}\right\} = \mathcal{L}^{-1}\left\{\dfrac{3}{(s-2)^2+3^2}\right\} = e^{2t}\sin 3t$, from (xii) of *Table 1*.

(b) $\mathcal{L}^{-1}\left\{\dfrac{2(s+1)}{s^2+2s+10}\right\} = \mathcal{L}^{-1}\left\{\dfrac{2(s+1)}{(s+1)^2+3^2}\right\} = 2e^{-t}\cos 3t$, from (xiii) of *Table 1*.

*Problem 6* Determine (a) $\mathcal{L}^{-1}\left\{\dfrac{5}{s^2+2s-3}\right\}$; (b) $\mathcal{L}^{-1}\left\{\dfrac{4s-3}{s^2-4s-5}\right\}$

(a) $\mathcal{L}^{-1}\left\{\dfrac{5}{s^2+2s-3}\right\} = \mathcal{L}^{-1}\left\{\dfrac{5}{(s+1)^2-2^2}\right\} = \mathcal{L}^{-1}\left\{\dfrac{\frac{5}{2}(2)}{(s+1)^2-2^2}\right\}$

$= \dfrac{5}{2}e^{-t}\sinh 2t$, from (xiv) of *Table 1*.

(b) $\mathcal{L}^{-1}\left\{\dfrac{4s-3}{s^2-4s-5}\right\} = \mathcal{L}^{-1}\left\{\dfrac{4s-3}{(s-2)^2-3^2}\right\} = \mathcal{L}^{-1}\left\{\dfrac{4(s-2)+5}{(s-2)^2-3^2}\right\}$

$= \mathcal{L}^{-1}\left\{\dfrac{4(s-2)}{(s-2)^2-3^2}\right\} + \mathcal{L}^{-1}\left\{\dfrac{5}{(s-2)^2-3^2}\right\}$

since $\mathcal{L}^{-1}\{\ \}$ is a linear operator,

$= 4e^{2t}\cosh 3t + \mathcal{L}^{-1}\left\{\dfrac{\frac{5}{3}(3)}{(s-2)^2-3^2}\right\}$, from (xv) of *Table 1*

$= 4e^{2t}\cosh 3t + \dfrac{5}{3}e^{2t}\sinh 3t$, from (xiv) of *Table 1*.

*Problem 7* Determine $\mathcal{L}^{-1}\left\{\dfrac{4s-5}{s^2-s-2}\right\}$

$\dfrac{4s-5}{s^2-s-2} \equiv \dfrac{4s-5}{(s-2)(s+1)} \equiv \dfrac{A}{(s-2)} + \dfrac{B}{(s+1)} \equiv \dfrac{A(s+1)+B(s-2)}{(s-2)(s+1)}$

Hence $4s-5 \equiv A(s+1)+B(s-2)$
When $s = 2$, $3 = 3A$, from which, $A = 1$
When $s = -1$, $-9 = -3B$, from which, $B = 3$

Hence $\mathcal{L}^{-1}\left\{\dfrac{4s-5}{s^2-s-2}\right\} \equiv \mathcal{L}^{-1}\left\{\dfrac{1}{s-2} + \dfrac{3}{s+1}\right\} = \mathcal{L}^{-1}\left\{\dfrac{1}{s-2}\right\} + \mathcal{L}^{-1}\left\{\dfrac{3}{s+1}\right\}$

$= e^{2t}+3e^{-t}$, from (iii) of *Table 1*

*Problem 8* Find $\mathcal{L}^{-1}\left\{\dfrac{9s^2+4s-10}{s(s-1)(s+2)}\right\}$.

$$\frac{9s^2+4s-10}{s(s-1)(s+2)} \equiv \frac{A}{s} + \frac{B}{s-1} + \frac{C}{s+2} \equiv \frac{A(s-1)(s+2)+B(s)(s+2)+C(s)(s-1)}{s(s-1)(s+2)}$$

Hence $9s^2+4s-10 \equiv A(s-1)(s+2)+B(s)(s+2)+C(s)(s-1)$

When $s = 0$, $-10 = -2A$, from which, $A = 5$

When $s = 1$, $3 = 3B$, from which, $B = 1$

When $s = -2$, $18 = 6C$, from which, $C = 3$

Hence $\mathcal{L}^{-1}\left\{\frac{9s^2+4s-10}{s(s-1)(s+2)}\right\} \equiv \mathcal{L}^{-1}\left\{\frac{5}{s} + \frac{1}{s-1} + \frac{3}{s+2}\right\}$

$\qquad\qquad\qquad = 5+e^t+3e^{-2t}$, from (ii) and (iii) of *Table 1*

*Problem 9* Find $\mathcal{L}^{-1}\left\{\frac{3s^3+s^2+12s+2}{(s-3)(s+1)^3}\right\}$

$$\frac{3s^3+s^2+12s+2}{(s-3)(s+1)^3} \equiv \frac{A}{s-3} + \frac{B}{s+1} + \frac{C}{(s+1)^2} + \frac{D}{(s+1)^3}$$

$$\equiv \frac{A(s+1)^3+B(s-3)(s+1)^2+C(s-3)(s+1)+D(s-3)}{(s-3)(s+1)^3}$$

Hence $3s^3+s^2+12s+2 \equiv A(s+1)^3+B(s-3)(s+1)^2+C(s-3)(s+1)+D(s-3)$

When $s = 3$, $128 = 64A$, from which, $A = 2$

When $s = -1$, $-12 = -4D$, from which, $D = 3$

Equating $s^3$ terms gives: $3 = A+B$, from which, $B = 1$

Equating $s^2$ terms gives: $1 = 3A-B+C$, from which, $C = -4$

Hence $\mathcal{L}^{-1}\left\{\frac{3s^3+s^2+12s+2}{(s-3)(s+1)^3}\right\} \equiv \mathcal{L}^{-1}\left\{\frac{2}{s-3} + \frac{1}{s+1} - \frac{4}{(s+1)^2} + \frac{3}{(s+1)^3}\right\}$

$\qquad = 2e^{3t}+e^{-t}-4e^{-t}t + \frac{3}{2}e^{-t}t^2$, from (iii) and (xi) of *Table 1*

*Problem 10* Determine $\mathcal{L}^{-1}\left\{\frac{5s^2+8s-1}{(s+3)(s^2+1)}\right\}$.

$$\frac{5s^2+8s-1}{(s+3)(s^2+1)} \equiv \frac{A}{s+3} + \frac{Bs+C}{s^2+1} \equiv \frac{A(s^2+1)+(Bs+C)(s+3)}{(s+3)(s^2+1)}$$

Hence $5s^2+8s-1 \equiv A(s^2+1)+(Bs+C)(s+3)$

When $s = -3$, $20 = 10A$, from which, $A = 2$

Equating $s^2$ terms gives: $5 = A+B$, from which, $B = 3$

Equating $s$ terms gives: $8 = 3B+C$, from which, $C = -1$

Hence $\mathcal{L}^{-1}\left\{\frac{5s^2+8s-1}{(s+3)(s^2+1)}\right\} \equiv \mathcal{L}^{-1}\left\{\frac{2}{s+3} + \frac{3s-1}{s^2+1}\right\}$

$\qquad\qquad = \mathcal{L}^{-1}\left\{\frac{2}{s+3}\right\}+\mathcal{L}^{-1}\left\{\frac{3s}{s^2+1}\right\} -\mathcal{L}^{-1}\left\{\frac{1}{s^2+1}\right\}$

$\qquad\qquad = 2e^{-3t}+3\cos t-\sin t$, from (iii), (v) and (iv) of *Table 1*.

*Problem 11* Find $\mathcal{L}^{-1}\left\{\frac{7s+13}{s(s^2+4s+13)}\right\}$.

$$\frac{7s+13}{s(s^2+4s+13)} \equiv \frac{A}{s} + \frac{Bs+C}{s^2+4s+13} \equiv \frac{A(s^2+4s+13)+(Bs+C)(s)}{s(s^2+4s+13)}$$

Hence $7s+13 \equiv A(s^2+4s+13)+(Bs+C)(s)$

When $s=0$, $13=13A$, from which, $A=1$

Equating $s^2$ terms gives: $0=A+B$, from which, $B=-1$

Equating $s$ terms gives: $7=4A+C$, from which, $C=3$

Hence $\mathcal{L}^{-1}\left\{\frac{7s+13}{s(s^2+4s+13)}\right\} \equiv \mathcal{L}^{-1}\left\{\frac{1}{s} + \frac{-s+3}{s^2+4s+13}\right\} = \mathcal{L}^{-1}\left\{\frac{1}{s}\right\} + \mathcal{L}^{-1}\left\{\frac{-s+3}{(s+2)^2+3^2}\right\}$

$$= \mathcal{L}^{-1}\left\{\frac{1}{s}\right\} + \mathcal{L}^{-1}\left\{\frac{-(s+2)+5}{(s+2)^2+3^2}\right\}$$

$$= \mathcal{L}^{-1}\left\{\frac{1}{s}\right\} - \mathcal{L}^{-1}\left\{\frac{s+2}{(s+2)^2+3^2}\right\} + \mathcal{L}^{-1}\left\{\frac{5}{(s+2)^2+3^2}\right\}$$

$$= 1 - e^{-2t}\cos 3t + \frac{5}{3}e^{-2t}\sin 3t, \text{ from (i), (xiii) and}$$
(xii) of *Table 1*

**Problem 12** Use Laplace transforms to solve the differential equation

$2\frac{d^2y}{dx^2} + 5\frac{dy}{dx} - 3y = 0$, given that when $x=0$, $y=4$ and $\frac{dy}{dx}=9$

This is the same problem as *Problem 1* of chapter 15, page 124 and a comparison of methods can be made. Using the procedure of para. 4:

(i) $2\mathcal{L}\left\{\frac{d^2y}{dx^2}\right\} + 5\mathcal{L}\left\{\frac{dy}{dx}\right\} - 3\mathcal{L}\{y\} = \mathcal{L}\{0\}$

$2[s^2\mathcal{L}\{y\}-sy(0)-y'(0)] + 5[s\mathcal{L}\{y\}-y(0)] - 3\mathcal{L}\{y\} = 0$, from equations (3) and (4) of chapter 23.

(ii) $y(0)=4$ and $y'(0)=9$.

Thus $2[s^2\mathcal{L}\{y\}-4s-9] + 5[s\mathcal{L}\{y\}-4] - 3\mathcal{L}\{y\} = 0$

i.e. $2s^2\mathcal{L}\{y\}-8s-18+5s\mathcal{L}\{y\}-20-3\mathcal{L}\{y\} = 0$

(iii) Rearranging gives: $(2s^2+5s-3)\mathcal{L}\{y\} = 8s + 38$

i.e., $\mathcal{L}\{y\} = \frac{8s+38}{2s^2+5s-3}$

(iv) $y = \mathcal{L}^{-1}\left\{\frac{8s+38}{2s^2+5s-3}\right\}$

$$\frac{8s+38}{2s^2+5s-3} \equiv \frac{8s+38}{(2s-1)(s+3)} \equiv \frac{A}{2s-1} + \frac{B}{s+3} \equiv \frac{A(s+3)+B(2s-1)}{(2s-1)(s+3)}$$

Hence $8s+38 \equiv A(s+3)+B(2s-1)$

When $s=\frac{1}{2}$, $42=3\frac{1}{2}A$, from which, $A=12$

When $s=-3$, $14=-7B$, from which, $B=-2$

Hence $y = \mathcal{L}^{-1}\left\{\frac{8s+38}{2s^2+5s-3}\right\} \equiv \mathcal{L}^{-1}\left\{\frac{12}{2s-1} - \frac{2}{s+3}\right\} = \mathcal{L}^{-1}\left\{\frac{12}{2(s-\frac{1}{2})}\right\} - \mathcal{L}^{-1}\left\{\frac{2}{s+3}\right\}$

Hence $y = 6e^{-\frac{1}{2}x} - 2e^{-3x}$, from (ii) of *Table 1*

**Problem 13** Use Laplace transforms to solve the differential equation:

$$\frac{d^2y}{dx^2} + 6\frac{dy}{dx} + 13y = 0, \text{ given that when } x = 0, y = 3 \text{ and } \frac{dy}{dx} = 7.$$

This is the same as *Problem 3* of chapter 15, page 125. Using the procedure of para. 4:

(i) $\mathcal{L}\left\{\dfrac{d^2y}{dx^2}\right\} + 6\mathcal{L}\left\{\dfrac{dy}{dx}\right\} + 13\mathcal{L}\{y\} = \mathcal{L}\{0\}$

    Hence $[s^2\mathcal{L}\{y\} - sy(0) - y'(0)] + 6[s\mathcal{L}\{y\} - y(0)] + 13\mathcal{L}\{y\} = 0$, from equations
(3) and (4) of chapter 23.

(ii) $y(0) = 3$ and $y'(0) = 7$.

    Thus $s^2\mathcal{L}\{y\} - 3s - 7 + 6s\mathcal{L}\{y\} - 18 + 13\mathcal{L}\{y\} = 0$

(iii) Rearranging gives: $(s^2 + 6s + 13)\mathcal{L}\{y\} = 3s + 25$

$$\text{i.e.} \quad \mathcal{L}\{y\} = \frac{3s + 25}{s^2 + 6s + 13}$$

(iv) $y = \mathcal{L}^{-1}\left\{\dfrac{3s+25}{s^2+6s+13}\right\} = \mathcal{L}^{-1}\left\{\dfrac{3s+25}{(s+3)^2+2^2}\right\} = \mathcal{L}^{-1}\left\{\dfrac{3(s+3)+16}{(s+3)^2+2^2}\right\}$

$$= \mathcal{L}^{-1}\left\{\frac{3(s+3)}{(s+3)^2+2^2}\right\} + \mathcal{L}^{-1}\left\{\frac{8(2)}{(s+3)^2+2^2}\right\}$$

$$= 3e^{-3t}\cos 2t + 8e^{-3t}\sin 2t, \text{ from (xiii) and (xii) of } Table\ 1.$$

**Hence $y = e^{-3t}(3\cos 2t + 8\sin 2t)$.**

---

**Problem 14** Use Laplace transforms to solve the differential equation

$$\frac{d^2y}{dx^2} - 3\frac{dy}{dx} = 9, \text{ given that when } x = 0, y = 0 \text{ and } \frac{dy}{dx} = 0.$$

This is the same problem as *Problem 2* of chapter 16, page 132. Using the procedure of para. 4:

(i) $\mathcal{L}\left\{\dfrac{d^2y}{dx^2}\right\} - 3\mathcal{L}\left\{\dfrac{dy}{dx}\right\} = \mathcal{L}\{9\}$

    Hence $[s^2\mathcal{L}\{y\} - sy(0) - y'(0)] - 3[s\mathcal{L}\{y\} - y(0)] = \dfrac{9}{s}$

(ii) $y(0) = 0$ and $y'(0) = 0$.

    Hence $s^2\mathcal{L}\{y\} - 3s\mathcal{L}\{y\} = \dfrac{9}{s}$

(iii) Rearranging gives: $(s^2 - 3s)\mathcal{L}\{y\} = \dfrac{9}{s}$

$$\text{i.e., } \mathcal{L}\{y\} = \frac{9}{s(s^2-3s)} = \frac{9}{s^2(s-3)}$$

(iv) $y = \mathcal{L}^{-1}\left\{\dfrac{9}{s^2(s-3)}\right\}$

$$\frac{9}{s^2(s-3)} \equiv \frac{A}{s} + \frac{B}{s^2} + \frac{C}{s-3} \equiv \frac{A(s)(s-3) + B(s-3) + Cs^2}{s^2(s-3)}$$

Hence $9 \equiv A(s)(s-3) + B(s-3) + Cs^2$

When $s = 0$, $9 = -3B$, from which, $B = -3$

When $s = 3$, $9 = 9C$, from which, $C = 1$

Equating $s^2$ terms gives: $0 = A+C$, from which, $A = -1$

Hence $\mathcal{L}^{-1}\left\{\dfrac{9}{s^2(s-3)}\right\} = -\dfrac{1}{s} - \dfrac{3}{s^2} + \dfrac{1}{s-3}$

$\qquad\qquad\qquad\qquad = -1-3x+e^{3x}$, from (i), (vi) and (iii) of *Table 1*.

i.e., $y = e^{3x} - 3x - 1$.

---

*Problem 15* Use Laplace transforms to solve the differential equation
$\dfrac{d^2 y}{dx^2} - 7\dfrac{dy}{dx} + 10y = e^{2x}+20$, given that when $x = 0$, $y = 0$ and $\dfrac{dy}{dx} = -\dfrac{1}{3}$.

Using the procedure of para. 4:

(i) $\mathcal{L}\left\{\dfrac{d^2 y}{dx^2}\right\} - 7\mathcal{L}\left\{\dfrac{dy}{dx}\right\} + 10\mathcal{L}\{y\} = \mathcal{L}\{e^{2x}+20\}$

Hence $[s^2\mathcal{L}\{y\}-sy(0)-y'(0)]-7[s\mathcal{L}\{y\}-y(0)]+10\mathcal{L}\{y\} = \dfrac{1}{s-2}+\dfrac{20}{s}$

(ii) $y(0) = 0$ and $y'(0) = -\dfrac{1}{3}$

Hence $s^2\mathcal{L}\{y\}-0-\left(-\dfrac{1}{3}\right)-7s\mathcal{L}\{y\}+0+10\mathcal{L}\{y\} = \dfrac{21s-40}{s(s-2)}$

(iii) $(s^2-7s+10)\mathcal{L}\{y\} = \dfrac{21s-40}{s(s-2)} - \dfrac{1}{3} = \dfrac{3(21s-40)-s(s-2)}{3s(s-2)} = \dfrac{-s^2+65s-120}{3s(s-2)}$

$\qquad\qquad \mathcal{L}\{y\} = \dfrac{-s^2+65s-120}{3s(s-2)(s^2-7s+10)} = \dfrac{1}{3}\left[\dfrac{-s^2+65s-120}{s(s-2)(s-2)(s-5)}\right]$

$\qquad\qquad\qquad\quad = \dfrac{1}{3}\left[\dfrac{-s+65s-120}{s(s-5)(s-2)^2}\right]$

(iv) $y = \dfrac{1}{3}\mathcal{L}^{-1}\left\{\dfrac{-s^2+65s-120}{s(s-5)(s-2)^2}\right\}$

$\dfrac{-s^2+65s-120}{s(s-5)(s-2)^2} \equiv \dfrac{A}{s} + \dfrac{B}{s-5} + \dfrac{C}{s-2} + \dfrac{D}{(s-2)^2}$

$\qquad\qquad\qquad = \dfrac{A(s-5)(s-2)^2+B(s)(s-2)^2+C(s)(s-5)(s-2)+D(s)(s-5)}{s(s-5)(s-2)^2}$

Hence $-s^2+65s-120 \equiv A(s-5)(s-2)^2+B(s)(s-2)^2+C(s)(s-5)(s-2)+D(s)(s-5)$
When $s = 0$, $-120 = -20A$, from which, $A = 6$
When $s = 5$, $180 = 45B$, from which, $B = 4$
When $s = 2$, $6 = -6D$, from which, $D = -1$

Equating $s^3$ terms gives: $0 = A+B+C$, from which, $C = -10$

Hence $\dfrac{1}{3}\mathcal{L}^{-1}\left\{\dfrac{-s^2+65s-120}{s(s-5)(s-2)^2}\right\} = \dfrac{1}{3}\mathcal{L}^{-1}\left\{\dfrac{6}{s} + \dfrac{4}{s-5} - \dfrac{10}{s-2} - \dfrac{1}{(s-2)^2}\right\}$

$\qquad\qquad\qquad\qquad\qquad\qquad = \dfrac{1}{3}[6+4e^{5x}-10e^{2x}-xe^{2x}]$

Thus $y = 2+\dfrac{4}{3}e^{5x} - \dfrac{10}{3}e^{2x} - \dfrac{x}{3}e^{2x}$

---

*Problem 16* The current flowing in an electrical circuit is given by the differential equation $Ri+L(di/dt)=E$, where $E, L$ and $R$ are constants. Use Laplace transforms to solve the equation for current $i$ given that when $t = 0$, $i = 0$.

Using the procedure of para. 4:

(i) $\mathcal{L}\{Ri\} + \mathcal{L}\left\{L\dfrac{di}{dt}\right\} = \mathcal{L}\{E\}$. $R\mathcal{L}\{i\} + L[s\mathcal{L}\{i\} - i(0)] = \dfrac{E}{s}$

(ii) $i(0) = 0$, hence $R\mathcal{L}\{i\} + LS\mathcal{L}\{i\} = \dfrac{E}{s}$

(iii) Rearranging gives: $(R + Ls)\mathcal{L}\{i\} = \dfrac{E}{s}$ i.e. $\mathcal{L}\{i\} = \dfrac{E}{s(R+Ls)}$

(iv) $i = \mathcal{L}^{-1}\left\{\dfrac{E}{s(R+Ls)}\right\}$

$\dfrac{E}{s(R+Ls)} \equiv \dfrac{A}{s} + \dfrac{B}{R+Ls} \equiv \dfrac{A(R+Ls)+Bs}{s(R+Ls)}$

Hence $E \equiv A(R+Ls) + Bs$

When $s = 0$, $E = AR$, from which, $A = \dfrac{E}{R}$

When $s = -\dfrac{R}{L}$, $E = B\left(-\dfrac{R}{L}\right)$, from which $B = \dfrac{-EL}{R}$

Hence $\mathcal{L}^{-1}\left\{\dfrac{E}{s(R+Ls)}\right\} = \mathcal{L}^{-1}\left\{\dfrac{E/R}{s} + \dfrac{-EL/R}{R+Ls}\right\} = \mathcal{L}^{-1}\left\{\dfrac{E}{Rs} - \dfrac{EL}{R(R+Ls)}\right\}$

$= \mathcal{L}^{-1}\left\{\dfrac{E}{R}\left(\dfrac{1}{s}\right) - \dfrac{E}{R}\left(\dfrac{1}{\dfrac{R}{L}+s}\right)\right\} = \dfrac{E}{R}\mathcal{L}^{-1}\left\{\dfrac{1}{s} - \dfrac{1}{(s+\dfrac{R}{L})}\right\}$

Hence current $i = \dfrac{E}{R}\left(1 - e^{-\frac{R}{L}t}\right)$

## C. FURTHER PROBLEMS ON INVERSE LAPLACE TRANSFORMS AND THE SOLUTION OF DIFFERENTIAL EQUATIONS

Determine the inverse Laplace transforms in *Problems 1 to 13*.

1 (a) $\dfrac{7}{s}$; (b) $\dfrac{2}{s-5}$     [(a) 7; (b) $2e^{5t}$]

2 (a) $\dfrac{3}{2s+1}$; (b) $\dfrac{2s}{s^2+4}$     [(a) $\dfrac{3}{2}e^{-\frac{1}{2}t}$; (b) $2\cos 2t$]

3 (a) $\dfrac{1}{s^2+25}$ ; (b) $\dfrac{4}{s^2+9}$     [(a) $\dfrac{1}{5}\sin 5t$; (b) $\dfrac{4}{3}\sin 3t$]

4 (a) $\dfrac{5s}{2s^2+18}$ ; (b) $\dfrac{6}{s^2}$     [(a) $\dfrac{5}{2}\cos 3t$; (b) $6t$]

5 (a) $\dfrac{5}{s^3}$ ; (b) $\dfrac{8}{s^4}$     [(a) $\dfrac{5}{2}t^2$; (b) $\dfrac{4}{3}t^3$]

6 (a) $\dfrac{10}{s^5}$ ; (b) $\dfrac{5s}{s^2-4}$     [(a) $\dfrac{5}{12}t^4$; (b) $5\cosh 2t$]

7 (a) $\dfrac{3s}{\frac{1}{2}s^2-8}$ ; (b) $\dfrac{7}{s^2-16}$     [(a) $6\cosh 4t$; (b) $\dfrac{7}{4}\sinh 4t$]

8 (a) $\dfrac{15}{3s^2-27}$ ; (b) $\dfrac{4}{(s-1)^3}$     [(a) $\dfrac{5}{3}\sinh 3t$; (b) $2e^t t^2$]

9 (a) $\dfrac{1}{(s+2)^4}$ ; (b) $\dfrac{3}{(s-3)^5}$     [(a) $\dfrac{1}{6}e^{-2t}t^3$; (b) $\dfrac{1}{8}e^{3t}t^4$]

10 (a) $\dfrac{3}{s^2-4s+13}$ ; (b) $\dfrac{4}{2s^2-8s+10}$     [(a) $e^{2t}\sin 3t$; (b) $2e^{2t}\sin t$]

11 (a) $\dfrac{s+1}{s^2+2s+10}$ ; (b) $\dfrac{3}{s^2+6s+13}$     [(a) $e^{-t}\cos 3t$; (b) $\dfrac{3}{2}e^{-3t}\sin 2t$]

12 (a) $\dfrac{2(s-3)}{s^2-6s+13}$ ; (b) $\dfrac{7}{s^2-8s+12}$     [(a) $2e^{3t}\cos 2t$; (b) $\dfrac{7}{2}e^{4t}\sinh 2t$]

13 (a) $\dfrac{2s+5}{s^2+4s-5}$ ; (b) $\dfrac{3s+2}{s^2-8s+25}$     $\left[\begin{array}{l}\text{(a) } 2e^{-2t}\cosh 3t + \dfrac{1}{3}e^{-2t}\sinh 3t \\[2mm] \text{(b) } 3e^{4t}\cos 3t + \dfrac{14}{3}e^{4t}\sin 3t\end{array}\right]$

Use partial fractions to find the inverse Laplace transforms of the functions in *Problems 14 to 21*.

14 $\dfrac{11-3s}{s^2+2s-3}$     $[2e^t-5e^{-3t}]$

15 $\dfrac{2s^2-9s-35}{(s+1)(s-2)(s+3)}$     $[4e^{-t}-3e^{2t}+e^{-3t}]$

16 $\dfrac{2s+3}{(s-2)^2}$     $[2e^{2t}+7te^{2t}]$

17 $\dfrac{5s^2-2s-19}{(s+3)(s-1)^2}$     $[2e^{-3t}+3e^t-4e^t t]$

18 $\dfrac{3s^2+16s+15}{(s+3)^3}$     $[e^{-3t}(3-2t-3t^2)]$

19 $\dfrac{7s^2+5s+13}{(s^2+2)(s+1)}$     $[2\cos\sqrt{2}t + \dfrac{3}{\sqrt{2}}\sin\sqrt{2}t+5e^{-t}]$

20 $\dfrac{3+6s+4s^2-2s^3}{s^2(s^2+3)}$     $[2+t+\sqrt{3}\sin\sqrt{3}t-4\cos\sqrt{3}t]$

21 $\dfrac{26-s^2}{s(s^2+4s+13)}$     $[(2-3e^{-2t}\cos 3t - \dfrac{2}{3}e^{-2t}\sin 3t]$

In *Problems 22 to 29*, use Laplace transforms to solve the given differential equations.

22 $9\dfrac{d^2y}{dt^2}-24\dfrac{dy}{dt}+16y = 0$, given $y(0)=0$ and $y'(0)=3$.     $\left[y=(3-t)e^{\frac{4}{3}t}\right]$

23 $\dfrac{d^2x}{dt^2}+100x = 0$, given $x(0)=2$ and $x'(0)=0$.     $[x=2\cos 10t]$

24 $\dfrac{d^2i}{dt^2}+1000\dfrac{di}{dt}+250\,000i$, given $i(0)=0$ and $i'(0)=100$.     $[i=100te^{-500t}]$

25 $\dfrac{d^2x}{dt^2}+6\dfrac{dx}{dt}+8x=0$, given $x(0)=4$ and $x'(0)=8$.     $[x=4(3e^{-2t}-2e^{-4t})]$

26 $\dfrac{d^2y}{dx^2}-2\dfrac{dy}{dx}+y=3e^{4x}$, given $y(0)=-\dfrac{2}{3}$ and $y'(0)=4\dfrac{1}{3}$.     $[y=(4x-1)e^x + \dfrac{1}{3}e^{4x}]$

27 $\dfrac{d^2y}{dx^2}+16y=10\cos 4x$, given $y(0)=3$ and $y'(0)=4$.

$[y=3\cos 4x+\sin 4x + \dfrac{5}{4}x\sin 4x]$

28 $\dfrac{d^2y}{dx^2}+\dfrac{dy}{dx}-2y=3\cos 3x-11\sin 3x$, given $y(0)=0$ and $y'(0)=6$.
$[y=e^x-e^{-2x}+\sin 3x]$

29 $\dfrac{d^2y}{dx^2}-2\dfrac{dy}{dx}+2y=3e^x\cos 2x$, given $y(0)=2$ and $y'(0)=5$.
$[y=e^x(3\cos x+\sin x)-e^x\cos 2x]$

30 Solve, using Laplace transforms, *Problems 7 to 18* of chapter 15, page 128.
31 Solve, using Laplace transforms, *Problems 19 to 27, 29 and 31* of chapter 16, page 139.

# 25 Linear correlation

## A. MAIN POINTS CONCERNING LINEAR CORRELATION

1 Correlation is a measure of the amount of association existing between two variables. For linear correlation, if points are plotted on a graph and all the points lie on a straight line, then **perfect linear correlation** is said to exist. When a straight line having a positive gradient can reasonably be drawn through points on a graph **positive or direct linear correlation** exists. Similarly, when a straight line having a negative gradient can reasonably be drawn through points on a graph, **negative or inverse linear correlation** exists. When there is no apparent relationship between co-ordinate values plotted on a graph then no correlation exists between the points.

2 The amount of linear correlation between two variables is expressed by a **coefficient of correlation**, given the symbol $r$. This is defined in terms of the deviations of the co-ordinates of two variables from their mean values and is given by the **product–moment formula** which states:

$$\text{coefficient of correlation, } r = \frac{\Sigma xy}{\sqrt{\{(\Sigma x^2)(\Sigma y^2)\}}} \tag{1}$$

where the $x$-values are the values of the deviations of co-ordinates $X$ from $\overline{X}$, their mean value and the $y$-values are the values of the deviations of co-ordinates $Y$ from $\overline{Y}$, their mean value. That is, $x = (X - \overline{X})$ and $y = (Y - \overline{Y})$. The results of this determination give values of $r$ lying between $+1$ and $-1$, where $+1$ indicates perfect direct correlation, $-1$ indicates perfect inverse correlation and $0$ indicates that no correlation exists. Between these values, the smaller the value of $r$, the less is the amount of correlation which exists. Generally, values of $r$ in the ranges 0.7 to 1 and $-0.7$ to $-1$ show that a fair amount of correlation exists.

3 When the value of the coefficient of correlation has been obtained from the product–moment formula, some care is needed before coming to conclusions based on this result. Checks should be made to ascertain the following two points:

(a) That a 'cause and effect' relationship exists between the variables; it is relatively easy, mathematically, to show that some correlation exists between, say, the number of ice creams sold in a given period of time and the number of chimneys swept in the same periods of time, although there is no relationship between these variables;

(b) that a linear relationship exists between the variables; the product–moment formula given in para. 2 is based on linear correlation. Perfect non-linear correlation may exist, (for example, the co-ordinates exactly following the curve $y = x^3$),

204

but this gives a low value of coefficient of correlation since the value of $r$ is determined using the product–moment formula, based on a linear relationship.

4 To test the significance of a correlation coefficient requires an understanding of such statistical techniques as hypothesis testing, the Student's t-distribution and significance testing theory, and has therefore been omitted from this text.

## B. WORKED PROBLEMS ON LINEAR CORRELATION

*Problem 1* In an experiment to determine the relationship between force on a wire and the resulting extension, the following data is obtained:

| Force (N) | 10 | 20 | 30 | 40 | 50 | 60 | 70 |
|---|---|---|---|---|---|---|---|
| Extension (mm) | 0.22 | 0.40 | 0.61 | 0.85 | 1.20 | 1.45 | 1.70 |

Determine the linear coefficient of correlation for this data.

Let $X$ be the variable force values and $Y$ be the dependent variable extension values. The coefficient of correlation is given by:

$$r = \frac{\Sigma xy}{\sqrt{\{(\Sigma x^2)(\Sigma y^2)\}}}$$

where $x = (X-\bar{X})$ and $y = (Y-\bar{Y})$, $\bar{X}$ and $\bar{Y}$ being the mean values of the $X$ and $Y$ values respectively. Using a tabular method to determine the quantities of this formula gives:

| $X$ | $Y$ | $x = (X-\bar{X})$ | $y = (Y-\bar{Y})$ | $xy$ | $x^2$ | $y^2$ |
|---|---|---|---|---|---|---|
| 10 | 0.22 | −30 | −0.699 | 20.97 | 900 | 0.489 |
| 20 | 0.40 | −20 | −0.519 | 10.38 | 400 | 0.269 |
| 30 | 0.61 | −10 | −0.309 | 3.09 | 100 | 0.095 |
| 40 | 0.85 | 0 | −0.069 | 0 | 0 | 0.005 |
| 50 | 1.20 | 10 | 0.281 | 2.81 | 100 | 0.079 |
| 60 | 1.45 | 20 | 0.531 | 10.62 | 400 | 0.282 |
| 70 | 1.70 | 30 | 0.781 | 23.43 | 900 | 0.610 |
| $\Sigma X = 280$ | $\Sigma Y = 6.43$ | | | | | |
| $\bar{X} = \dfrac{280}{7}$ $= 40$ | $\bar{Y} = \dfrac{6.43}{7}$ $= 0.919$ | | | $\Sigma xy =$ 71.30 | $\Sigma x^2 =$ 2800 | $\Sigma y^2 =$ 1.829 |

Thus $r = \dfrac{71.3}{\sqrt{[2800 \times 1.829]}} = 0.996$

This shows that a very good direct correlation exists between the values of force and extension.

*Problem 2* The relationship between expenditure on welfare services and absenteeism for similar periods of time is shown below, for a small company.

| Expenditure (£'000) | 3.5 | 5.0 | 7.0 | 10 | 12 | 15 | 18 |
|---|---|---|---|---|---|---|---|
| Days lost | 241 | 318 | 174 | 110 | 147 | 122 | 86 |

Determine the coefficient of linear correlation for this data.

Let $X$ be the expenditure in thousands of pounds and $Y$ be the days lost.

The coefficient of correlation, $r = \dfrac{\Sigma xy}{\sqrt{\{(\Sigma x^2)(\Sigma y^2)\}}}$

where $x = (X-\bar{X})$ and $y = (Y-\bar{Y})$, $\bar{X}$ and $\bar{Y}$ being the mean values of $X$ and $Y$ values respectively. Using a tabular approach:

| $X$ | $Y$ | $x = (X-\bar{X})$ | $y = (Y-\bar{Y})$ | $xy$ | $x^2$ | $y^2$ |
|---|---|---|---|---|---|---|
| 3.5 | 241 | −6.57 | 69.9 | −459.2 | 43.2 | 4886 |
| 5.0 | 318 | −5.07 | 146.9 | −744.8 | 25.7 | 21580 |
| 7.0 | 174 | −3.07 | 2.9 | −8.9 | 9.4 | 8 |
| 10 | 110 | −0.07 | −61.1 | 4.3 | 0 | 3733 |
| 12 | 147 | 1.93 | −24.1 | −46.5 | 3.7 | 581 |
| 15 | 122 | 4.93 | −49.1 | −242.1 | 24.3 | 2411 |
| 18 | 86 | 7.93 | −85.1 | −674.8 | 62.9 | 7242 |
| $\Sigma X = 70.5$ | $\Sigma Y = 1198$ | | | | | |
| $\bar{X} = \dfrac{70.5}{7}$ | $\bar{Y} = \dfrac{1198}{7}$ | | | $\Sigma xy =$ −2172 | $\Sigma x^2 =$ 169.2 | $\Sigma y^2 =$ 40 441 |
| $= 10.07$ | $= 171.1$ | | | | | |

Thus $r = \dfrac{-2\,172}{\sqrt{[169.2 \times 40\,441]}} = -0.830$

This shows that there is fairly good inverse correlation between the expenditure on welfare and days lost due to absenteeism.

*Problem 3* The relationship between monthly car sales and income from the sale of petrol for a garage is as shown:

| Cars sold | 2 | 5 | 3 | 12 | 14 | 7 | 3 | 28 | 14 | 7 | 3 | 13 |
|---|---|---|---|---|---|---|---|---|---|---|---|---|
| Income from petrol sales (£'000) | 12 | 9 | 13 | 21 | 17 | 22 | 31 | 47 | 17 | 10 | 9 | 11 |

Determine the linear coefficient of correlation between these quantities.

Let $X$ represent the number of cars sold and $Y$ the income, in thousands of pounds, from petrol sales. Using the tabular approach (see opposite page).

The coefficient of correlation,

$r = \dfrac{\Sigma xy}{\sqrt{\{(\Sigma x^2)(\Sigma y^2)\}}} = \dfrac{613.4}{\sqrt{\{(616.7)(1372.7)\}}} = 0.667$

Thus, there is no appreciable correlation between petrol and car sales.

| $X$ | $Y$ | $x = (X-\bar{X})$ | $y = (Y-\bar{Y})$ | $xy$ | $x^2$ | $y^2$ |
|-----|-----|------|------|------|------|------|
| 2 | 12 | −7.25 | −6.25 | 45.3 | 52.6 | 39.1 |
| 5 | 9 | −4.25 | −9.25 | 39.3 | 18.1 | 85.6 |
| 3 | 13 | −6.25 | −5.25 | 32.8 | 39.1 | 27.6 |
| 12 | 21 | 2.75 | 2.75 | 7.6 | 7.6 | 7.6 |
| 14 | 17 | 4.75 | −1.25 | −5.9 | 22.6 | 1.6 |
| 7 | 22 | −2.25 | 3.75 | −8.4 | 5.1 | 14.1 |
| 3 | 31 | −6.25 | 12.75 | −79.7 | 39.1 | 162.6 |
| 28 | 47 | 18.75 | 28.75 | 539.1 | 351.6 | 826.6 |
| 14 | 17 | 4.75 | −1.25 | −5.9 | 22.6 | 1.6 |
| 7 | 10 | −2.25 | −8.25 | 18.6 | 5.1 | 68.1 |
| 3 | 9 | −6.25 | −9.25 | 57.8 | 39.1 | 85.6 |
| 13 | 11 | 3.75 | −7.25 | −27.2 | 14.1 | 52.6 |
| $\Sigma X = 111$ $\bar{X} = \dfrac{111}{12}$ $= 9.25$ | $\Sigma Y = 219$ $\bar{Y} = \dfrac{219}{12}$ $= 18.25$ | | | $\Sigma xy =$ 613.4 | $\Sigma x^2 =$ 616.7 | $\Sigma y^2 =$ 1372.7 |

## C. FURTHER PROBLEMS ON LINEAR CORRELATION

In *Problems 1 to 5*, determine the coefficient of correlation for the data given, correct to 3 decimal places.

1

| $X$ | 14 | 18 | 23 | 30 | 50 |
|-----|-----|-----|-----|-----|-----|
| $Y$ | 900 | 1200 | 1600 | 2100 | 3800 |

[0.999]

2

| $X$ | 2.7 | 4.3 | 1.2 | 1.4 | 4.9 |
|-----|-----|-----|-----|-----|-----|
| $Y$ | 11.9 | 7.10 | 33.8 | 25.0 | 7.50 |

[−0.916]

3

| $X$ | 24 | 41 | 9 | 18 | 73 |
|-----|-----|-----|-----|-----|-----|
| $Y$ | 39 | 46 | 90 | 30 | 98 |

[0.422]

4

| $X$ | 6 | 3 | 9 | 15 | 2 | 14 | 21 | 13 |
|-----|-----|-----|-----|-----|-----|-----|-----|-----|
| $Y$ | 1.3 | 0.7 | 2.0 | 3.7 | 0.5 | 2.9 | 4.5 | 2.7 |

[0.992]

5

| $X$ | 1142 | 1870 | 1171 | 1234 | 1471 | 1882 | 1960 |
|-----|-----|-----|-----|-----|-----|-----|-----|
| $Y$ | 0.56 | 0.35 | 0.55 | 0.52 | 0.43 | 0.34 | 0.31 |

[−0.993]

6  In an experiment to determine the relationship between the current flowing in an electrical circuit and the applied voltage, the results obtained are:

| Current (mA) | 5 | 11 | 15 | 19 | 24 | 28 | 33 |
|-----|-----|-----|-----|-----|-----|-----|-----|
| Applied voltage (V) | 2 | 4 | 6 | 8 | 10 | 12 | 14 |

Determine, using the product–moment formula, the coefficient of correlation for these results. [0.999]

7 A gas is being compressed in a closed cylinder and the values of pressures and corresponding volumes at constant temperature are as shown:

| Pressure (kPa) | 160 | 180 | 200 | 220 | 240 | 260 | 280 | 300 |
|---|---|---|---|---|---|---|---|---|
| Volume (m³) | 0.034 | 0.036 | 0.030 | 0.027 | 0.024 | 0.025 | 0.020 | 0.019 |

Find the coefficient of correlation for these values. [−0.962]

8 The relationship between the number of miles travelled by a group of salesmen in ten equal time periods and the corresponding value of orders taken, is given below. Calculate the coefficient of correlation using the product–moment formula for these values.

| Miles travelled | 1370 | 1050 | 980 | 1770 | 1340 | 1560 | 2110 | 1540 | 1480 | 1670 |
|---|---|---|---|---|---|---|---|---|---|---|
| Orders taken (£'000) | 23 | 17 | 19 | 22 | 27 | 23 | 30 | 23 | 25 | 19 |

[0.632]

9 The weight of applies harvested from a small orchard and the number of nights on which the temperature fell below 4°C after the blossom appeared, for equal time periods, are as shown:

| Number of nights | 10 | 13 | 25 | 22 | 5 | 18 | 12 |
|---|---|---|---|---|---|---|---|
| Apple crop (kg) | 670 | 417 | 183 | 274 | 770 | 216 | 549 |

Determine the coefficient of correlation for this data. [−0.942]

10 The data shown below refers to the number of times machine tools had to be taken out of service, in equal time periods, due to faults occurring and the number of hours worked by maintenance teams. Calculate the coefficient of correlation for this data.

| Machines out of service: | 4 | 13 | 2 | 9 | 16 | 8 | 7 |
|---|---|---|---|---|---|---|---|
| Maintenance hours: | 400 | 515 | 360 | 440 | 570 | 380 | 415 |

[0.937]

# 26 Linear regression

## A. MAIN POINTS CONCERNING LINEAR REGRESSION

1 Regression analysis, usually termed **regression**, is used to draw the line of 'best fit' through co-ordinates on a graph. The techniques used enable a mathematical equation of the straight line form $y = mx + c$ to be deduced for a given set of co-ordinate values, the line being such that the sum of the deviations of the co-ordinate values from the line is a minimum, i.e. it is the line of 'best fit'.

2 When a regression analysis is made, it is possible to obtain two lines of best fit, depending on which variable is selected as the dependent variable and which variable is the independent variable. For example, in a resistive electrical circuit, the current flowing is directly proportional to the voltage applied to the circuit. There are two ways of obtaining experimental values relating the current and voltage. Either, certain voltages are applied to the circuit and the current values are measured, in which case, the voltage is the independent variable and the current is the dependent variable. Alternatively, the voltage can be adjusted until a desired value of current is flowing and the value of voltage is measured, in which case, the current is the independent value and the voltage is the dependent value.

3 **The least-squares regression line**
For a given set of co-ordinate values, $(X_1, Y_1), (X_2, Y_2), \ldots, (X_N, Y_N)$ let the $X$-values be the independent variables and the $Y$-values be the dependent values. Also let $D_1, D_2, \ldots D_N$ be the vertical distances between the line shown as PQ in *Fig 1* and the points representing the co-ordinate values. The least-squares regression line, i.e. the line of best fit, is the line which makes the value of $D_1^2 + D_2^2 + \ldots + D_N^2$ a minimum value.

4 The equation of the least-squares regression line is usually written as $Y = a_0 + a_1 X$, where $a_0$ is the $Y$-axis intercept value and $a_1$ is the gradient of the line (analogous to $c$ and $m$ in the equation $y = mx + c$). The values of $a_0$ and $a_1$ to make the sum of the 'deviations squared' a minimum can be obtained from the two equations:

$$\Sigma Y = a_0 N + a_1 \Sigma X \tag{1}$$
$$\Sigma(XY) = a_0 \Sigma X + a_1 \Sigma X^2 \tag{2}$$

where $X$ and $Y$ are the co-ordinate values, $N$ is the number of co-ordinates and $a_0$ and $a_1$ are called the **regression coefficients** of $Y$ on $X$. Equations (1) and (2) are called the **normal equations** of the regression line of $Y$ on $X$. The regression line of $Y$ on $X$ is used to estimate values of $Y$ for given values of $X$.

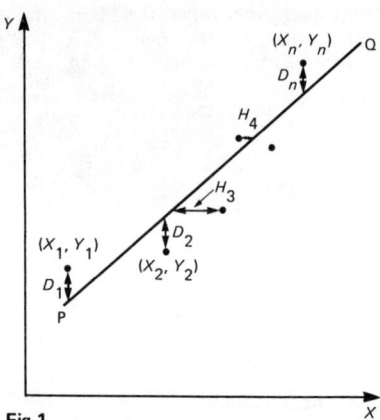

**Fig 1**

5   If the $Y$-values, (vertical-axis), are selected as the independent variables, the horizontal distances between the line shown as PQ in *Fig 1* and the co-ordinate values, $(H_3, H_4,$ etc.) are taken as the deviations. The equation of the regression line is of the form: $X = b_0 + b_1 Y$ and the normal equations become:

$$\Sigma X = b_0 N + b_1 \Sigma Y \tag{3}$$
$$\Sigma(XY) = b_0 \Sigma Y + b_1 \Sigma Y^2 \tag{4}$$

where $X$ and $Y$ are the co-ordinate values, $b_0$ and $b_1$ are the regression coefficients of $X$ on $Y$ and $N$ is the number of co-ordinates. These normal equations are of the regression line of $X$ on $Y$, which is slightly different to the regression line of $Y$ on $X$. The regression line of $X$ on $Y$ is used to estimate values of $X$ for given values of $Y$.

6   The regression line of $Y$ on $X$ is used to determine any value of $Y$ corresponding to a given value of $X$. If the value of $Y$ lies within the range of $Y$-values of the extreme co-ordinates, the process of finding the corresponding value of $X$ is called **linear interpolation**. If it lies outside of the range of $Y$-values of the extreme co-ordinates then the process is called **linear extrapolation** and the assumption must be made that the line of best fit extends outside of the range of the co-ordinate values given. By using the regression line of $X$ on $Y$, values of $X$ corresponding to given values of $Y$ may be found by either interpolation or extrapolation.

## B. WORKED PROBLEMS ON LINEAR REGRESSION

*Problem 1*  In an experiment to determine the relationship between frequency and the inductive reactance of an electrical circuit, the following results were obtained:

| Frequency (Hz) | 50 | 100 | 150 | 200 | 250 | 300 | 350 |
|---|---|---|---|---|---|---|---|
| Inductive reactance (ohms) | 30 | 65 | 90 | 130 | 150 | 190 | 200 |

Determine the equation of the regression line of inductive reactance on frequency, assuming a linear relationship.

Since the regression line of inductive reactance on frequency is required, the frequency is the independent variable, $X$, and the inductive reactance is the dependent variable, $Y$. The equation of the regression line of $Y$ on $X$ is:

$Y = a_0 + a_1 X$, and the regression coefficients $a_0$ and $a_1$ are obtained by using the normal equations

$$\Sigma Y = a_0 N + a_1 \Sigma X$$
and $\quad \Sigma XY = a_0 \Sigma X + a_1 \Sigma X^2$, (see para. 4)

A tabular approach is used to determine the summed quantities.

| Frequency, X | Inductive reactance, Y | $X^2$ | XY | $Y^2$ |
|---|---|---|---|---|
| 50 | 30 | 2 500 | 1 500 | 900 |
| 100 | 65 | 10 000 | 6 500 | 4 225 |
| 150 | 90 | 22 500 | 13 500 | 8 100 |
| 200 | 130 | 40 000 | 26 000 | 16 900 |
| 250 | 150 | 62 500 | 37 500 | 22 500 |
| 300 | 190 | 90 000 | 57 000 | 36 100 |
| 350 | 200 | 122 500 | 70 000 | 40 000 |
| $\Sigma X = 1\,400$ | $\Sigma Y = 855$ | $\Sigma X^2 =$ 350 000 | $\Sigma XY =$ 212 000 | $\Sigma Y^2 =$ 128 725 |

The number of co-ordinate values given, $N$ is 7. Substituting in the normal equations, gives:

$$855 = 7a_0 + 1\,400 a_1$$
$$212\,000 = 1400 a_0 + 350\,000 a_1$$

Solving these simultaneous equations gives $a_0 = 5.00$ and $a_1 = 0.586$, correct to 3 significant figures. Thus the equation of the regression line of inductive reactance on frequency is

$$Y = 5.00 + 0.586X$$

*Problem 2* For the data given in *Problem 1*, determine the equation of the regression line of frequency on inductive reactance, assuming a linear relationship.

In this case, the inductive reactance is the independent variable $X$ and the frequency is the dependent variable $Y$. From para. 5, the equation of the regression line of $X$ on $Y$ is:

$X = b_0 + b_1 Y$, and the normal equations are
$\Sigma X = b_0 N + b_1 \Sigma Y$ and $\Sigma XY = b_0 \Sigma Y + b_1 \Sigma Y^2$.
From the table shown in *Problem 1*, the simultaneous equations are:

$$1\,400 = 7b_0 + 855 b_1$$
$$212\,000 = 855 b_0 + 128\,725 b_1$$

Solving these equations gives $b_0 = -6.15$ and $b_1 = 1.69$, correct to 3 significant figures. Thus the equation of the regression line of frequency on inductive reactance is

$$X = -6.15 + 1.69Y$$

*Problem 3* Use the regression equations calculated in *Problems 1 and 2* to find
(a) the value of inductive reactance when the frequency is 175 Hz and (b) the value
of frequency when the inductive reactance is 250 ohms, assuming the line of best
fit extends outside of the given co-ordinate values. Draw a graph showing the two
regression lines.

(a) From *Problem 1* the regression equation of inductive reactance on frequency
is $Y = 5.00 + 0.586X$. When the frequency ($X$) is 175 Hz, $Y = 5.00 + 0.586(175)$
$= 107.6$, correct to 4 significant figures, i.e. the inductive reactance is **107.6 ohms**
when the frequency is 175 Hz.
(b) From *Problem 2*, the regression equation of frequency on inductive reactance is
$X = -6.15 + 1.69Y$. When the inductive reactance, ($Y$), is 250 ohms,
$X = -6.15 + 1.69(250) = 416.4$ Hz, correct to 4 significant figures, i.e., the
frequency is **416.4 Hz** when the inductive reactance is 250 ohms.

The graph depicting the two regression lines is shown in *Fig. 2*. To obtain the
regression line of inductive reactance on frequency the regression line equation

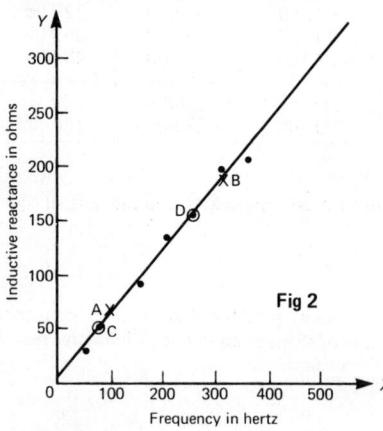

$Y = 5.00 + 0.586X$ is used, and $X$
(frequency) values of 100 and 300
have been selected in order to
find the corresponding $Y$ values.
These values gave the
co-ordinates as (100, 63.6) and
(300, 180.8), shown as points A
and B in *Fig. 2*. Two co-ordinates
for the regression line of
frequency on inductive reactance
are calculated using the equation
$X = -6.15 + 1.69Y$, the values of
inductive reactance of 50 and 150
being used to obtain the
co-ordinate values. These values
gave co-ordinates (78.4, 50) and
(274.4, 150), shown as points C
and D in *Fig. 2*.

**Fig 2**

It can be seen from *Fig 2* that to the scale drawn, the two regression lines coincide.
Although it is not necessary to do so, the co-ordinate values are also shown to indi-
cate that the regression lines do appear to be the lines of best fit. A graph showing
co-ordinate values is called a **scatter diagram** in statistics.

---

*Problem 4* The experimental values relating centripetal force and radius, for a mass
travelling at constant velocity in a circle, are as shown:

| Force (N) | 5 | 10 | 15 | 20 | 25 | 30 | 35 | 40 |
|-----------|-----|-----|-----|-----|-----|-----|-----|-----|
| Radius (cm) | 55 | 30 | 16 | 12 | 11 | 9 | 7 | 5 |

Determine the equations of (a) the regression line of force on radius and (b) the
regression line of radius on force. Hence, calculate the force at a radius of 40 cm
and the radius corresponding to a force of 32 newtons.

Let the radius be the independent variable $X$, and the force be the dependent variable $Y$. (This decision is usually based on a 'cause' corresponding to $X$ and an 'effect' corresponding to $Y$.)

(a) The equation of the regression line of force on radius is of the form $Y = a_0 + a_1 X$ and the constants $a_0$ and $a_1$ are determined from the normal equations:

$$\Sigma Y = a_0 N + a_1 \Sigma X \text{ and } \Sigma XY = a_0 \Sigma + a_1 \Sigma X^2 \text{ (see para. 4)}$$

Using a tabular approach to determine the values of the summations, gives:

| Radius, $X$ | Force, $Y$ | $X^2$ | $XY$ | $Y^2$ |
|---|---|---|---|---|
| 55 | 5 | 3025 | 275 | 25 |
| 30 | 10 | 900 | 300 | 100 |
| 16 | 15 | 256 | 240 | 225 |
| 12 | 20 | 144 | 240 | 400 |
| 11 | 25 | 121 | 275 | 625 |
| 9 | 30 | 81 | 270 | 900 |
| 7 | 35 | 49 | 245 | 1225 |
| 5 | 40 | 25 | 200 | 1600 |
| $\Sigma X = 145$ | $\Sigma Y = 180$ | $\Sigma X^2 =$ 4601 | $\Sigma XY =$ 2045 | $\Sigma Y^2 =$ 5100 |

Thus $180 = 8a_0 + 145a_1$ and $2045 = 145a_0 + 4601a_1$

Solving these simultaneous equations gives $a_0 = 33.7$ and $a_1 = -0.617$, correct to 3 significant figures. Thus the equation of the regression line of force on radius is: $Y = 33.7 - 0.617X$.

(b) The equation of the regression line of radius on force is of the form $X = b_0 + b_1 Y$ and the constants $b_0$ and $b_1$ are determined from the normal equations:

$$\Sigma X = b_0 N + b_1 \Sigma Y \text{ and } \Sigma XY = b_0 \Sigma Y + b_1 \Sigma Y^2, \text{ (see para. 5)}$$

The values of the summations have been obtained in part (a) giving:

$145 = 8b_0 + 180b_1$ and $2045 = 180b_0 + 5100b_1$

Solving these simultaneous equations gives $b_0 = 44.2$ and $b_1 = -1.16$, correct to 3 significant figures. Thus the equation of the regression line of radius on force is:

$X = 44.2 - 1.16Y$

The force, $Y$, at a radius of 40 cm, is obtained from the regression line of force on radius, i.e.

$Y = 33.7 - 0.617(40) = 9.02$, i.e. **the force at a radius of 40 cm is 9.02N.**

The radius, $X$, when the force is 32 newtons is obtained from the regression line of radius on force, i.e.

$X = 44.2 - 1.16(32) = 7.08$, i.e. **the radius when the force is 32N is 7.08 cm.**

## C. FURTHER PROBLEMS ON LINEAR REGRESSION

In *Problems 1 to 3*, determine the equation of the regression line of $Y$ on $X$, correct to 3 significant figures.

**1**

| X | 14 | 18 | 23 | 30 | 50 |
|---|----|----|----|----|----|
| Y | 900 | 1200 | 1600 | 2100 | 3800 |

$[Y = -256 + 80.6X]$

**2**

| X | 2.7 | 4.3 | 1.2 | 1.4 | 4.9 |
|---|-----|-----|-----|-----|-----|
| Y | 11.9 | 7.10 | 33.8 | 25.0 | 7.50 |

$[Y = 35.9 - 6.50X]$

**3**

| X | 6 | 3 | 9 | 15 | 2 | 14 | 21 | 13 |
|---|---|---|---|----|---|----|----|----|
| Y | 1.3 | 0.7 | 2.0 | 3.7 | 0.5 | 2.9 | 4.5 | 2.7 |

$[Y = 0.0477 + 0.216X]$

In *Problems 4 to 6*, determine the equations of the regression lines of $X$ on $Y$ for the data stated, correct to 3 significant figures.

4  The data given in *Problem 1*.      $[X = 3.20 + 0.0124Y]$

5  The data given in *Problem 2*.      $[X = 5.10 - 0.129Y]$

6  The data given in *Problem 3*.      $[X = -0.0472 + 4.56Y]$

7  The relationship between the voltage applied to an electrical circuit and the current flowing is as shown:

| Current (mA) | 2 | 4 | 6 | 8 | 10 | 12 | 14 |
|--------------|---|---|---|---|----|----|----|
| Applied voltage (V) | 5 | 11 | 15 | 19 | 24 | 28 | 33 |

Assuming a linear relationship, determine the equation of the regression line of applied voltage, $(Y)$, on current $(X)$, correct to 4 significant figures.

$[Y = 1.117 + 2.268X]$

8  For the data given in *Problem 7*, determine the equation of the regression line of current on applied voltage, correct to 3 significant figures.   $[X = -0.483 + 0.440Y]$

9  Draw the scatter diagram for the data given in *Problem 7* and show the regression lines of applied voltage on current and current on applied voltage. Hence determine the values of (a) the applied voltage needed to give a current of 3 mA and (b) the current flowing when the applied voltage is 40 volts, assuming the regression lines are still true outside of the range of values given.      [(a) 7.92 V; (b) 17.1 V]

10  In an experiment to determine the relationship between force and momentum, a force $(X)$ is applied to a mass, by placing the mass on an inclined plane, and the time, $(Y)$, for the velocity to change from $u$ m/s to $v$ m/s is measured. The results obtained are as follows:

| Force (N) | 11.4 | 18.7 | 11.7 | 12.3 | 14.7 | 18.8 | 19.6 |
|-----------|------|------|------|------|------|------|------|
| Time (s) | 0.56 | 0.35 | 0.55 | 0.52 | 0.43 | 0.34 | 0.31 |

Determine the equation of the regression line of time on force, assuming a linear relationship between the quantities, correct to 3 significant figures.

$[Y = 0.881 - 0.0290X]$

11  Find the equation for the regression line of force on time for the data given in *Problem 10*, correct to 3 decimal places.      $[X = 30.187 - 34.041Y]$

12  Draw a scatter diagram for the data given in *Problem 10* and show the regression lines of time on force and force on time. Hence find (a) the time corresponding to a force of 16 N, and (b) the force at a time of 0.25 s, assuming the relationship is linear outside of the range of values given.      [(a) 0.417 s; (b) 21.7 N]

# Index